T0137399

Springer Water

The book series Springer Water comprises a broad portfolio of multi- and interdisciplinary scientific books, aiming at researchers, students, and everyone interested in water-related science. The series includes peer-reviewed monographs, edited volumes, textbooks, and conference proceedings. Its volumes combine all kinds of water-related research areas, such as: the movement, distribution and quality of freshwater; water resources; the quality and pollution of water and its influence on health; the water industry including drinking water, wastewater, and desalination services and technologies; water history; as well as water management and the governmental, political, developmental, and ethical aspects of water.

More information about this series at http://www.springer.com/series/13419

El-Sayed E. Omran · Abdelazim M. Negm
Editors

Technological and Modern Irrigation Environment in Egypt

Best Management Practices & Evaluation

Springer

Editors
El-Sayed E. Omran
Faculty of Agriculture
Suez Canal University
Ismailia, Egypt

Abdelazim M. Negm
Department of Water and Water Structures
Engineering, Faculty of Engineering
Zagazig University
Zagazig, Egypt

ISSN 2364-6934 ISSN 2364-8198 (electronic)
Springer Water
ISBN 978-3-030-30377-8 ISBN 978-3-030-30375-4 (eBook)
https://doi.org/10.1007/978-3-030-30375-4

This Springer imprint is published by the registered company Springer Nature Switzerland AG
The registered company address is: Gewerbestrasse 11, 6330 Cham, Switzerland

Preface

The idea of writing this book was conceived in late 2016. The demand for writing this book is also increasing due to the great challenge facing Egypt as a poor country in water resources. The purpose of this book is to shed light on advanced technology and modern irrigation environment in Egypt, which can solve part of the problem. New irrigation technologies might be a significant contribution toward helping us reach that goal. Emerging technologies are to improve agricultural procedures and the use of modern irrigation techniques. Modernization of an irrigation system is defined as the act of upgrading or improving the system capacity to enable it to respond appropriately to the water service demands of the current times, keeping in future perspective needs, or as a process of technical and managerial upgrading (as opposed to mere rehabilitation) of irrigation schemes with the objective to improve resource utilization (labor, water, economics, environment) and water delivery service to farms.

This book consists of 16 chapters and contributed by more than 25 scientists, specialists, and researchers from Egypt. Keeping in mind the philosophy of "The Art of Irrigation," the first chapter introduces this book by providing the salient features of each chapter under its theme. The 14 main chapters of this book were grouped into five main parts, which sequentially examine the technological and modern irrigation environment in Egypt. A brief description of each is given here.

The first part is organized to present a comprehensive overview of basic irrigation practice: problems and evaluation. This part consists of three chapters. Three approaches shed light on the main problems and irrigation evaluation in Egypt. The first approach deals with water resources, types, and common problems in Egypt. The second approach calls for an evaluation of irrigation schemes and irrigation systems. The third approach is the evolution of irrigation system, tools, and technologies.

The second part, which consists of three chapters, describes the smart irrigation technology concepts. This part covers three potential approaches, which are identified for using new and modern technology to increase sustainable agriculture in Egypt. A first potential approach is the smart sensing system for precision agriculture. The second potential way is the development of recent information and data on irrigation technology and management. The third potential approach is the medicinal plants in a hydroponic system under water deficit conditions as a way to save water.

The third part, which involves three chapters, reviews the various irrigation management. This topic is covered in detail in three approaches, which were identified for managing irrigation as a potential way for sustainable agriculture—first, accurate estimation of crop coefficients for better irrigation water management in Egypt. Second, vermicomposting influences agriculture in Egypt. There is a nice prospect for vermicomposting adaptation by municipal waste systems in country operation—finally, irrigation water use efficiency and economic water productivity of different plants under Egyptian conditions.

The fourth part comprises three chapters. It discusses the different methods and approaches for irrigation system design. Three distinct technologies are identified for irrigation system design. First, improving the performance of surface irrigation system by designing pipes for water conveyance and on-farm distribution. The second technology is the micro-sprinkler irrigation of orchard. The orchard sprinkler is a small spinner or impact sprinkler designed to cover the inter-space between adjacent trees; there is little or no overlap between sprinklers. The third is the drip irrigation technology. The drip irrigation system is suitable for a wide variety of horticultural and agronomic crops, and in many respects, it applies to those crops presently under surface drip irrigation.

Part five consists of three chapters to provide a comprehensive overview of water reuse and treatment. Two techniques were identified to treat water irrigation with a magnetic field. The first technique is to use irrigation with magnetically treated water induces antioxidative responses of *Vicia faba* L. to Ni and Pb stress at the harvest stage. The second technique is to use irrigation with magnetically treated water to enhance plant growth and rehabilitates the toxicity of nickel and lead. The management of wastewater as a new resource of water is also referred to enlighten the readers on the important items of reusing treated water. Selection of the most effective and proper wastewater treatment is an essential part of generating a new water resource as well as protecting the discharge environments.

This book ends with the sixteenth chapter, which briefly summarizes the most significant findings and recommendations of this book. The concluding chapter highlights major challenges to achieving equitable and sustainable water security in Egypt and offers cautious prospects for the future.

This book on the technological and modern irrigation environment in Egypt emerged at this point. There are various books written worldwide on the subject; however, this book aims to put forth and focus on the recent advances of technological and modern irrigation environment in Egypt. A group of experts has come together to write this book who target to pass recent available knowledge and information to the readers. One of the important features of this book is that it does not have a textbook structure when the chapters, in order to be understood, need to be read in the sequence given. You can start the journey from any part, based on your interests and preferences. The readers and beneficiaries vary from academicians, professionals, and scientists, to undergraduate and graduate students.

This book is intended to be of interest to all stakeholders of the irrigation sector: irrigation agencies, financing institutions, water users' associations, planners, designers, training, and research institutions. We wish it would act as a handbook for those interested readers on the technological and modern irrigation. We believe that the information presented here will be most helpful to policy makers, managers, and researchers interested in a broad perspective of urgent water issues in Egypt and hope that this book will contribute to some real if modest, progress toward the beneficial management of the irrigated agriculture. It is hoped that it will stimulate and bring a useful contribution to the debate on irrigation sector reform and modernization and to the success of efforts to improve the performance of irrigation and to provide a better service to the farmers, by increasing the awareness of the critical importance of proper modernization procedures and design criteria.

Advances in this book would not have been possible without the great efforts paid by all the authors, and we are sure their valuable contributions increase the significance of this book. Without their patience and significant efforts in writing and revising the different versions of the chapters to satisfy the high-quality demand of Springer, it would not have been possible to produce this book and make it a reality. All appreciation and thanks must be extended to include all members of Springer team who have worked long and hard to produce this volume and make it a reality for the researchers, graduate students, and scientists around the world. We must thank all the experts who contributed to the review processes of the book chapters. We hope this book is widely read. Now, we can avoid the blunders of the past by changing the direction and start benefiting from the knowledge base created by the scientists. We did not have this chance a decade ago. Now, it is the right time during the sustainability era.

Abdelazim M. Negm acknowledges the partial support of the Science and Technology Development Fund (STDF) of Egypt in the framework of the grant no. 30771 for the project titled "A Novel Standalone Solar-Driven Agriculture Greenhouse—Desalination System: That Grows Its Energy And Irrigation Water" via the Newton-Mosharafa funding scheme.

The editors welcome any comment which might contribute to enriching the next editions of this book during the next editions.

Ismailia, Egypt
Zagazig, Egypt
May 2019

El-Sayed E. Omran
Abdelazim M. Negm

Contents

Introduction

Introduction to "Technological and Modern Irrigation Environment in Egypt: Best Management Practices and Evaluation"

El-Sayed E. Omran and Abdelazim M. Negm

1 Background/Overview

Water is certainly the only widespread and global concern that interests all of the world's living bodies. It is well documented that water resources are not evenly distributed among countries based on population densities. Some countries have abundant water resources, whereas some others suffer from inadequate water, and even face severe water scarcity problems. By 2025, Egypt was ranked among the top ten countries with water shortages, partly due to the rapidly growing population [1]. On the one hand, climate change is likely to increase resource stress, particularly water, which is the most important and central factor in the development of agriculture. The current situation in Egypt is framed by water shortages, which are under serious pressure from growing populations, rapid development, intensification of agriculture, and the degrading environment in Egypt [2]. On the other hand, in 2011, the Government of Ethiopia announced a plan to build a hydroelectric dam on the Blue Nile River called the Grand Ethiopian Renaissance Dam (GERD). GERD construction is believed to affect the water quota in Egypt by decreasing discharge from the Aswan High Dam (AHD) [1]. The GERD will put additional pressure on the Egyptian situation [3]. As our water resources come under increasing pressure, hard decisions should be made so that resources are not degraded or tipping points reached.

E.-S. E. Omran
Soil and Water Department, Faculty of Agriculture, Suez Canal University, 41522 Ismailia, Egypt
e-mail: ee.omran@gmail.com

Institute of African Research and Studies and Nile Basin Countries, Aswan University, Aswan, Egypt

A. M. Negm (✉)
Water and Water Structures Engineering Department, Faculty of Engineering, Zagazig University, Zagazig 44519, Egypt
e-mail: amnegm85@yahoo.com; amnegm@zu.edu.eg

E.-S. E. Omran and A. M. Negm (eds.), *Technological and Modern Irrigation Environment in Egypt*, Springer Water,
https://doi.org/10.1007/978-3-030-30375-4_1

Agricultural production in Egypt depends on irrigation. Although irrigation is in a quiet crisis, irrigated agriculture remains essential for future food security. Regarding undeniable previous achievements in donating to food production, irrigation expansion has significantly lost momentum due to a significant slowdown in new investment, loss of irrigated areas due to waterlogging, salinization, overdrafting of aquifers, and urban intrusion.

Technology and innovative solutions to irrigation must address the problems of water scarcity affecting Egypt. The type of irrigation system is important, and the availability of appropriate irrigation systems hardly meets the agricultural expansion needs. Irrigation water is rapidly becoming the primary limiting factor in crop production. Water is threatened by the way we treat it. Water is a resource that cannot be replaced and can only be renewed if it is well managed. Egypt faces water scarcity, an issue that can be partially mitigated by improving efficiency in the use of agricultural water. Efficiency in water use is of vital importance in parts of the world where water resources are limited. Water efficiency has now become a strategic goal in Egypt. By law, with pressurized irrigation systems, newly reclaimed land must be irrigated. Over the past several decades, pressurized irrigation systems have played an important role in improving the efficiency of irrigation and water application uniformity.

2 Themes of the Book

Therefore, this book intends to improve and address the following main theme:

- Irrigation practice: problems and evaluation.
- Smart irrigation technology.
- Irrigation management.
- Irrigation system design.
- Water reuse and treatment.

The next sections present briefly the main technical elements of each chapter under its related theme.

3 Chapters' Summary

3.1 Irrigation Practice: Problems and Evaluation

This theme is covered in three chapters. Chapters deal with Water Resources, Types and Common Problems in Egypt which are connected to Irrigation. The chapter illustrates several important points of water resources in Egypt as follows:

- The research draws attention to the wide gap between water resources and requirements in Egypt.
- The agricultural sector consumes about 85% of the available water.
- Overpopulation, poor water management as well as establishing of Grand Ethiopian Renaissance Dam (GERD) are expected to maximize the water gab and hence increase the water scarcity in Egypt.
- Increasing the available water resources by using groundwater, reuse of agricultural drainage water, and the industrial wastewater seems not quite enough to overcome water scarcity in Egypt.
- Therefore, alternative options are available to solve the problem through:

(i) Desalination of seawater which requires high energy through establishing nuclear power station might contribute, to a large extent, to overcome the problem of water scarcity in Egypt.
(ii) More advanced techniques of irrigation water management should be followed.
(iii) Attention should be paid toward cultivation of crops characterized by low water requirements instead of those of high water consumptive uses.

In Chapter 3 titled "Need for Evaluation of Irrigation Schemes and Irrigation Systems," the authors explain terms of indicators and a basic concept to track and evaluate the technical and agronomic performance of smallholder irrigation projects. They describe soil moisture depletion, management allowed depletion, irrigation methods, efficiency, and uniformity of irrigation. Furthermore, it includes essential deficit irrigation between limited and stress irrigation and high-frequency irrigation.

Basic concepts and terms of indicators the technical and agronomic performance of smallholder irrigation schemes used are very significant for monitoring and evaluating. Essential deficit irrigation between limited and stress irrigation and high-frequency irrigation depends on soil moisture depletion, management allowed depletion, irrigation methods, efficiency, and uniformity of irrigation.

There are seven basic techniques or methods of irrigation, most of which have several variations. Each technique and variation has characteristics that are adaptable to different locations and crops. The basic component and operation for each of the seven techniques are basin, basin check, border strip, furrow or corrugation, sprinkler, and trickle (or drip) emitter.

Concerning an irrigation scheme evaluation, the values of performance indicators should be applied in order to evaluate the performance level concerning the irrigation scheme. For their calculation, performance indicators call upon a certain number of parameters that have to be measured in the areas of irrigation schemes for the progressive elaboration.

A detailed study is needed to optimize profit which would be beyond the scope of the following evaluation procedures described here. In addition to the evaluation of system performance in the field, which indicates the location and magnitude of water losses, such a study would require a thorough knowledge of system costs, plus the relation between water and crop production in the area studied.

The first theme ends with an interesting chapter on the "Evolution of Irrigation System, Tools and Technologies" where in Egypt, irrigation is dependent upon the

Nile River and has been systematically utilized since the predynastic era. In the time of the political unification of Egypt under the Pharaohs, basin irrigation was supplemented with the growth of crops year-round, specifically cereal grains. Although the river floods in summer, it carries a much volume of water for the rest of the year. Egypt's dry and warm climate supports more than one crop every year, so farmers started growing multiple crops. The Aswan Dam was built in 1902 to be completed twice in 1912 and the second in 1933. In order to better regulate the extra water, the government built a series of other barrages: Asyut, Zifta, Esna, Edfina, and Nag Hammadi, as well as the structure of the Delta Bridge, which replaced Al-Qanatir Al-Khairiya.

First utilizing buckets and then shadoof after the Middle Kingdom, Egyptians lifted water from the river to irrigate their crops. The Archimedes' screw (tanbour) and waterwheel (saqia) were introduced to Egypt by the second century. These water-lifting devices were by animal-powered and meant a drastic increase in the water amount that could be poured onto the fields during the summer when the water volume in the Nile was at its lowest.

3.2 Smart Irrigation Technology

The "Smart Sensing System for Precision Irrigation" is the title of the first chapter in the second theme. The chapter focuses on two objectives. First objective answers the question "What is the status of smart sensors, cloud computing, and IoT development to support PA." This chapter presented the state-of-the-art survey of the research literature on how emerging technology used to solve agricultural problems specifically related to precision agriculture (PA). Proximal sensing allows measuring many soil and plant properties in situ. These include portable X-ray, spectroscopy, digital camera, smartphone, multistripe laser triangulation scanning, ground-penetrating radar, and electromagnetic induction sensor.

Second objective is to propose a system architecture and key technical elements of new technology for sustainable agricultural management. Four layers are proposed in the system, which are sensor layer, transmission layer, Cloud services layer, and application layer. Benefits and possible limitations of the proposed system are identified. Environmental sensors have been utilized in applications according to the need to construct smart PA. The Cloud is a gathering of platforms and infrastructures on which data is stored and processed, enabling farmers to recover and transfer their data for a particular mobile application, at any site with Internet access. Joining the Cloud, IoT, and sensors are fundamental, with the goal that the sensing data can be stored or handled. The proposed system comprises the sensor layer, the transmission layer, the Cloud services layer, and the application layer. At last, the advantages and the possible limitations of the system are talked about.

Moreover, the second chapter in this section is titled "Development of Recent Information and Data on Irrigation Technology and Management." It describes the required knowledge, information and data for irrigation, evapotranspiration concepts, and computer software to calculate evapotranspiration of crop and irrigation requirements. It focuses on data records associated with irrigators, irrigation management advisors, agricultural extension workers, experts, and external evaluators. Besides, a detailed assessment and specific information for the experimental areas for irrigation schemes is necessary to update information and data on irrigation technology, and to promote intensive and sustainable irrigated agriculture and improve food security and support decision making in Egypt.

The aim of this chapter is to develop and manage the recent information and data on irrigation technology to promote intensive and sustainable irrigated agriculture and improve food security in Egypt.

Development and recent data provide Egypt with a planning tool for rational exploitation of developing and manage recent information and data concerning soil, crops, and water resources. This planning is intended to lead to an increase in crop production for local consumption, as well as promote the production of high-value crops.

The planning tool will support decision making with direct regard to establishing agro-ecological zones maps; identifying water availability and selection of potential irrigation areas; identifying criteria for crop selection and estimated water requirements; identification of the most favorable regions to develop irrigation management practices; giving priority to irrigation water distribution; organization and control of irrigation supply management; developing irrigated agriculture in small-, medium-, and large-scale projects on old and new lands. Further, decision making will be made with regard to supporting the national policy options for irrigation water distribution; upgrading the Egyptian agricultural production; recommending options for water harvesting and storage in the coastal regions; suggesting solutions for mismanagement; and identifying energy requirements for irrigation systems

On the other hand, the chapter titled "Medicinal Plants in Hydroponic System under Water Deficit Conditions—A Way to Save Water" focuses on saving and investing water. The importance of the production of medicinal plants to meet the human and pharmaceutical requirements in the case of water shortage, scientific studies contributed to the management of this problem through some low-water production systems. When using deficit irrigation systems, marginal or appropriate production with the least amount of irrigation water can be achieved. Under a deficient irrigation system, as a system for providing water with a degree of potential stress that produces minimal effects on the crop and occasionally increases the active substances in medicinal plants. Deficit irrigation can achieve and attain greater economic benefit by increasing the yield per unit of water for a particular medicinal plant. In fact, the choice of the type of hydroponics, as a technique for deficit irrigation, was facilitated by the availability of plant monitor technology and mobile apps to aid farmers in "when," "where," "how," or "what" to plant and correctness agriculture. There are numerous ways in which medicinal plants can be produced with the lowest consumption of water. The choice of method is associated with cultivated species, quality of irrigation water, and the objective of production.

In general, biostimulants can help plants tolerate water deficit stresses. These include microbial inoculants, biochemicals, amino acids, humic acids, fulvic acids, plant and seaweed extracts, and more. Deficit irrigation techniques can be the hydroponic systems, where plants grow in nutrient solutions. Nutrients, along with growth promoters and elicitors, are available in the water. In these systems, water is reclaimed and the water consumption decreases to produce the final crop unit and more active compounds.

3.3 Irrigation Management

The "Accurate Estimation of Crop Coefficients for Better Irrigation Water Management in Egypt" is an interesting chapter particularly for irrigated agriculture which is prevailing in Egypt to provide the best level of production. However, water is becoming a scarce natural resource, and agriculture represents the major water consumption (about 85% of water resources are allocated to agriculture). Therefore, proper irrigation scheduling has to be employed by the farmers to attain water-saving measures. Thus, knowledge of crops water requirements is crucial for water resource management and planning. Crop water requirements consist of two components, namely evapotranspiration (ETo) and crop coefficients (Kc). Irrigation scheduling models can calculate both ETo and crop Kc. Example of these models is BISm model developed by Snyder et al. [4]. To do that, the weather elements in the five agro-climatic zones in 2016 were input in the BISm model to calculate ETo in the five agro-climatic zones of Egypt. Using planting and harvest dates of 13 vegetable crops, 14 field crops, and 6 fruit crops, the model calculated the Kc values and dates of these crops. The model was also used to calculate the value of water consumptive use of these crops in the five agro-climatic zones of Egypt. The validation of the model was done by comparing the calculating values of water consumptive use with the measured values from field experiments implemented in previous years, where the estimated values lay in a reasonable range.

Also, the chapter titled "Irrigation Water Use Efficiency and Economic Water Productivity of Different Plants under Egyptian Conditions" is of special interest as Egypt as an arid country with fixed water assets which is a limiting factor in reclamation processes. Improving IWUE and EWP is most important than increasing yield for each land unit. So, decision makers in government, farmers, and scientific communities' efforts should come together to improve the agricultural management that reflects on the increase both of IWUE and EWP especially in arid and semi-arid regions then decreasing the demand for water in the agricultural sector. EWP values should be considered at the selection between strategic crops during putting the agricultural-economic politics, so more efficient agronomic management must be adopted. Irrigation management (supplemental irrigation, irrigation systems, rates, and frequencies), tillage methods, organic amendments, mulch technique, hydrogel application, grafting technique, selection of tolerant genotypes, crop rotation, as well as spraying of abscisic acid, proline, silicate, salicylic, auxin, and cytokinin, etc. are the fundamental practices have a high potential for maximizing the benefits of water unit. As well as, the unit of measuring IWUE kg/m^3 not suitable, it should determine the weighted or the economic part (seed, straw or biological yield, etc.) to become kg seeds/m^3 that more accurate than the first one.

On the other hand, "Vermicomposting Impacts on Agriculture in Egypt" are presented in the last chapter of this section. The solid waste is upsurging in an abrupt way in Egypt. The solid waste like agricultural and municipal solid waste generation increases by urban and rural individuals, respectively. Vermicomposting is one of the best methods to overcome the waste generation and GHG emissions that can be produced by the waste burning. Vermicomposting converts the organic waste into value-added products which play an important role to maintain the circular economy by making money from organic waste. Vermicomposting method comprised of many aspects such as solid waste reduction, as biofertilizers have positive effects on plant growth. It also decreases GHG emissions, so, from every point of view vermicomposting is a better option to manage Egypt waste generation. And, with these above points vermicomposting promotions needed further. And, more aspects need to be considered for vermicomposting adoption like substrate composition, earthworm species, and precomposting.

3.4 Update Irrigation System Design

In the chapter titled "Improving Performance of Surface Irrigation System by Designing Pipes for Water Conveyance and On-Farm Distribution," the authors provide introduction to primary theories of water flow by pipes and static and dynamic head calculation of head losses in pipe flow consequently, designing pipe sizes for irrigation water flow. Discussing considerations for designing a surface irrigation system by most variables in hydraulics of surface irrigation system and specify three phases of water-front in a surface irrigation system are as follows: advance phase, wetting phase (or ponding) and recession phase. Presenting how audit evaluating irrigation efficiencies, application efficiency storage efficiency/water requirement efficiency, irrigation distribution uniformity, uniformity coefficient, and low-quarter distribution uniformity. Consequently, describing how to determine the performance of basin/furrow irrigation systems. Because performance terms measure how close an irrigation event is an ideal one in order to the modernization of an irrigated area must start with a diagnosis of its current situation, through the indicators of performance, engineering, and on-farm water use. In some cases, cost–benefit or social uplift and social acceptance aspects are measured via socioeconomic indicators and summarized the field (soil) and crop condition representing the ideal/normal field condition during the evaluation of an irrigation system. Case studies from Egypt to improve irrigation efficiency were presented, and also, some of the obstacles and constraints were shown in the chapter.

Additionally, the "Micro-sprinkler Irrigation of Orchard" is dealt in this second chapter in this part. Micro-spray irrigation system, also known as micro-jet or spray emitter irrigation system, is very similar to a drip irrigation system except that each emitter (sprayer) can wet larger areas compare to emitters of drip irrigation systems.

Spray emitters are normally used on coarse-textured soils (sandy soil), where wetting sufficiently large areas would require a large number of drippers. This is because coarse-textured soils have inherently vertical wetting compared to do fine textures soil (clayey soil). They are also used for trees where one to two sprayers per tree can go over the wetted area requirements.

The uniformity of the watering pattern produced by over-tree sprinklers, useful for frost protection and climate control as well as for irrigation, can be evaluated only at the top of the tree canopy level. Interference of the catch pattern by the trees makes soil surface measurements meaningless. However, ground-level distribution is of most importance to irrigation. Observations indicate how much soil is dry, and probing can indicate uniformity of application. Under-tree systems requiring overlap from adjacent sprinklers to obtain uniformity can be evaluated by the standard technique for open-field evaluation.

The irrigation objectives must be known before the operation of the system can be evaluated intelligently. Uniformity of application and the efficiency of storing water for plant use are the two most important points to be considered. For evaluating orchard sprinkler systems, uniformity and efficiency must be qualified, for often it is not practical to try to have complete coverage. Fortunately, mature trees have such extensive root systems that they can extract soil moisture wherever it is available. Therefore, the roots may absorb any available stored water.

Good management of micro-sprinkler irrigation system provides great water and soil conservation and reduces applied water requirement. Using micro-sprinkler irrigation system improves the existing water productivity; water use; efficiency, economic returns, increase yield and great control of applied water through improving distribution characteristic, and storage efficiency for field crops growing under orchard area.

Furthermore, the “Drip Irrigation Technology: Principles, Design, and Evaluation” is needed to be widely used in Egypt these days to increase the productivity of water as it saves water, improves crop yields and quality, and facilitates fertilizer application. The most modern drip irrigation systems in Egypt are ultra-low drip irrigation system (ULDI) and mobile drip irrigation system (MDIS). Although drip irrigation systems generally performance is dependent upon skilled management, they are growing in popularity in Egypt for many reasons. One reason is that all drip irrigation systems maintain a relatively constant level of moisture content in the root zone, which increases production of many crops. A second reason for the increasing popularity of drip irrigation systems in Egypt is that the potential efficiency (90%) is greater than for surface irrigation systems and sprinkler irrigation systems. For subsurface drip irrigation, evaporation from the soil surface is dramatically reduced or may be zero. Also, drip irrigation systems are effective chemigation systems because they apply water directly to the root zone of plant.

3.5 Water Reuse and Treatment

As the demand to the water resources in Egypt is higher than the supply, Egypt is trying to close the gap through the reuse of wastewater in irrigation. Therefore, the "Irrigation with Magnetically Treated Water Induces Antioxidative Responses of *Vicia faba* L. to Ni and Pb Stress at Harvest Stage" is the first chapter in the fifth theme. However, in Egypt, agriculture sustainability is still under the persistent threat of trace elements (TEs). So far, most researches are still circulating around finding ameliorating non-expensive method to eliminate the danger of translocating TEs into food chain. Therefore, in this research we are exploring and highlighting the outcomes of changing the quality of irrigation water in enhancing crop behavior under one of the abiotic stresses that are referred to. Therefore, the effect of two different concentrations of two different heavy metals (Ni, Pb) had been chosen with/and without irrigation with MTW. The choice of broad bean plants comes from its economic importance as one of the daily consumed legumes in Egypt. Ni and Pb lead to significant oxidative stress damage. Consequently, most of the antioxidants had increased aggressively with normal tap water irrigation. Meanwhile, irrigation with MTW had cooled plants' temper by decreasing contents of H_2O_2, lipid peroxidation, and electrolyte leakage. However, surprisingly irrigation with MTW had a double negative effect on plants treated with the high concentration of the heavy metals. That later effect was evident from increasing proline and total protein contents comparing to respective controls.

Also, the second chapter (Chapter 15 in the book) is under the fifth theme which is titled "Irrigation with Magnetically Treated Water Enhances Growth and Defense Mechanisms of Broad Bean (*Vicia faba* L.) and Rehabilitates the Toxicity of Nickel and Lead." The authors present their experiences where in a cultivated pot experiment. Heavy metals (Ni and Pb) showed altered growth and antioxidant enzyme activities of broad bean (*Vicia faba* L.). However, irrigation with magnetically treated water (MTW) succeeded in alleviating the negative effect of heavy metals. Increasing Anthocyanins, proline, and total protein contents were the most highlighted stress markers under heavy metals stress. Meanwhile referring to oxidative stress, there was a significant increase in hydrogen peroxide (H_2O_2) level and lipid peroxidation. However, great repression of glutathione (GSH), catalase (CAT), ascorbate peroxidase (APX), peroxidase (POX), and superoxide dismutase (SOD) has been observed. Enhancing the morphological parameters and decreasing proline and total protein content in all treatments could be achieved under irrigation with MTW. Moreover, MTW caused a significant increase in most antioxidants activities in plants grown in soils with/without heavy metals. Water passed through 1000 gauss magnet revealed an innovative role of irrigation with MTW in the adaptation of broad bean plants to detrimental effects of heavy metals based on plant growth, physiological, and biochemical criteria.

This book ends with the conclusions and recommendations chapter numbered 16.

Acknowledgements The editors who wrote this chapter would like to acknowledge the authors of the chapters for their efforts during the different phases of the book including their inputs in this chapter.

Abdelazim Negm acknowledges the partial support of the Science and Technology Development Fund (STDF) of Egypt in the framework of the grant no. 30771 for the project titled "A Novel Standalone Solar-Driven Agriculture Greenhouse—Desalination System: That Grows Its Energy and Irrigation Water" via the Newton-Mosharafa funding scheme.

References

1. Omran E-SE, Negm A (2018) Environmental impacts of the GERD project on Egypt's Aswan high dam lake and mitigation and adaptation options. In: Negm A, Abdel-Fattah S (eds) Grand Ethiopian Renaissance Dam Versus Aswan High Dam. The Handbook of Environmental Chemistry, vol 79. Springer, Cham
2. Omran ESE (2017a) Cloud-based non-conventional land and water resources for sustainable development in Sinai Peninsula, Egypt. In: The handbook of environmental chemistry. Springer, Berlin. https://doi.org/10.1007/698_2017_63
3. Omran ESE (2017b) Land and groundwater resources in the Egypt's Nile Valley, Delta and its Fringes, Egypt. In: The handbook of environmental chemistry. Springer, Berlin. https://doi.org/10.1007/698_2017_64
4. Snyder RL, Orang M, Bali K, Eching S (2004) Basic irrigation scheduling (BISm). http://www.waterplan.water.ca.gov/landwateruse/wateruse/Ag/CUP/Californi/Climate_Data_010804.xls

Irrigation Practice: Problems and Evaluation

Irrigation: Water Resources, Types and Common Problems in Egypt

Ahmed A. Abdelhafez, Sh. M. Metwalley and H. H. Abbas

1 Introduction

Water is the most important substance for human being and living organisms. Water scarcity is the lack of sufficient available water resources to meet the demands of water usage within a region. Water scarcity can occurr through two main mechanisms as follows: physical water scarcity, which occurs due to inadequate natural water resources to supply a region's demand and economic water scarcity, which occurs as a result of poor management of the available water resources. In Africa, there are several water resources, i.e., Congo, Nile, Zambezi and Niger rivers and Lake Victoria. Therefore, the scarcity of water in Africa is mainly economic due to the poor management of water resources. In Egypt, the development of water resources is very poor; on the other hand, there is an excessive growth of population. Furthermore, agricultural activities in Egypt consumes about 80% of The Nile water budget. Therefore, potential scarcity might occur in Egypt; especially, there is a critical argument due to the buildup of the Grand Ethiopian Renaissance Dam. Clearly, the construction of the Grand Ethiopian Renaissance Dam will negatively affect the recent situation of natural water resources in Egypt. It is worth mentioning that filling the Grand Ethiopian Renaissance Dam by 74 billion m^3 of Nile water may lead to change the demographic map of Egypt. In this chapter, we will address

A. A. Abdelhafez (✉)
Department of Soils and Water, Faculty of Agriculture, New Valley University, New Valley, Egypt
e-mail: ahmed.aziz@aun.edu.eg

Sh. M. Metwalley
Department of Soils and Water, Faculty of Technology and Development, Zagazig University, Zagazig, Egypt
e-mail: metwallysh@yahoo.com

H. H. Abbas
Department of Soils and Water, Faculty of Agriculture, Benha University, Benha, Egypt
e-mail: hharsalem@yahoo.com

E.-S. E. Omran and A. M. Negm (eds.), *Technological and Modern Irrigation Environment in Egypt*, Springer Water,
https://doi.org/10.1007/978-3-030-30375-4_2

the water resources in Egypt. In addition, we will try to present the major problems of irrigation and the potential hazards that might occur due to the construction of the Grand Ethiopian Renaissance Dam.

2 Egyptian Water Resources Management History

In ancient times, the great Greek historian Herodotus said:

> Egypt is the Gift of The Nile.

He meant that The Nile is the one which gives life and existence to Egypt (Fig. 1). Nowadays, we can say:

> Just a River, it is Lifeblood for the Egyptians.

Management of irrigation water sources in Egypt is as old as the country itself. For thousands of years, agriculture in Egypt depended on the annual floodwaters of The Nile River to irrigate the arable lands. Historically, water level gauges or "Nilometers" were used to measure the height of The Nile floods and set taxes accordingly [1, 2]. The expected annual flood of The Nile River provided both fertile sediments and irrigation water for Egyptian agriculture. The Islamic conquests have developed these policies by introducing agricultural land levels, according to the ease by which irrigation from The Nile water is performed; in the calculation of taxes and revenues imposed on farmers [3].

Napoleon Bonaparte, commander of the French expedition to Egypt in 1798, was the first to consider the introduction of theory to The Nile water storage. Muhammad Ali Pasha, who ruled Egypt from 1805 to 1845, was a genius leader. He took Napoleon's idea and considered that the management of The Nile water is one of the pillars of the agricultural and economic renaissance in Egypt. He established the barrages and summer (*Seifi*) canals across The Nile River as a means of surface storage in The Nile and its branches during flooding periods, as well as allowing the recharge of groundwater reservoirs that are used when water is scarce. It started with the construction of the Esna Barrage with a total length of 900 m, and this barrage was rehabilitated by a new one. Esna Barrage was considered to be the main control unit for regulating Nile water and supplemented with a hydropower station [4]. Naga Hammadi Barrage was also established in the early 1900s with a total length of 330 m and reconstructed by the Egyptian governorate to maintain continued supply of water to the large irrigation areas downstream. Assiut Barrage was designed by the famous British engineer Sir William Willcocks in the year of 1898 with a total length of 1200 m. Thereafter, Delta Barrages were established in the year of 1939, followed by Zifta and Damietta reservoirs on the branch of Damietta and Edfina on Rosetta's branch (Fig. 2) [4]. A point to note, the main objectives of constructing barrages and dams in the past were to control The Nile water and to ensure continuous supply of The Nile water to agricultural areas.

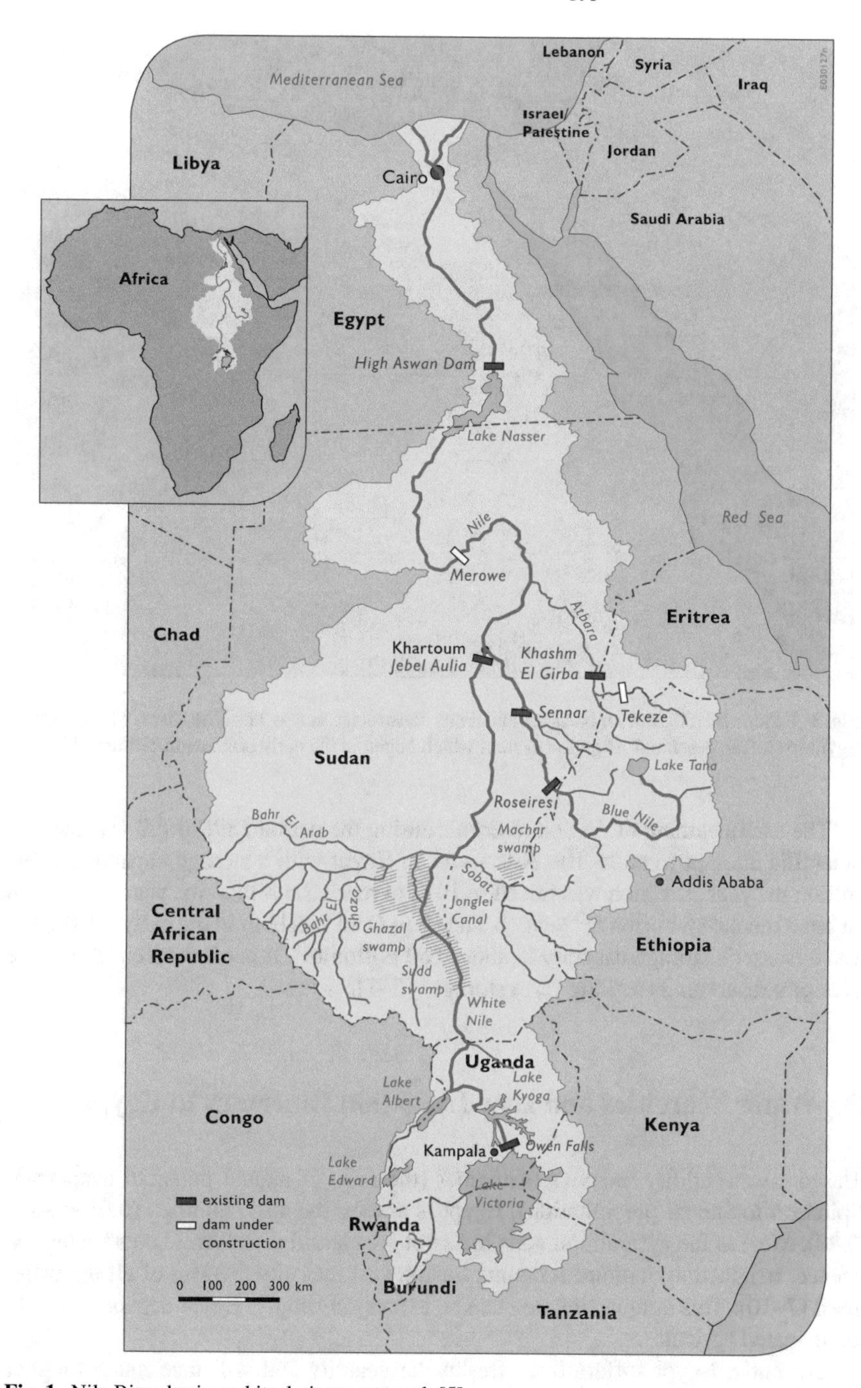

Fig. 1 Nile River basin and its drainage network [5]

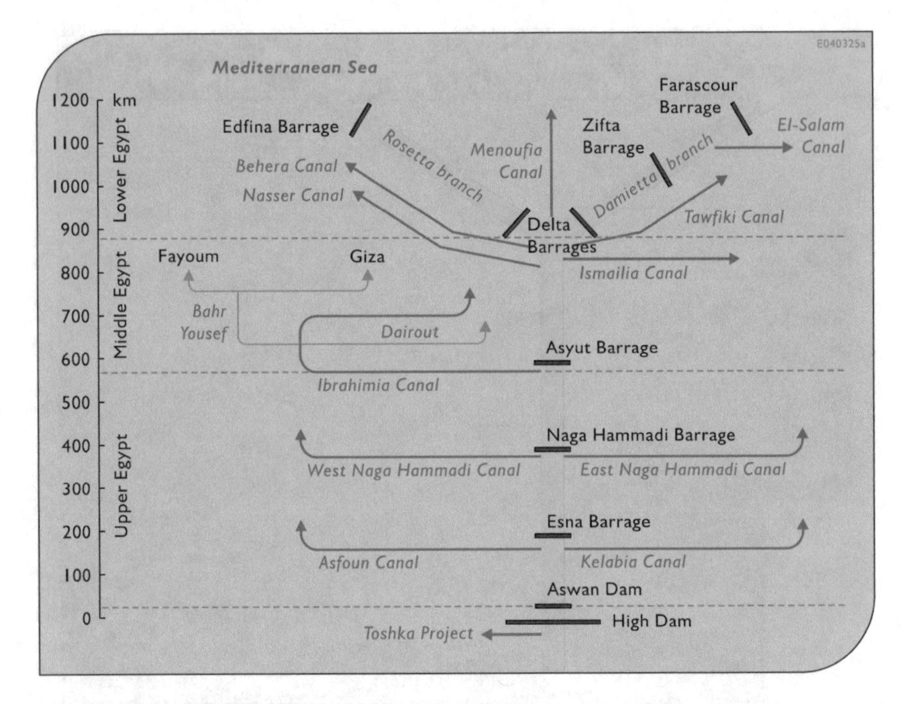

Fig. 2 Schematic diagram for barrages and dams constructed across The Nile River, which enabled Egypt to shift to perennial irrigation system which began in the early nineteenth century [5]

The establishment of Aswan reservoir during the period 1898–1902 was the first scientific attempt to store The Nile water in Egypt with a storage capacity of one billion m^3 $year^{-1}$, which was raised in 1912 to reach 2.5 billion m^3 $year^{-1}$ and then in 1932 reached 5 billion m^3 $year^{-1}$. The High Dam was built in the early 1960s, and Lake Nasser's storage capacity is about 160 billion m^3. It is often referred to Lake Nasser's reservoir as a "long-term storage" [5–7].

3 Water Scarcities and Low Irrigation Efficiency in Egypt

Based on the aridity index (AI) number (the ratio of annual potential evapotranspiration to annual precipitation), Egypt is among the arid countries ($0.05 < AI < 0.20$). Most of the agricultural soils in Egypt are classified as "Arid lands" category. Hence, irrigated agriculture accounts for the vast majority (>85%) of all the water used [7–10]. This unique situation can be expressed simply: Egypt depends entirely on irrigated agriculture.

Currently, Egypt suffers from freshwater scarcity and will face much tougher situations due to increasing the present and future water demands where the country has no enough alternative water resources. The Ministry of Water Resources and

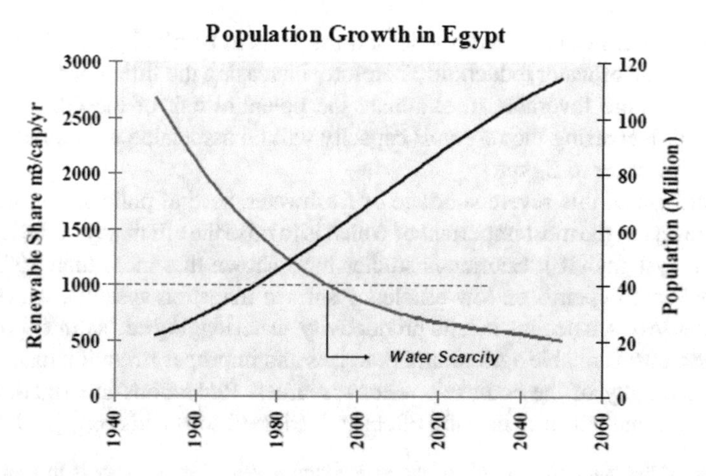

Fig. 3 Projected annual per capita share of renewable water resources in Egypt [12]

Irrigation (MWRI) announced in May 2016 that Egypt's water resources produce 62 billion m^3 annually, while the consumption could reach 80 billion m^3 [5]. Thus, the annual per capita of water supply in Egypt is about 600 m^3 only, while the international average is 1000 m^3 [11]. In this context, MWRI [5] reported that average per capita share of about 800 m^3 cap^{-1} y^{-1} as of year 2004, while projections forecast a share of about 600 m^3 cap^{-1} y^{-1} by the year 2025 as depicted in Fig. 3. This indicates a sharp decline in water availability per capita more severe than predicted by the previous studies [2, 6, 9].

Consequently, water availability per capita rate has become already one of the lowest in the world. This is suggested for further declines due to the excessive increase in population on one hand and climate change and global warming on the other one. Accordingly, the gap between the limited water resources and the increasing demands for water is getting wider over time [13], and Egypt will soon face water scarcity.

3.1 *Grand Ethiopian Renaissance Dam (GERD): View Point*

The water shortage crisis is compounded owing to the fact that the main source of water, The Nile, originates from outside the borders of Egypt, making its discharge rates controlled by many other factors. Furthermore, in April 2011, Ethiopia has launched the construction of the Grand Ethiopian Renaissance Dam (GERD), with water storage capacity of 74 billion m^3. Egyptian experts predict a water reduction of about 20–34% owing to the construction of that Ethiopian dam. "This is estimated to be 11–19 billion m^3 on average over the Dam's filling period" [13, 14]; consequently, about one-third of the total agricultural lands might be subjected to drought. It is worth mentioning that the filling duration of GERD reservoir is of a major concern. According to the annual rate of precipitation in Ethiopia whose average is about

817 mm, GERD reservoir needs at least three years to be filled [15], which poses a potential threat of water reduction. Therefore, increasing the filling period of GERD reservoir is more favorable to eliminate the potential risk of the GERD. On the other hand, increasing the reservoir capacity will be associated with more risk and reduction of water to Egypt.

In the light of this severe shortage of freshwater, several policies and measures should be taken; the most important of which is to raise the efficiency of the Egyptian irrigation systems [16]. Enormous studies have shown that more than 70% of the cultivated area depends on low-efficiency surface irrigation systems, which cause high water losses, a decline in land productivity, waterlogging and salinity problems. Moreover, unsustainable agricultural practices and improper irrigation management affect the quality of the country's water resources. Reductions in irrigation water quality have, in their turn, harmful effects on irrigated soils and crops [2, 9, 10, 12].

> … one of the main components of the agricultural development strategy is to achieve a gradual improvement of the efficiency of irrigation systems to reach 80 per cent in an area of 3.36 million ha, and to reduce the areas planted with rice from 702.66 thousands ha (2007) to 546 thousands ha by 2030 in order to save an estimated 12 400 million cubic meters of water. [17]

However, attention should be paid regarding the reduction of rice agricultural soils due to the potential salinization of agriculture soils in the delta regions.

4 Water Resources of Egypt

Water resources in Egypt are confined to the withdrawal quota from The Nile water; the limited amount of rainfall; the shallow and renewable groundwater reservoirs in The Nile Valley, The Nile Delta and the coastal strip; and the deep groundwater in the Eastern desert, the Western desert and Sinai, which are almost nonrenewable. The nontraditional water resources include reuse of agricultural drainage water and treated wastewater, as well as the desalination of seawater and brackish groundwater [6, 7, 13, 18].

4.1 Conventional Water Resources

4.1.1 The Nile Water

Egypt reliese on The Nile water to provide nearly 97% of freshwater needs [7, 13]. This clearly shows that Egypt's national security depends on The Nile River. Currently, more than 90 million capita live around the valley and delta of The Nile. The Nile Valley is a narrow strip that expands from Aswan in the south to Cairo in the north while The Nile Delta starts from the south of Cairo and stretches up to the

Table 1 Egypt water resources in 2017 (unit: billion m^3 y^{-1}) [19]

Source	billion m^3 y^{-1}	Usage	billion m^3 y^{-1}
The share of the River Nile water	55.50	Drinking	10.70
Underground water (surface and deep wells)	9.60	Industries	5.40
Reuse of drainage waters	13.5	Agriculture	61.65
Rains and floods	1.30	Evaporation	2.50
Seawater desalinization	0.35		
Total	80.25		80.25

Mediterranean Sea in the form of a triangle. Both the valley and the delta depend entirely on The Nile River for drinking, farming, industry and the other economic activities. Therefore, any changes that might occur in the net water budget of The Nile River will affect the Egyptian demographic map.

The Nile water represents the main source of water in Egypt, amounted to be 55.5 billion m^3 annually according to the agreement of Egypt and Sudan in 1959. The Nile water source mainly serves irrigated soils in The Nile Valley and Nile Delta which together constitute 85% of total irrigated land in Egypt [7, 13] and consume about 77% of the total water resources in Egypt (Table 1).

The Nile River is the second longest river in the world, with a length of about 6700 km. The Nile basin stretches over 35 latitudes from latitude 4° south of the equator at its sources near Lake Tanganyika and reaches the line 31° north of the equator. The Nile basin extends its authority over more than nine longitudes, from the longitude of 29° at its roots in the tropical plateau to a longitude of ~38° at its springs in the Abyssinian plateau as shown in Fig. 1. The area of The Nile basin is about 2.9 million km^2, which includes parts of ten African countries, namely Ethiopia, Eritrea, Uganda, Burundi, Tanzania, Rwanda, Sudan, Congo, Kenya and Egypt.

The total area of these ten countries is about 8.7 million km^2. The Nile River passes through its long travel in different regions, languages and civilizations. It passes through several climatic regions, from the equatorial region with an average annual rainfall of about 800 mm at its source, and even the desert region that is very dry in northern Sudan and south Egypt.

The Nile River income, like most rivers, varies from a year to another while the lowest was 42 billion m^3 $year^{-1}$, and the highest was 150 billion m^3 $year^{-1}$ measured in Aswan. The average natural Nile River revenues during the twentieth century—estimated at Aswan—were about 84 billion m^3. The Nile collects water from three main basins, i.e., the Ethiopian plateau, tropical lakes' plateau and Bahr el Ghazal.

4.1.2 Underground Water

Renewable groundwater reservoirs are distributed between The Nile Valley (with a stock of 200 billion m^3) and the delta region (with a stock of approximately 400 billion m^3). This water is considered a part of The Nile water resources. The water withdrawn from these reservoirs is estimated by 6.5 billion m^3 $year^{-1}$ since 2006. This is within the safe withdrawn level, which has a maximum limit of about 7.5 billion m^3 $year^{-1}$ according to the Groundwater Research Institute. It is also characterized by a good quality where salinity range is about 0.47–1.25 dS m^{-1} in the south delta regions. The depletion of these reservoirs is allowed only when drought occurs for a long period. It is estimated that the drawdown of these reservoirs is around 9.6 billion m^3 in 2017 [2].

Nonrenewable aquifers are located under the Eastern and Western Desert and the Sinai Peninsula. The most important of these is the Nubian sandstone reservoir in the Western Desert, which is estimated by about 40 thousand billion m^3. It extends in the regions of northeast Africa and includes the lands of Egypt, Sudan, Libya and Chad. This reservoir is one of the most important sources of fresh groundwater; however, it is not readily available for use in Egypt due to its deep depths. It is therefore estimated that only about 0.6 billion m^3 of these waters are withdrawn annually that is enough to irrigate about 63,000 ha in Owainat area.

The annual withdrawal rate is expected to increase to about 2.5–3 billion m^3 $year^{-1}$ as a safe and economic withdrawal. In general, the effects of the expected decline in the level of the groundwater reservoir should be avoided by shifting from the extensive plantation system to the specific plantation system of scattered areas (840–2100 ha) to conserve these aquifers for extended periods.

4.1.3 Rainfall and Floods Waters

Egypt is almost rainless except for the northern coast where rainfalls at an annual rate of 50–250 mm. On the northwestern coast, rainfall ranges from 50 to 150 mm $year^{-1}$. In the years of relatively heavy rain, the area cultivated with barley may exceed 42,000 ha. On the northeast coast, rainfall increases as we head eastward. The rate at El-Arish is 150 mm, while in Rafah, it is about 250 mm.

In the light of normal winter rainfall, the amount of rainwater falling over the northern parts of Egypt (about 200,000 km^2) is estimated at 5–10 billion m^3 $year^{-1}$. Of this amount, about 1.5–3 billion m^3 are flowing over the surface; much of it returns to atmosphere via evaporation and transpiration, and the rest leaks into the layers to be added to groundwater recharge. It is noted that the surface runoff water from the valley's ravine is lost into the sea or in the coastal saline ponds.

When winter rains are relatively high, a phenomenon that is repeated once every 4–5 years, the amount of surface runoff water may reach 2 billion m^3, and its impact extends to wider areas of Egyptian deserts. When the Egyptian lands are exposed to seasonal rains, a phenomenon that is repeated every 10 years, the amount of rain that flows above the surface may reach 5 billion m^3 and has significant impacts on the regions of Red Sea, South Sinai and The Nile River basin and often cause extensive environmental damage.

The amount of rains falling on the Sinai Peninsula and distributed over its various water basins as well as the amount of rains that flow on the surface and out of the water basins toward the sea are about 131.67 million m^3 annually and represent 5.25% of the total rainfalls.

It is worth mentioning that the average annual rainfall on the entire Egyptian territory is about 8 billion m^3 and that the flow is in the area of 1.3 billion m^3, and this helps to attract and harvest about 200–300 million m^3 $year^{-1}$ of these rains in Sinai and the northern coast and the eastern Red Sea.

4.2 *Nonconventional Water Resources*

4.2.1 Reuse of Agricultural Drainage Water

Agricultural drainage water includes the water used for leaching salts from soils (leaching requirements), the leakage from the irrigation and drainage systems, the unused canal endings and the sewage and industrial wastewaters that pour into agricultural drains. Therefore, this water is of low quality due to its high salinity and is mixed with water from drains that are often contaminated with chemicals used in the agriculture and industry. Salinity in this type of water ranges from 700 to more than 3000 mg L^{-1} [20], which are not suitable for irrigation for many crops.

Reuse of agricultural wastewater is a significant source of water resources development in the future. The improvement of the agricultural drainage water quality should be taken into consideration by directly treating the water of sub-drains or main drains before mixing them with freshwater and avoiding mixing them with sewage or industrial water to avoid the environmental risks of reusing such water without treatment. It is very important to discharge at least 50% of the total amount of agricultural drainage water to the sea to maintain the water and salt balance of The Nile Delta and to prevent the increased impact of deep-sea water interference with the North Delta aquifer [21].

4.2.2 Reuse of Municipal and Industrial Wastewater

Treated wastewater is one of the water resources that can be used for irrigation purposes if it meets the appropriate health conditions. The amount of treated water increased annually from 0.26 billion m^3 $year^{-1}$ in the early 1990s to about 0.6 billion m^3 $year^{-1}$ in 2000 and about 4 billion m^3 in 2017. Since the amount of collected wastewater exceeds 5 billion m^3 [22], the future policy is to increase the use of wastewater up to 4.5 billion m^3 [23]. It is used to irrigate nonfood crops for humans or animals and to plant forests in the desert for wood production. It is very crucial to focus on the treatments of municipal and industrial wastewater before use and their separation far away from agricultural drainage waters to avoid the risks of chemical waste on public health and the environment [22, 24].

4.2.3 Desalination of Sea Water

Owing to the length of Egypt's coastline both on the Mediterranean Sea and on the Red Sea, and the effective government action during the last two decades and the present to maximize tourism and industrial development of coastal areas, the provision of water resources for this development is a guarantee of its existence and sustainability.

One of the most important sources of water available in coastal areas is desalination, whether for seawater or brackish water. The challenge for adopting desalination technique has been and is still in the development of commercially viable methods. The extensive experience gained over the past 40 years and the improvements in desalination technology have made desalination very technically acceptable and provide high-quality water to arid areas that have been deprived of water sources for sustainable economic and social development. In the mid-1960s, the idea of desalinization was illusory, and most of the activity in that field was experimental, and many of the first projects failed to meet the expectations on which they were adopted. At present, it is a highly reliable technology, and many countries rely on it for financial capacity such as the Gulf countries, but costs remain relatively high compared to other water resources. The Middle East and North African countries contribute to 41.3% of the world's desalinated water [25].

For Egypt, desalinization of seawater can be used as an untraditional water source for the development of desert and coastal communities. Solar, wind and nuclear energies can be used in the desalination process instead of transferring electricity or oil to these regions to raise the economies of exploitation of this source [25].

A previous study indicated that the cost of desalting 1 m^3 of seawater ranged between 5.40 and 18.88 LE depending on the capacity of the desalination plant, which makes the use of this water for agricultural purposes economically inefficient at present [26]. Nowadays, some studies are conducted aiming at desalination of brackish water in the groundwater reservoir near the northern coast of the Mediterranean Sea and North Sinai, which are relatively less saline than seawater, thus reducing the cost of desalination technique [25, 26].

5 Irrigation Systems and Their Common Problems

Methods of irrigation in ancient Egypt are thought to be the basis of many agricultural technologies followed worldwide [27]. According to de Feo et al. [28], the ancient Egyptians used various hydraulic technologies to lift water from the near river banks to water their crops, i.e., Shadouf, Tanbur and Saqiya. The Shadouf device consists of a bucket attached from one side to a wooden arm and from the other side to a counterbalance [29]. It can lift water from the near river banks up to 1–6 m [30]. In case of Tanbur (or Archimedes' screw), it consists of a watertight cylinder enclosing a chamber called off by spiral divisions running from end to end [31], and the water is lifted in the turn of cylinder around its axis from the immersed bottom of the spiral chamber in the water [32, 33]. Saqiya (or waterwheel) consists of a row of pots attached to the rim of a revolving wheel so that turning wheels by oxen allow the pots to be filled with water when dipped into an irrigation canal and then lifted on the wheel to a height of 3–6 m [31].

Recently, increasing global population has placed major stresses on the available water resources especially with declining groundwater availability and decreasing water quality [34] to supply fruits, vegetables and cereal foods consumed by humans [35]. Thus, irrigated agriculture is thought to play major roles for satisfying the basic needs and improve life standards [36]. Increasing water use efficiency is the key to overcome water shortage problems and, on the other hand, reduce the environmental problems [37] through optimizing source to sink balance and avoid excessive vigor [38]. This might take place through adopting low-volume irrigation technologies, especially in arid soils [39]. The techniques of water distribution within the field differ according to the quantity of water available for irrigation, other natural resources, system cost and irrigation efficiency in addition to a number of institutional, financial and production inputs [36]. Generally, there are many types of irrigation systems, e.g., surface irrigation, sprinkler irrigation and drip irrigation.

5.1 Surface Irrigation

It is often called *flood irrigation* in which water moves by gravity surface flow [40] either across the entire field (basin irrigation) or the water is fed into small channels (furrows) or strips of land (borders) [41]. Surface irrigation was and is still the common irrigation system in many regions. This method is applied on lower lands of heavy textured soils in the presence of suitable drainage system that may lead to over logging conditions [42]. On the other hand, several problems might be associated with this system first of which, the high water lose due to evaporation and seepage, in addition, the phenomenon of water shortage at canal tail end (CT); that commonly occurs due to several reasons, i.e., widening and cleaning of the canal pathway that may lead to decrease the velocity of water, loses of water through seepage along the canal path and the inadequate amount of water at the beginning of irrigation way [43, 44]. Surface irrigation consisted of several types as follows:

5.1.1 Basin Irrigation

Water is supplied in small level field ground leveled to zero slope in both directions [45] enclosed by a dyke to prevent runoff [42] to form small bounded units [46]. In case of large areas, uneven distribution of irrigation water might occur [47]. The first artificial basin irrigation was established by deliberating flooding and draining using sluice gates and water contained by longitudinal and transverse dikes [31]. Although this type of surface irrigation is the least expensive to operate and manage, it is considered to be the most expensive to develop and maintain [48]. Basin irrigation is suitable for many crops, especially paddy rice. However, soil type and surface morphology are important factors controlling the use of basin irrigation. The size of basin should be small if the soil is sand and had steep slope [41]. Shortage of water occurs in some places in basin irrigation system where the water was stood for a short time affecting the growing plants due to the lack of moisture. As a result, farmers used to build up dicks and dams to maintain water on a soil surface for a constant time [49].

5.1.2 Furrow Irrigation

Furrow irrigation is directed by small parallel channels conveying water down or along the slope of the field under the influence of gravity [50]. Water infiltrates through the soil and spreads along the primary directions of the field vertically and horizontally [42]. Crops are grown on ridges between furrows [41]. However, there might exist higher erosion and pollution risks with the concentrated water flow in the furrows [51, 52]. In this concern, applying water-soluble polyacrylamide to the irrigation water can minimize water runoff and consequently soil erosion [53]. It is worthy to mention that the water use efficiency under furrow irrigation might be low. Wide-spaced furrow irrigation or skipped crop rows can lessen water evaporation from the soil surface, as is the case for drip irrigation [54]. Moreover, alternate irrigation of one of the two neighboring furrows probably saves in arid areas [54]. The furrow irrigation is more favorable for many crops, especially crops that would be damaged if water covered their stem or crown [41]. A point to note that flat soil is more suitable for furrow irrigation, and slope should not exceed 0.5% [41, 55].

Since the furrow irrigation system wets as little as 20% of the field surface, extra time might be needed to ensure the complete wetting of fine particle soils [56].

5.1.3 Border Irrigation

It is the extension of basin irrigation to suit larger mechanized farms in which land is divided into wide, level rectangular strips [57] "with a minimum slope of 0.05% to provide adequate drainage and a maximum slope of 2% to limit problems of soil erosion" [41] and its drain exists at the lower end [42]. Length and width of the strips and basins depend on field slope, soil texture and the amount of water received in each strip. In case of sandy and silty loam soils, shorter strips are required while the clayey soils require longer strips [57]. The border irrigation system is suitable for many crops, except those that require prolonged ponding [58].

5.2 *Irrigation Using Sprinkler Systems*

It is the method of irrigation through which water is sprayed in the air in the form of small water droplets similar to natural rainfall [41] to optimize water application on sloping fields [59], also to overcome the waterlogging problems [60]. However, "most of the water losses take place from the water that leaves the nozzle until it reaches soil rhizosphere" [61] probably through wind drift and evaporation and transpiration losses [62]. Farmers should carefully select the right wind conditions for irrigation [63]. Also, they can avoid further evaporation and transpiration losses by nighttime irrigations which might bring such losses to almost negligible levels [64]. Furthermore, matching the sprinkler rate can guarantee more uniform entry of irrigation water into the root zone [65]. The area wetted is circular, and sprinklers should be operated close to each other to attain an overlap of at least 65% of the wetted diameter [41]. Some sprinklers are designed to optimize the use of various inputs for improving or enhancing economic crop production using precision agriculture (PA) technologies [60]. Others are supplied by remote sensors to simultaneously monitor water status in the field for improving the efficiency of water use [66].

5.3 *Drip Irrigation*

Drip (or micro) irrigation, also known as trickle irrigation, is a system of irrigation through which water falls drop by drop at or near the root zone of plants [67]. It is then considered the most efficient method of water irrigation since evaporation and runoff are thought to be low [68]; also, this method can effectively minimize evapotranspiration rates from plants grown in tropical climate regions, with dry and hot summers [69]. Under proper management, the efficiency of drip irrigation is typically within the range of 80–90% [70]. This method is the ideal for wastewater reuse [71] and for irrigation with low-quality water [72] to protect the environment on one hand and to sustain economic vitality on the other hand [73]. Micro-irrigation with fertilization also provides an effective and cost-efficient way to supply water and nutrients to crops [74] in a process known by fertigation [75, 76]; however, using low water quality in fertigation can accumulate salts within the root zone to reach toxic levels [77]. Drip irrigation is also used as a method of partial wetting of the root zone (controlled alternate partial root-zone irrigation (CAPRI), also called partial root-zone drying (PRD) [78] through simultaneously exposed to both wet and dry zones [79]. Water quality plays an important role for sustainable performance of irrigation system, for example, in New Valley, Egypt; groundwater is the major source of different applications. The groundwater in New Valley contains high concentrations of salts and iron metal ions; consequently, when iron ions are oxidized they lead to corrupt the dripping units (Fig. 4). The wetted zone supplies water and nutrients to plants while, at the same time, the non-wetted zone reduces loss of water by evaporation from soil surface [80] probably because the plants that receive fully

Fig. 4 Precipitation of iron metal ions and salts on the soil surface in the New Valley, Egypt (Image was taken by the corresponding author from the Faculty of Agriculture farm, New Valley University in October, 2017)

irrigations usually have widely opened stomata; therefore, the partially dried part of the roots stimulates partial closure of the stomata to reduce water loss through transpiration) [78].

5.4 Subsurface Irrigation

It is used in areas of high water tables in which the water table is artificially raised to moisture below the plants' root zone, yet the moisture rises past the roots [81]. This type of irrigation can also be applied on the seepage of the nearby water sources (e.g., canals or rivers) [82]. Subsurface drip irrigation (SDI) can further substantially improve irrigation water use efficiency (IWUE) than did the surface drip irrigation through reducing evaporative loss while maximizing capture of in-season rainfall by the soil profile [69]. Furthermore, roots concentrate at a depth of the irrigation tubes [83]. However, SDI can eliminate soil air (including oxygen) within the rhizosphere during and following irrigation events [84]. Such reduced condition of root respiration hinders plant growth by decreasing transpiration while improving the unselective flow of ions and salt ingress into the plant to reach toxic levels [69]. Thus, injection of air into the drip lines improves the aeration of the root zone more generally and therefore affects crop growth [85]. Generally, the life of a SDI system should be at least 10–15 years for economic competitiveness [86].

6 Conclusions and Recommendations

It could be concluded clearly that Egypt may suffer from water shortage through the next few years due to increasing the present and future water demands beside of the expected shortage in Egypt share of The Nile water due to the building of the Grand Ethiopian Renaissance Dam (GERD). Egyptian experts predict a water reduction of about 11–19 billion m^3 owing to the construction of that Ethiopian dam. Consequently, about one-third of the total agricultural lands in Egypt might be subjected for drought. Therefore, attention should be paid to overcome these problems through development of irrigation systems, reuse of agricultural wastewater, desalination of seawater and managing the discharge and usage of groundwater. Furthermore, applications of advanced irrigation systems should be preceded by studying the major characteristics of water resource to ensure its safe use.

References

1. Collins B, Montgomery DR, Hass AD (2002) Historical changes in the distribution and functions of large wood in Puget Lowland rivers. Can J Fish Aquat Sci (No. 59):66–76
2. Ashour MA, El Attar ST, Rafaat YM, Mohamed MN (2009) Water resources management in Egypt. J Eng Sci Assiut Univ 37(2):269–279
3. Shahin M (2002) Hydrology and water resources of Africa. Springer, Berlin, pp 286–287
4. El Gamal F, Hesham M, Shalaby AR (2007) Country paper on harmonization and integration of water saving options in Egypt. In: Karam F, Karaa K, Lamaddalena N, Bogliotti C (eds) Harmonization and integration of water saving options. Convention and promotion of water saving policies and guidelines, CIHEAM/EU DG Research, Bari, pp 81–90
5. Ministry of Water Resources and Irrigation (MWRI) (2005) Water for the future, national water resources plan 2017, January 2005
6. Allam MN, Allam GI (2007) Water resources in Egypt: future challenges and opportunities. Water Int 32(2):205–218
7. FAO (2016) AQUASTAT-FAO's information, Regional report. Egypt, FAO, Rome
8. UNCOD (United Nations Conference on Desertification) (1977) Desertification: its causes and consequences. Pergamon Press, 448 p
9. Abu-Zeid M (1992) Water resources assessment for Egypt. Canadian J Dev Stud (Rev can d'études dév) 13(4):173–194
10. Abdrabbo MA, Farag A, Abul-Soud M, El-Mola MM, Moursy FS, Sadek II, Hashem FA, Taqi MO, El-Desoky WMS, Shawki HH (2012) Utilization of satellite imagery for drought monitoring in Egypt. World Rural Obs 4(3)
11. Central Agency of Public Mobilization and Statistics (CAPMAS) (2016) Statistical of Egyptian water resources. Annually Statistics Book
12. Wagdy A (2008) Progress in water resources management: Egypt. In: Proceedings of the 1st technical meeting of Muslim Water Researchers Cooperation (MUWAREC), December 2008 (Malaysia), vol 1, p 13
13. Ferrari E, McDonald S, Osman R (2014) Water scarcity and irrigation efficiency in Egypt. In: The 17th annual conference on global economic analysis "New challenges in food policy, trade and economic vulnerability", Dakar, Senegal, 1–28. https://www.gtap.agecon.purdue.edu/resources/download/7118.pdf
14. Ibrahim AIR, Impact of Ethiopian Renaissance Dam and population on future Egypt water needs. Am J Eng Res (AJER) 6(5):160–171
15. Fazzini M, Bisic C, Billi P (2015) Landscapes and landforms of Ethiopia: the climate of Ethiopia. Springer, Netherlands, pp 65–87. http://dx.doi.org/10.1007/978-94-017-8026-1_3
16. Goher AA, Ward FA (2011) Gains from improved irrigation water use efficiency in Egypt. Water Resour Dev, 1–22
17. FAO (2013) Country Programming Framework (CPF) Government of Egypt 2012–2017. FAO, Rome
18. State Information Service (SIS), Egypt (2017) Annually statistics book. Egypt Fig 2017:172
19. Ministry of Water Resources and Irrigation (MWRI) (2018) https://www.mwri.gov.eg/index.php/ministry-7/2017-02-13-15-05-27
20. Abdel Azim RA (1999) Agricultural drainage water reuse in Egypt: current practices and a vision for future development (Ph.D. thesis, Faculty of Engineering, Cairo University, Cairo)
21. Attia BB (2018) Unconventional water resources and agriculture in Egypt. Part V Securing water resources in Egypt: Securing water resources for Egypt: a major challenge for policy planners, 485–506
22. Elbana TA, Bakr N, Elbana M (2018) Unconventional water resources and agriculture in Egypt. Part V Securing water resources in Egypt: Reuse of treated wastewater in Egypt: challenges and opportunities, 429–453
23. Aboulroos S, Satoh M (2017) Irrigated agriculture in Egypt, past, present and future. Chapter 11: Challenges in exploiting resources—general conclusion, 267–283

24. Sherif AEEA, Rabie AR, Abdelhafez AA (2017) Accumulation trends of heavy metals in *Cupressus sempervirens* and *Eucalyptus camaldulensis* trees grown in treated wastewater irrigated soil. Alex Sci Exch J 38(2):220–230
25. Fath HES (2018) Unconventional water resources and agriculture in Egypt. Part V Securing water resources in Egypt: Desalination and greenhouses, 455–483
26. Lamei A, van der Zaag P, von Münch E (2008) Impact of solar energy cost on water production cost of seawater desalination plants in Egypt. Energy Policy 36(2008):1748–1756. https://doi.org/10.1016/j.enpol.2007.12.026
27. Janick J (2002) Ancient Egyptian agriculture and the origins of Horticulture. In: ISHS Acta Horticulturae 582: international symposium on mediterranean horticulture: issues and prospects. https://doi.org/10.17660/ActaHortic.2002.582.1
28. de Feo G, Angelakis AN, Antoniou GP, El-Gohary F, Haut B, Passchier CW, Zheng XY (2013) Historical and technical notes on aqueducts from prehistoric to medieval times. Water 5(4):1996–2025
29. Mays LW (2008) A very brief history of hydraulic technology during antiquity. Environ Fluid Mech (Dordr) 8(5–6):471–484. https://doi.org/10.1007/s10652-008-9095-2
30. Mirti TH, Wallender WW, Chancellor WJ, Grismer ME (1999) Performance characteristics of the shaduf: a manual water-lifting device. Appl Eng Agric 15(3):225–231
31. Mays LW (2010) Water technology in ancient Egypt. In Mays L (ed) Ancient water technologies. Springer Netherlands, Dordrecht, pp 53–65. https://doi.org/10.1007/978-90-481-8632-7_3
32. Rorres C (2000) The turn of the screw: optimum design of an Archimedes screw. J Hydraul Eng 126(1). http://dx.doi.org/10.1061/(ASCE)0733-9429(2000)126:1(72)
33. Koutsoyiannis D, Angelakis AN (2003) Hydrologic and hydraulic science and technology in ancient Greece. In: Stewart BA, Howell T (eds) Encyclopedia of water science. Dekker, New York, pp 415–417. http://dx.doi.org/10.13140/RG.2.1.1333.5282
34. Evans RG, LaRue J, Stone KC, King BA (2013) Adoption of site-specific variable rate sprinkler irrigation systems. Irrig Sci 31:871–887. https://doi.org/10.1007/s00271-012-0365-x
35. Howell TA (2001) Enhancing water use efficiency in irrigated agriculture contrib. from the USDA-ARS, Southern Plains Area, Conserv. and Production Res. Lab., Bushland, TX 79012. Mention of trade or commercial names is made for information only and does not imply an endorsement, recommendation, or exclusion by USDA-ARS. Agron J 93:281–289. https://doi.org/10.2134/agronj2001.932281x
36. Gilley JR (1996) Sprinkler irrigation systems. In: Pereira LS, Feddes RA, Gilley JR, Lesaffre B (eds) Sustainability of irrigated agriculture. NATO ASI Series (Series E: Applied Sciences), vol 312. Springer, Dordrecht, pp 291–307. https://doi.org/10.1007/978-94-015-8700-6_17
37. Deng XP, Shan L, Zhang H, Turner NC (2006) Improving agricultural water use efficiency in arid and semiarid areas of China. Agric Water Manag 80(1):23–40. https://doi.org/10.1016/j.agwat.2005.07.021
38. Chaves MM, Santos TP, Souza CR, Ortuño MF, Rodrigues ML, Lopes CM, Maroco JP, Pereira JS (2007) Deficit irrigation in grapevine improves water-use efficiency while controlling vigour and production quality. Ann Appl Biol 150:237–252. https://doi.org/10.1111/j.1744-7348.2006.00123.x
39. Caswell MF (1991) Irrigation technology adoption decisions: empirical evidence. In: Dinar A, Zilberman D (eds) The economics and management of water and drainage in agriculture. Springer US, Boston, MA, pp 295–312. https://doi.org/10.1007/978-1-4615-4028-1_15
40. USDA (1997) Irrigation guide. National engineering handbook. Part 652. The United States Department of Agriculture (USDA). Available at: https://policy.nrcs.usda.gov/17837.wba
41. Brouwer C (1985) Irrigation water management: irrigation methods. Training manual no 5. FAO—Food and Agricultural Organization of the United Nations. http://www.fao.org/agl/aglw/fwm/Manual5.pdf
42. Verheye WH (2009) Management of agricultural land: climatic and water aspects. In: Verheye WH (ed) Encyclopedia of land use, land cover and soil sciences: land use management and case studies. Volume IV, Encyclopedia of life support systems, Eolss Publishers Co. Ltd., Oxford, UK, pp 61–97

43. El-Nashar WY, Elyamany AH (2017) Value engineering for canal tail irrigation water problem. Ain Shams Eng J. https://doi.org/10.1016/j.asej.2017.02.004
44. El-Kilani RMM, Sugita M (2017) (Chapter 6) Irrigation methods and water requirements in the Nile Delta 125-151. Irrigated agriculture in Egypt, past, present and future
45. Khanna M, Malano HM (2006) Modelling of basin irrigation systems: a review. Agric Water Manag 83(1):87–99. https://doi.org/10.1016/j.agwat.2005.10.003
46. Skogerboe GV, Bandaragoda DJ (1998) Managing irrigation for environmentally sustainable agriculture in Pakistan. Report No. R-77. Towards environmentally sustainable agriculture in the Indus basin irrigation system. Pakistan National Program, International Irrigation Management Institute (IIMI), Lahore
47. Erie LJ, Dedrick AR (1979) Level-basin irrigation: a method for conserving water and labor. USDA, Farmers' Bulletn Number 2261, Washington, DC
48. USDA (2012) Part 623 Irrigation, national engineering handbook. Chapter 4, Surface irrigation, 1–131. https://directives.sc.egov.usda.gov/opennonwebcontent.aspx?content=32940.wba
49. Norris PK (1934) Cotton production in Egypt. Tech Bull 451:1–42. United State, Department of Agriculture. Washington DC
50. USDA (1983) Section 15: Irrigation-Chapter 5: Furrow irrigation. National engineering handbook. United States Department of Agriculture. https://www.wcc.nrcs.usda.gov/ftpref/wntsc/waterMgt/irrigation/NEH15/ch5.pdf
51. Mailapalli DR, Raghuwanshi NS, Singh R (2009) Sediment transport in furrow irrigation. Irrig Sci 27:449–456. https://doi.org/10.1007/s00271-009-0160-5
52. Dibal JM, Igbadun HE, Ramalan AA, Mudiare OJ (2014) Modelling furrow irrigation-induced erosion on a sandy loam soil in Samaru, Northern Nigeria. International Scholarly Research Notices 2014(8). https://doi.org/10.1155/2014/982136
53. Lentz RD, Sojka RE (2000) Applying polymers to irrigation water: evaluating strategies for furrow erosion control. Am Soc Agric Eng 43(6):1561–1568. http://dx.doi.org/10.13031/2013.3056
54. Kang S, Shi P, Pan Y, Liang ZS, Hu XT, Zhang J (2000) Soil water distribution, uniformity and water-use efficiency under alternate furrow irrigation in arid areas. Irrig Sci 19:181–190. https://doi.org/10.1007/s002710000019
55. El Gamal F (2000) Irrigation in Egypt and role of national water research center. In: Lamaddalena N (ed) Annual meeting of the mediterranean network on collective irrigation systems (CIS_Net). Bari, CIHEAM, pp 33–44
56. USDA (2012) Chapter 4: Surface irrigation. National engineering handbook. Part 623 Irrigation. Available at https://directives.sc.egov.usda.gov/opennonwebcontent.aspx?content=32940.wba
57. Finey S (2016) Sustainable water management in smallholder farming: theory and practice. CABI, Oxfordshire, UK. https://doi.org/10.1079/9781780646862.0000
58. Walker WR (1989) Guidelines for designing and evaluating surface irrigation systems. Food and Agriculture Organization of the United Nations, Rome. http://www.fao.org/docrep/T0231E/t0231e00.htm#Contents
59. Mateos L (1998) Assessing whole-field uniformity of stationary sprinkler irrigation systems. Assessing whole-field uniformity of stationary sprinkler irrigation systems. Irrig Sci 18:73–81. https://doi.org/10.1007/s002710050047
60. Evans RG, King BA (2012) Site-specific sprinkler irrigation in a water-limited future. Am Soc Agric Biol Eng 55(2):493–504
61. McLean RK, Sri Ranjan R, Klassen G (2000) Spray evaporation losses from sprinkler irrigation systems. Can Agric Eng 32(1):1–8
62. Playán E, Salvador R, Faci JM, Zapata N, Martínez-Cob A, Sánchez I (2005) Day and night wind drift and evaporation losses in sprinkler solid-sets and moving laterals. Agric Water Manag 76(3):139–159. https://doi.org/10.1016/j.agwat.2005.01.015
63. Dechmi F, Playán E, Cavero J, Facl JM, Martínez-Cob A (2003) Wind effects on solid set sprinkler irrigation depth and yield of maize (Zea mays). Irrig Sci 22:67–77. https://doi.org/10.1007/s00271-003-0071-9

64. Martínez-Cob A, Playán E, Zapata N, Cavero J (2008) Contribution of evapotranspiration reduction during sprinkler irrigation to application efficiency. J Irrig Drain Eng 134(6):745–756. http://dx.doi.org/10.1061/(ASCE)0733-9437(2008)134:6(745)
65. Clothier BE, Green SR (1994) Root zone processes and the efficient use of irrigation water. Agric Water Manag 25(1):1–12. https://doi.org/10.1016/0378-3774(94)90048-5
66. Colaizzi PD, Barnes EM, Clarke TR, Choi CY, Waller PM, Haberland J, Kostrzewski M (2003) Water stress detection under high frequency sprinkler irrigation with water deficit index. J Irrig Drain Eng 129(1):36–43. http://dx.doi.org/10.1061(ASCE)0733-9437(2003)129:1(36)
67. Henkel M (2015) 21st century homestead: sustainable agriculture III: agricultural practices. First edition
68. Brouwer C, Prins K, Heibloem M (1989) Irrigation water management-Training manual no. 4: Irrigation scheduling. FAO Land and Water Development Division. www.fao.org/docrep/t7202e/
69. Bhattarai SP, Midmore DJ, Pendergast L (2008) Yield, water-use efficiencies and root distribution of soybean, chickpea and pumpkin under different subsurface drip irrigation depths and oxygation treatments in vertisols. Irrig Sci 26:439. https://doi.org/10.1007/s00271-008-0112-5
70. Subramani T, Prabakaran DJ (2015) Uniformity studies and performance of sprinkler and drip irrigation. Int J Appl Innov Eng Manag (IJAIEM) 4(5):284–293
71. Capra A, Scicolone B (2007) Recycling of poor quality urban wastewater by drip irrigation systems. J Clean Prod 15(16):1529–1534. https://doi.org/10.1016/j.jclepro.2006.07.032
72. Rajak D, Manjunatha MV, Rajkumar GR, Hebbara M, Minhas PS (2006) Comparative effects of drip and furrow irrigation on the yield and water productivity of cotton (L.) in a saline and waterlogged vertisol. Agric Water Manag 83(1):30–36. http://dx.doi.org/10.1016/j.agwat.2005.11.005
73. Dinar A, Yaron D (1990) Influence of quality and scarcity of inputs on the adoption of modern irrigation technologies. Western J. Agr. Econ. 15(2):224–233
74. Hanson BR, Šimůnek J, Hopmans JW (2006) Evaluation of urea–ammonium–nitrate fertigation with drip irrigation using numerical modeling. Agric Water Manag 86(1):102–113. https://doi.org/10.1016/j.agwat.2006.06.013
75. Thompson TL, Doerge TA, Godin RE (2002) Subsurface drip irrigation and fertigation of broccoli. Soil Sci Soc Am J 66:186–192. https://doi.org/10.2136/sssaj2002.1860
76. Singandhupe RB, Rao GGSN, Patil NG, Brahmanand PS (2003) Fertigation studies and irrigation scheduling in drip irrigation system in tomato crop (L.). Eur J Agron 19(2):327–340. http://dx.doi.org/10.1016/S1161-0301(02)00077-1
77. Mmolawa K, Or D (2000) Root zone solute dynamics under drip irrigation: a review. Plant Soil 222(1–2):163–190. https://doi.org/10.1023/A:1004756832038
78. Kang S, Zhang J (2004) Controlled alternate partial root-zone irrigation: its physiological consequences and impact on water use efficiency. J Exp Bot 55(407):2437–2446. https://doi.org/10.1093/jxb/erh249
79. Loveys BR, Dry PR, Stoll M, McCarthy MG (2000) Using plant physiology to improve the water use efficiency of horticultural crops. ISHS Acta Horticulturae 537: III international symposium on irrigation of horticultural crop. https://doi.org/10.17660/ActaHortic.2000.537.19
80. Bielorai H (1982) The effect of partial wetting of the root zone on yield and water use efficiency in a drip-and sprinkler-irrigated mature grapefruit grove. Irrig Sci 3:89–100. https://doi.org/10.1007/BF00264852
81. Crossley PL (2004) Sub-irrigation in wetland agriculture. Agric Hum Values 21(23):191–205. https://doi.org/10.1023/B:AHUM.0000029395.84972.5e
82. Zuo Q, Shi J, Li Y, Zhang R (2006) Root length density and water uptake distributions of winter wheat under sub-irrigation. Plant Soil 285:45. https://doi.org/10.1007/s11104-005-4827-2
83. Machado RM, do Rosário M, Oliveira G, Portas CAM (2003) Tomato root distribution, yield and fruit quality under subsurface drip irrigation. Plant Soil 255:333–341. https://doi.org/10.1023/A:1026191326168
84. Bhattarai SP, Ninghu Su, David J (2005) Midmore, oxygation unlocks yield potentials of crops in oxygen-limited soil environments. Adv Agron 88:313–377. https://doi.org/10.1016/S0065-2113(05)88008-3

85. Goorahoo D, Carstensen G, Zoldoske DF, Norum E, Mazzei A (2002) Using air in sub-surface drip irrigation (SDI) to increase yields in Bell peppers. In: Proceedings of Californial plant and soil conference. Energy and agriculture: now and the future, Feb 5–6, Frenco, CA, pp 22–52
86. Lamm FR, Trooien TP (2003) Subsurface drip irrigation for corn production: a review of 10 years of research in Kansas. Irrig Sci 22(3–4):195–200. https://doi.org/10.1007/s00271-003-0085-3

Need for Evaluation of Irrigation Schemes and Irrigation Systems

S. A. Abd El-Hafez, M. A. Mahmoud and A. Z. El-Bably

1 Introduction

Evaluation is a process of determining systematically and objectively the relevance, effectiveness, efficiency, and effect of activities in light of their objectives. It is an organizational process for improving activities still in progress and for aiding management in future planning, programming, and decision-making [1]. Evaluation in the rural development programs is associated with the assessment of effects, benefits or disbenefits, and impacts, on the beneficiaries.

Evaluation focus on: who or which group has benefited (or has been adversely affected), by how much (compared to the situation before the activity), in what manner (directly or indirectly), and why (establishing causal relationships between activities and results to the extent possible). While monitoring is a continuous or regular activity, evaluation is a management task that takes place at critical times of the life of a scheme or program. Evaluation can be executed [2]:

- during project planning (ex-ante): to assess the potential impact
- during project implementation (ongoing): to evaluate the performance and quality
- at project end (ex-post): to define the successful completion
- some years after completing (impact): to evaluate its final impact on the development

S. A. Abd El-Hafez · M. A. Mahmoud · A. Z. El-Bably (✉)
Water Requirements and Field Irrigation Research Department, Soil, Water and Environment Research Institute (SWERI), Agriculture Research Center (ARC), 9th Cairo Univ. Street, Giza, Egypt
e-mail: elbabli@gmail.com

S. A. Abd El-Hafez
e-mail: s.a.abdelhafez@gmail.com

M. A. Mahmoud
e-mail: mahmoud_abdalla96@yahoo.com

E.-S. E. Omran and A. M. Negm (eds.), *Technological and Modern Irrigation Environment in Egypt*, Springer Water,
https://doi.org/10.1007/978-3-030-30375-4_3

- The main objective of monitoring and evaluation (M&E) design and process is to ensure that the program or project fulfills the stated goals and objectives within the financial supports that are set at the beginning.

The objectives of an irrigation scheme can be grouped into six categories [3, 4]:

- Production and productivity
- Profitability
- Equity
- Rational utilization of the resource
- Sustainability
- Non-agricultural objectives.

Irrigation systems may or may not be well designed and properly used. The basic concepts and terms for system evaluation described in this chapter are specified for evaluating actual operation and management and for determining the potential for more efficient and economical operation. This is vital to provide direction to manage in deciding whether to continue existing practices or to improve them [5].

Improvement of water management on the farm may conserve water, soil, and labor and may increase crop yields. A system evaluation should show and measure the existing irrigation practice effectiveness. Careful study of the system evaluation will indicate whether improvements can be made and will provide information for management with a reasoned basis for selecting possible modifications that may be both economical and practical.

Most modifications suggested for irrigation systems improvement require only simple changes in management practices. Evaluations frequently indicate the need for estimates of soil moisture deficiency and for better maintenance practices for systems. These often save water, labor, and working hours. Sometimes it is worthwhile to invest the capital necessary to mechanize or even automate an irrigation system.

Operation of sprinkler irrigation systems may be improved greatly by such simple changes as altering operating pressures, nozzle sizes, heights of risers, and water application durations; operating at different pressures at alternate irrigations; using alternate set sequencing; obtaining larger sized lateral pipes; and by tipping risers along the edge of the field.

For border and strip furrow irrigation systems, any of the following simple changes may greatly improve performance: use of larger, smaller, or cutback streams; irrigation at a different soil moisture deficiency; using different spacing or shape of furrows; revising strip width or length; using supplemental pipelines and portable gated pipe; and using return-flow systems to recover runoff water. Capital investment for such projects as grading the land to provide a smoother surface or more uniform slope and soil conditions, constructing reservoirs, increasing capacity for water delivery, and automation or semi-automation often proves profitable where it improves labor and water efficiency.

Basin irrigation systems may be improved greatly by relocating a dike conforming to changes in the surface texture of the soil; grading land more carefully to achieve, as nearly as possible, a level surface and uniform intake; or changing the basin area so that it more nearly matches the volume of water from the available stream.

Trickle irrigation systems may require a different duration of application, a different frequency of irrigation, additional infiltration, or a higher density of emitters.

Possibilities for saving water and labor usually are best when the water supply is flexible in frequency, rate, and duration. Flexibility in frequency means that the water is available on or near the day when it is needed to match the moisture demands of the crop. Flexibility in the rate means that the rate of supply can be changed to match different sizes of fields, to cutback sizes of streams, to accommodate varied rates of infiltration, and to smooth out the irrigators workload. Flexibility in duration means that the water can be turned off as soon as the soil moisture deficiency has been supplied and requirements for leaching have been satisfied. These types of flexibility are necessary for achieving efficient use of water.

A principal cause of low efficiency is over-irrigation. When either furrow or border-strip irrigation is used, a chief part of any excess water is runoff, which may be recovered by using a return-flow system. Most excess water, used in the basin, basin check, sprinkler, and trickle systems, infiltrates and adds to the groundwater supply. Such water may be recovered from wells, but it may cause a drainage problem if the subsurface flow is restricted at a shallow depth.

2 Basic Concepts and Terms

Certain concepts are implicit in the design and operation of every irrigation system and irrigation schemes. Likewise, certain terms and their definitions are basic in describing these systems and in evaluating their operation. Some of the most frequently used terms are listed and briefly explained here and are explained in detail.

Evaluation is the analysis of any irrigation system based on measurements that have been taken in the field under the conditions and practices normally used. It also includes on-site studies of possible modifications such as changing sprinkler pressures, having larger or smaller streams in furrows, and changing the duration of application. Measurements needed for analysis include soil moisture depletion before irrigation, the rate of inflow, uniformity of application and infiltration, duration of application, the rate of advance, soil conditions, rates of infiltration, and irrigation adequacy [6].

As for indicators concerning irrigation schemes, indicators are the tools that measure the substantial progress toward the goal achievement such as the targets or standards to be met at each stage. They provide an objective basis for monitoring progress and evaluation of final achievements. A good indicator should define the level of achievement, specifically: how much? (quantity), how well? (quality), by when? (time). This can be demonstrated in the steps below [7]:

Step 1: **Identify indicator**: Small farmers increase rice yields

Step 2: **Add quantity**: A total of *5000* farmers *with landholdings of 1 acre or less* increase their rice yields *by 30%*

Step 3: **Add quality**: A total of *5000* farmers with landholdings of 1 *acre* or less increase their rice yields by 30% *while maintaining the same rice quality existing in the 2000 harvest*

Step 4: **Specify time**: A total of *5000* farmers with landholdings of *1 acre* or less increase their rice yields by 30% *between June 2000 and June 2008* while maintaining the same rice quality existing in the 2000 harvest. One set of indicators needs to be formulated to monitor and evaluate the process.

These indicators might be, for example, credit repaid amount, the cultivated crops, rate of the farmers' participation, attendance of the training, etc. Another set of indicators needs to be formulated to monitor and evaluate the effect of the program activities. These indicators could be, for example, yield increase, income gains, environmental effects, changes in workload, relationship between benefits and investment, etc. A set of indicators can of course also include both of the above at the same time.

Indicators should disaggregate the information by gender and various socio-economic groups. This means that instead of monitoring the number of farmers, data need to be gathered on the number of male and the number of female farmers from the different socio-economic groups participating. Equally, information on yield increases should be distinguished based on the gender of the household head, large versus small farmers, etc. The purpose of collecting gender-disaggregated monitoring data is that it may yield valuable information that can lead to measures to improve the program, especially the performance of specific groups of farmers [2].

Because of the difficulties in collecting information in the field, and because of the related costs, the number of indicators should be kept to the minimum required. A few key indicators should be selected that will adequately fulfill the objective of assessing the conditions of the scheme and identifying causes for failure or success. In this chapter, some common indicators are given for each type of performance, from which key indicators can be selected.

For the calculation of indicators, a certain number of parameters have to be measured in the irrigation scheme. The choice of these parameters has to be judicious. They should be easily measurable and remeasurable, at low cost, preferably by the farmers themselves. Some examples of indicators associated with related parameters are given in Table 1.

Examples of indicators to monitor and evaluate the technical and agronomic performance of smallholder irrigation scheme are listed down [2].

To these objectives, so-called performance indicators are mentioned down. For an irrigation scheme, the values of these obtained indicators should be compared to comparative values in order to evaluate the performance level concerning the irrigation scheme. For their calculation, performance indicators call upon a certain number of parameters that have to be measured in the irrigation improvement project (IIP) [8].

The first objective is to intensify and increase agricultural production on irrigated land.

Table 1 Examples of indicators associated with related parameters [2]

Indicators	Parameters	Expression
• Yield Y	• Harvest per season H (kg) • Area cultivated A (acre)	• $Y = H/A$ (kg/ha)
• Gross or net production per quantity of water applied P_{gIr} or P_{nIr}	• Harvest H (kg) • Volume of water applied W (m^3)	• P_{gIr} or $P_{\mathrm{nIr}} = H/W$ (kg/m^3)
• Cropping intensity CI	• Area harvested per year AH (= sum of the areas harvested per season) (acre) • Area cultivable CA (acre)	• CI = AH/CA × 100 (%)
• Overall project efficiency E_{p}	• Quantity of water entering the conveyance canal V (m^3) • Net irrigation requirements IR_n (m) • Actual irrigated area AIA (acre)	• $E_{\mathrm{p}} = 100 \times (\mathrm{AIA} \times 4000 \times \mathrm{IR}_n)/V$ (%)

The first indicator: Increase in average production

This indicator will measure the average increase that is being obtained in the demonstration phase as compared to the national averages and/or the production averages in the project area before the demonstration phase. The required data for its application are:

The average percentage of increase or decrease in production (CP) for all the crops is the indicator proposed for agricultural production:

$$\mathrm{CP} = 100 \times \sum_{1}^{N} \left[\frac{P(N) - A(N)}{A(N)} \right]^{\frac{1}{N}} \tag{1}$$

where

CP = Crop production increase or decrease (percentage)
P = Project crop production average
A = National crop production average
N = Number of crops

The second indicator: Cropping intensity

This indicator will provide an evaluation as to what extent second and third crops may take place in a year. The indicator (CI) is defined as follows:

$$\mathrm{CI} = \frac{A(C1) + B(C2) + C(C3)}{\mathrm{CA}} \tag{2}$$

where

$A(C1)$ = Total area harvested in the first season
$B(C2)$ = Total area harvested in the second season
$C(C3)$ = Total area harvested in the third season
CA = Cultivable area

The third indicator: Increase in planted area

The intensive use of irrigation water is a good indication that the change toward an intensive agriculture is taking place in an effective manner. Therefore, this indicator aims at evaluating to what extent this change is taking place. For this purpose, the increase in planted area from one season to the next (expressed in percentage) is a relevant indicator (IPA):

$$\mathrm{IPA} = 100 \times \frac{[\mathrm{AP}(S1) - \mathrm{AP}(S2)]}{\mathrm{AP}(S2)} \tag{3}$$

where

IPA = Increase in planted area (percentage)
AP($S1$) = Area planted during the current season
AP($S2$) = Area planted during the past season

The second objective is to improve the performance of existing schemes through on-farm irrigation technology.

The fourth indicator: Overall irrigation efficiency

Overall irrigation efficiency is a value that constantly varies through the year and is affected by the efficiency of the actual water distribution and farmers' ability to apply water effectively. Still, it is always a good reference for how efficiently irrigation water is utilized.

The following indicator is proposed:

$$\mathrm{OIE} = 100 \times \frac{(\mathrm{AIA} \times 4000 \times \mathrm{CWR})}{(\mathrm{FI} \times 3600 \times 30 \times N)} \tag{4}$$

where

OIE = Overall irrigation efficiency (percentage)
AIA = Actually irrigated area during peak month (acre)
CWR = Crop water or net irrigation requirement for the peak month (mm/month)
FI = Average flow of main intake in the peak month (l/s)
N = Number of irrigation hours per day

The above indicator will give the efficiency of the water use in the peak month. It is desirable to determine it for every month of the year in order to indicate the variations of the OIE along the year. This indicator will be particularly relevant when rehabilitation and improvements works have been undertaken, as the greater physical efficiency of the system must be reflected in higher values of OIE.

The fifth indicator: Costs of operation and maintenance

Operation and maintenance costs referred to the irrigated hectares are themselves already a good indicator of how efficiently the financial resources are being utilized:

$$\mathrm{OM} = \frac{\mathrm{TC}}{\mathrm{AIA}} \tag{5}$$

where

OM = Costs of operation and maintenance per acre
TC = Total annual costs incurred in O&M
AIA = Actually irrigated area (acre)

Once operation and maintenance costs have been determined, one can get an indication of the farmers' capacity to pay them by referring these costs to the farmers' income through the following equation:

$$\mathrm{IFI} = 100 \times \frac{\mathrm{TC}}{\mathrm{FI}} \tag{5.1}$$

where

IFI = Impact of costs operation and maintenance in farmer's income (percentage)
TC = Total annual costs incurred in O&M
FI = Farmers' income (assessed on the bases of a representative sample). For values of IFI greater than 10%, difficulties can be expected in the collection of fees

The third objective is to demonstrate technologies and methods of irrigation expansion.

The sixth indicator: Percentage of farmers that adopted the irrigation technology

A simple indicator is the percentage of farmers over the total participants in the demonstration area that have adopted the technological package:

$$\mathrm{AT} = 100 \times \frac{\mathrm{FAT}}{\mathrm{TNF}} \tag{6}$$

where

AT = Farmers that adopted the technology (percentage)
FAT = Number of farmers that adopted proposed technology
TNF = Total number of farmers of the demonstration area

The apparent simplicity of this indicator is constrained by the fact that is not so simple to clear whether or not a farmer has adopted a technology. As the technological packages will likely be different in each country or demonstration area, the criteria for determining the adoption by farmers must be developed locally.

The seventh indicator: Water use at farm level

One important aspect of the demonstration phase is the efficient application of water at the farm level. By this term, we mean that water is applied at suitable intervals (which will depend on the technology used) and the amounts necessary to satisfy the crop water requirements. If irrigation water is not applied with a minimum of technical bases, it is clear that the intended increases in crop production will not be reached. Therefore, it is of great importance to document how irrigation water is applied.

As the number of farmers participating in a scheme can be relatively large, it will be practically impossible to monitor the water use by every farmer as this will be time-consuming and costly. The only feasible way will be to do it on sample bases. The sample should be statistically representative, but this is again costly when the number of farmers is large.

The eighth indicator: Farm irrigation efficiency

The determination of the irrigation schedules mentioned for the seventh indicator implies the application of the farmer's efficiency in applying the irrigation water. The tendency is often to apply this figure based on empirical or personal experience. In the field, it can be carried out following standard procedures [2, 9]. It will be useful to determine these efficiencies yearly and monitor any progress made by farmers. However, as with the previous indicator it is an expensive indicator to be determined. More information on irrigation efficiencies is given in this chapter.

The fourth objective is to improve the capacity of staff and local community for self-management and develop institutional base for irrigation expansion.

The ninth indicator: Self-management

The aim of this indicator will be to assess the degree of self-management that has been achieved. The underlying assumption is that an effort was made to establish a WUA, and through the criteria proposed below, the degree of self-management is assessed as given in Table 2.

The tenth indicator: Training activities carried out

The number of training activities that have been done, the type of activity, its duration, and number of participants should be reported here. The number of participants should be related to this potential number to have an indication of what percentage has been covered.

As for soil moisture depletion, it (hereafter called SMD) is expressed numerically as a depth (in cm) indicating the dryness of the root zone at the time of measurement. This depth is identical to the water depth to be replaced by irrigation under normal management. For this reason, the idea of moisture deficit in the root zone is preferable

Table 2 Assessment of self-management degree

Self-management	Degree
The WUA functions satisfactorily and 80–90% of the water rates are collected	Fully independent
The WUA is established, the water distribution is effected by farmers at tertiary level but secondary canals and upward are operated by government staff, only minor maintenance works are carried out by farmers, and 65–80% of water rates are collected	Semi-independent
The WUA has been established but acts mainly as a consultative and information body. Decisions are still made by government officials, and 50–65% of water rates are effectively collected	Low degree of independence
The WUA has been established on paper, but none of its tasks are carried out in practice	Dependent, it needs explanation
The WUA has not been established	Needs justification

to the commonly used concept of water depth currently in the soil. Knowledge is needed of how dry the soil should be before irrigation and is related to the soil moisture tension at that SMD and to how well the crop will grow under that stress. Some plants produce better when they are kept moist by frequent irrigations, but they may be more subject to diseases and insect pests under such a regime. Other plants may produce more efficiently when the soil is allowed to become quite dry. Infrequent irrigating also reduces costs of labor and generally increases efficiency.

Management allowed depletion (hereafter called MAD) is the desired SMD at the time of irrigation. MAD is an expression of the degree of dryness that the manager believes the plants in a given area can tolerate and still produce the desired yield. The MAD is related to SMD and resulting in crop stress. It may be expressed as the percent of the total available soil moisture in the root zone or the corresponding depth of water that can be extracted from the root zone between irrigations to produce the best economic balance between crop returns and costs of irrigation.

Evaluation of furrow and border-strip irrigation systems should be made at about MAD, since infiltration rate, water movement, and duration of the irrigation are greatly affected by soil moisture deficit because the MAD appreciably affects all these factors, and small variations in the MAD become a useful management tool for improving the operation of certain surface irrigation systems, especially the border-strip system.

Efficient operation of an irrigation system depends as much or more on the capability of the irrigator as on the quality of the system. Any system may be properly used or misused. To determine what is the best use requires a thorough evaluation of the system or appreciable experience combined with shortcut evaluation procedures. The two following questions must always be considered to obtain the maximum efficiency from any given system:

- Is the soil dry enough to start irrigating?
- Is the soil wet enough to stop irrigating?

The irrigator must carefully estimate the SMD; if it is the same as MAD or greater, the soil is dry enough to start irrigating. The simplest method for evaluating SMD is field observation of the soil. This requires comparing soil samples taken from several depths in the root zone (preferably to the full rooting depth) with Table 3 This chart indicates the approximate relationship between field capacity and wilting point. For more accurate information, the soil must be checked by drying samples of it. The descriptions at the top of each textural column correspond to the condition of zero soil moisture deficiency, i.e., field capacity. Those descriptions at the bottom of a column describe a soil having the maximum deficiency, i.e., wilting point. The soil moisture deficiency at this condition is numerically equal to the available moisture range of the soil.

Intermediate soil moisture deficiency descriptions occur opposite corresponding numerical values of inches of water per foot of depth at which the soil is deficient. This chart describes a specific group of soils, and though it has been found to have general application, it may not apply to many other groups. Where this is the case, new descriptions will need to be prepared corresponding to particular soil moisture deficiency, feel, and appearance relationships.

Other methods for estimating SMD include the use of tensiometers when MAD values are low (high moisture situation) and resistance blocks or similar equipment when MAD values are high (low moisture content). Weighing and drying soil samples are precise, but slow and cumbersome and neutron soil moisture probes are expensive.

Water budgets based on the depth of evaporation from a pan and other methods for estimating the water consumed by the plants (potential evapotranspiration) are also satisfactory for estimating SMD. The SMD estimated from water budgets should occasionally be checked by field observations of the lower part of the root zone to see that SMD is not accumulating. Such checks show deficient irrigation, but unfortunately do riot reveal over-irrigation [10].

The second question, namely, when is soil wet enough to stop irrigating, is equally important because all water applied to the root zone after the SMD and leaching requirements have been satisfied is completely wasted. A probe, typically a 15/40-cm or 10/20-cm steel rod about 1.2 m long having a somewhat bulbous (not pointed) tip and a tee handle, can be used in most soils to quickly check the depth of penetration of irrigation at numerous points throughout the field. Such a probe easily penetrates to a moderate depth (about 90 cm) through the nearly saturated soil being irrigated, but it encounters considerable resistance when it meets plow pans or drier soil below the wetted soil. The proper depth of probe penetration is appreciably less than the desired final depth of water penetration because water continues to percolate deeper after the irrigation stops. This requires that the depth to which the probe penetrates during irrigation be calibrated later with depth penetrated after an adequate irrigation.

Alternately, to anticipate when the soil will be wet enough to stop dividing the SMD by the minimum rate of application at the soil surface. This will give the duration of irrigation needed to replace the SMD.

Table 3 Soil moisture and appearance relationship chart [2]

Soil texture					
Available soil moisture	Soil moisture conditions	Coarse fine sand; loamy fine sand	Moderate coarse sandy loam; fine sandy loam	Medium sandy clay loam; loam; silt loam	Fine clay loam; silty clay loam
0–25	Dry	Loose. Will hold together if not disturbed. Loose sand grains on fingers	Forms a very weak ball[a]. Aggregated soil grains break away easily from ball	Soil aggregations break away easily. No moisture staining on fingers. Clods crumble with applied pressure	Soil aggregations easily separate. Clods are hard to crumble with applied pressure
25–50	Slightly moist	Forms a very weak ball with well-defined marks. Light coating of loose and aggregated sand grains remains on fingers	Forms a weak ball with defined finger marks. Darkened. color. No water staining on fingers	Forms a weak ball with rough surfaces. No water staining on fingers. Few aggregated soil grains break away	Forms a weak ball. Very few soil aggregations break away. No water stains. Clods flatten with applied pressure
50–75	Moist	Forms a weak ball with loose and aggregated sand grains remaining on fingers. Darkened color. Heavy water staining on fingers Will not form into a ribbon[b]	Forms a ball with defined finger marks. Very light soil water staining on fingers. Darkened color. Will not slick	Forms a ball. Very light water staining. Darkened color. Pliable. Forms a weak ribbon between thumb and forefinger	Forms a smooth ball with defined finger marks. Light soil water staining on fingers. Ribbons form with thumb and forefinger

(continued)

Table 3 (continued)

Soil texture					
Available soil moisture	Soil moisture conditions	Coarse fine sand; loamy fine sand	Moderate coarse sandy loam; fine sandy loam	Medium sandy clay loam; loam; silt loam	Fine clay loam; silty clay loam
75–100	Wet	Forms a weak ball. Loose and aggregated sand grains remain on fingers. Darkened color. Heavy water staining on fingers. Will not ribbon	Forms a ball with wet outline left on hand. Light to medium water staining on fingers. Makes a weak ribbon between thumb and forefinger	Forms a ball with well-defined finger marks. Light to heavy soil water coating on fingers. Ribbons form	Forms a ball. Uneven medium to heavy soil water coating on fingers. Ribbon forms easily between thumb and forefinger
Field capacity (100)	Wet	Forms a weak ball. Light to heavy soil–water coating on fingers. Wet outline of soft ball remains on hand	Forms a soft ball. Free water appears briefly on surface after squeezing or shaking. Medium to heavy soil–water coating on fingers	Forms a soft ball. Free water appears briefly on soil surface after squeezing or shaking. Medium to heavy soil–water coating on fingers	Forms a soft ball. Free water appears on soil surface after squeezing or shaking. Thick soil water coating on fingers. Slick and sticky

[a] A “ball” is formed by squeezing a soil sample firmly in one’s hand
[b] A “ribbon” is formed by squeezing soil between one’s thumb and forefinger

Several devices for sensing soil moisture can indicate when to start and stop irrigating, but none is less expensive and easier to understand and use than the auger and simple probe described above. Some electrical or mechanical sensing devices may be connected to turn the irrigation system on and off automatically. However, their operation must be correlated with soil moisture values at the sensing point, which, in turn, must be related to values representative of the entire field under control.

The rate or volume of the application by sprinkler and trickle irrigation systems is usually known. When the application is reasonably uniform, the depth of application can be controlled easily by controlling the duration of the irrigation. However, under all the methods of irrigation field conditions must be checked to assure that the desired depth of application has been reached and that no excess water is being applied.

Information about soils and crops is fundamental to all planning for irrigation. Optimum MAD depends on the specific soil, crop, and depth of root zone, climate, and system of irrigation. The MAD should be established because it affects the depth, duration, and frequency of irrigation.

The available moisture, rate of infiltration, adaptability of method, and choice of crop are all related to soil texture; but depth of root zone, rate of intake, lateral wetting, perched water tables, and adaptability to land grading are mostly affected by soil profile and structure. The uniformity of soil in a field is important because it affects the uniformity of infiltration and therefore the choice of method of irrigation. Field surveys must thoroughly investigate soil uniformity. For all methods of irrigation in fields having more than one type of soil, the frequency and depth of irrigation should be governed by the soil that permits the lowest MAD.

Sprinkler or trickle irrigation is best for fields that have varied soils and topography because the depth of application of the water is independent of surface variations. For the areas where the rate of intake is slowest, the rate of application should be less than the basic rate of infiltration to prevent runoff.

Reasonable uniformity of soil surface is important to assure efficiency of furrow, border strip, or basin irrigation. It must be fully appreciated that the basic objective of land grading is to improve irrigation, not merely to produce a plane surface. The possibility of improving the uniformity of the soil within each field should not be overlooked during land grading. In basin and basin-check irrigation, uniformity of the intake rate is even more important than in-furrow and border-strip irrigations. However, uniformity of intake often can be improved by making boundaries of the basin conform to boundaries of areas having uniform soil texture. Low ridges can be formed over or temporarily removed as needed, and the shapes or sizes of basins may be varied as required [11].

To avoid confusion with certain similar but more general terms, three important terms used have been renamed. Irrigation System Efficiency is now called Potential Application Efficiency is now called Application Efficiency of the Low Quarter, and Distribution Efficiency has been changed to Distribution Uniformity.

3 Irrigation Methods

There are seven basic techniques or methods of irrigation, most of which have several variations. Each technique and variation has characteristics that are adaptable to different locations and crops [2, 6]. The basic component and operation for each of the seven techniques are:

3.1 Basin

A level area of any size or shape bounded by borders or ridges retains all the applied water until it infiltrates. Any loss of water results from either deep percolation or surface evaporation.

3.2 Basin Check

A fairly level area of any size or shape bounded by borders and with no depressions which cannot be readily drained. The borders (or ridges) retain all the applied water for a sufficient time to obtain a relatively uniform depth of infiltration over the area, and then, the remaining water is drained off the surface and used to irrigate an adjacent border check. Water is lost chiefly by deep percolation and evaporation.

3.3 Border Strip

A sloping area, usually rectangular, is bounded by borders or ridges that guide a moving sheet of water as it flows down the bordered strip. There should be little or no slope at right angles to the direction of flow. The on-flow of water is usually cut off when the advancing sheet has flowed six- to nine-tenths of the distance down the strip. Water is lost chiefly by runoff and deep percolation.

3.4 Furrow or Corrugation

A small sloping channel is scraped out of or pressed into the soil surface. For high uniformity of wetting, the irrigation stream should reach the end of the channel in about one-fourth of the time allotted for the irrigation; but the stream is not shut off until the root zone soil at the lower end of the furrow is adequately irrigated. Water in the soil moves both laterally and downward from the channel. Water is lost chiefly by deep percolation and runoff.

3.5 Sprinkler

Water discharged from a sprinkler should infiltrate the soil where it falls, but it should not wet the soil surface. For high wetting uniformity, the spray patterns from adjacent sprinklers must be properly overlapped. Evaporation, wind drift, and deep percolation are chief causes of loss of water.

3.6 Trickle (or Drip) Emitter

A device used in a trickle (or drip) irrigation for discharging water at some very low rate (less than 69 L per hour) through small holes in tubing placed near the soil surface. Water moves through the soil both sideways and downward away from the point of application to form a "bulb" of wet soil. Typically, only a portion of the soil mass is kept quite moist by very frequent or continuous application. Water loss is mainly by deep percolation.

Table 4 summarizes and compares the major physical characteristics that affect the adaptability of each of the six basic irrigation techniques. It also evaluates the probable Potential Application Efficiency of Low Quarter (PELQ) of a well designed and properly used system, employing each technique where appropriate. Most systems can be mechanized or even automated in order to reduce labor. This table leaves no allowance for such items as salinity and control of microclimate and takes no account of the costs or personal preferences of the irrigator.

4 Efficiency and Uniformity of Irrigation

The infiltrated water, evaporation from the plant and free water surfaces, wind drift, and runoff water must equal the total depth of applied (rain or irrigation) water. Furthermore, the sum of the transient and stored water, deep percolation, transpiration, and evaporation from the soil surface must equal the depth of infiltrated water. A growing crop may transpire transient water in the soil root zone before it is lost to deep percolation. However, some deep percolation is usually necessary to maintain a satisfactory salt balance since evaporation and transpiration (the only other ways to remove water from the root zone) leave the dissolved salts in the root zone. Transpiration and evaporation are interrelated and depend on atmospheric, plant, and soil moisture conditions.

Terms used to designate or rate the efficiency with which irrigation water is applied by a given system have been widely defined. To avoid confusion, the three primary terms that are used in field evaluation procedures (Distribution Uniformity, Application Efficiency of Low Quarter, and Potential Application Efficiency of Low Quarter) are defined below. These terms differ from those used in the first edition of this work and in some other publications; they should help avoid confusion with other terms and their definitions. The numerators and denominators of the definitions are expressed in equivalent depths of free water (volumes per unit area) for surface and most sprinkler-irrigated fields. However, water volume may be a more appropriate measure for trickle and sprinkler systems, which give only partial coverage.

High efficiency in the operation of an irrigation system is not necessarily economical, but a manager must evaluate the efficiency of any system in order to rationally decide whether he should merely modify his operation or adopt a different system. Efficiencies computed from ordinary field data are seldom more accurate than to

Table 4 Physical requirements and potential application efficiencies of the low quarter for the basic irrigation techniques [5]

Irrigation method	Physical requirements at site					PELQ
	Soil uniformity	Infiltration rate	Ground slope	Water supply	Labor intensity	
Basin	• Uniform within each basin	• Any	• Level, or graded to level	• Large intermittent	• High or infrequent intervals	• 60–85
Basin check	• Uniform within each basin	• All but extreme	• Fairly smooth with no depressions	• Large intermittent	• High or infrequent intervals	• 60–80[a]
Border strip	• Uniform within each strip	• All but extreme	• Mild and smooth	• Large intermittent	• High or infrequent intervals	• 70–85[a]
Furrow or corrugation	• Uniform along each furrow	• All but very rapid	• Mild or contour	• Medium to large intermittent	• High or infrequent intervals	• 70–75[a]
Sprinkler	• Soils may be intermixed	• All but very slow[b]	• Any farmable slope	• Small continuous	• High to very low daily[c]	• 65–85 • Depending on variance
Trickles (drip or subsurface)	• Soils may be intermixed	• Any	• Any farmable slope	• Small continuous	• Very low daily	• 75–90

[a]Values of 90% can be attained under ideal conditions if runoff water is reused
[b]Except for the center pivot and traveling sprinklers, which are best suited to use on soils that have medium and high infiltration rates?
[c]Labor inputs range from high intensity for hand move, moderate for the mechanical move, to low for automatic sprinkler irrigation systems

the nearest 5%. Therefore, variations of less than 5% in computed efficiency values are not significant except where identical data are being used for comparisons of alternative operational procedures [12, 13].

Distribution Uniformity (hereafter called DU) indicates the infiltration uniformity throughout the field.

$$\text{DU} = \frac{\text{depth infiltrated in the lowest one quarter of area} \times 100}{\text{average depth of water infiltrated}} \quad (7)$$

The average low quarter depth of water infiltrated is the lowest one-quarter of the measured or estimated values where each value represents an equal area. For sprinkler and trickle irrigation, the depth infiltrated is presumed equal to the depth applied or caught on the soil surface if there is no runoff.

The DU is a useful indicator of the magnitude of distribution problems. A low DU value indicates that losses due to deep percolation are excessive (and that the water table is likely to be too high) if adequate irrigation is applied to all areas. Although the concept of a low DU is relative, values less than 67% are generally considered as unacceptable. For example, if the desired depth of infiltrated water is 10 cm and the DU is 67%, the average depth infiltrated must be 15 cm and the deep percolation loss will be 5 cm. However, if deep percolation is limited by reducing the applied depth and the DU value is low, any area that receives the low quarter depth of irrigation will be seriously under-irrigated.

Application Efficiency of Low Quarter (hereafter called AELQ) achieved in the field indicates how well a system is being used.

$$\text{AELQ} = \frac{\text{average low quarter depth of water stored in the root zone} \times 100}{\text{average depth of water applied}} \quad (8)$$

When the average low quarter depth of irrigation water infiltrated exceeds the SMD, which is the storage capacity of the root zone, AELQ can be expressed as follows:

$$\text{AELQ} = \frac{\text{SMD}}{\text{Average depth of water applied}} \times 100 \quad (9)$$

The average low quarter depth of water infiltrated and stored in the root zone is the average of the lowest one-fourth of the measured or estimated values where each value represents an equal area of the field. Thus, about one-eighth of the irrigated area receives less than the average of the low quarter. "Irrigated area" means the area receiving water; for most systems, this is the entire field. However, where, a limited area is being wetted, the term refers only to that part of the area receiving water.

Implicit in AELQ is a measure of uniformity, but it does not indicate adequacy of the irrigation. It merely shows that, for any value greater than zero, all the area is receiving water. Low values for AELQ indicate problems in management and/or use of the system. Additional factors, which will be presented later, must be considered when any field is intentionally under-irrigated.

Potential Application Efficiency of Low Quarter (hereafter called PELQ) indicates a measure of system performance attainable under reasonably good management when the desired irrigation is being applied.

$$\text{PELQ} = \frac{\text{average low quarter depth infiltrated when equal to MAD} \times 100}{\text{average depth of water applied when MAD just satisfied}} \tag{10}$$

The PELQ is the precise value of AELQ when the low quarter depth of water infiltrated is just sufficient to satisfy the SMD when SMD = MAD in all parts of the field. Low PELQ usually is associated with inefficient system design, but may be intentional for economic reasons. The difference between PELQ and AELQ is a measure of management problems, whereas low values for AELQ merely indicate the possible existence of such problems.

Modifications of systems or methods can be compared meaningfully only by comparing values of PELQ. Such comparisons must be made when applying similar MAD depths. Economic comparisons should include costs of irrigation and crop production as well as expected returns.

$DU_{a,}$ AELA, and PELA may be used in place of DU, AELQ, and PELQ, respectively, to denote the use of absolute minimum depth instead of the average low quarter infiltrated. For convenience in the evaluation of surface irrigation systems, the depth of infiltration at the downstream end of the furrow (or borders) is often used in place of the average low quarter depth. This depth would be the absolute minimum depth infiltrated if the soil infiltration and furrow (or border) characteristics were uniform throughout the field. The absolute minimum should not be used for method comparisons [14, 15].

5 Essential Deficit Irrigation

Irrigation systems are usually managed to fill the SMD throughout the root zone at each irrigation; however, this should not always be the objective. Sometimes the interval between irrigations is extended to reduce the rate of water use below peak volumes by using a high MAD. This practice is used to aid other agricultural practices, to reduce requirements for system capacity, and/or to obtain maximum crop yields per unit of water or per unit of capital cost and is called stress irrigation. Another variation is to replace less than the SMD leaving the bottom portion of the root zone somewhat drier and is called limited irrigation. This type of intentional under-irrigation may be imposed rather uniformly throughout the field, or only in areas receiving minimum infiltration, or selectively. Intentional under-irrigation also enables better utilization of rainfall than full irrigation.

Limited irrigation is any of a group of procedures which result in under-irrigation to conserve water but do not reduce yields. If the root zone is full of moisture at the beginning of the period of peak water use, limited under-irrigation by not fully replacing SMD on the whole area can improve the efficiency of water use without reducing crop yields. However, yields can be maintained only if the period of peak use is relatively short and is followed by either a period of less use or by harvest. Moisture stored deep in the root zone from early or off-season irrigation, and rainwater is consumed during periods of under-irrigation. This plus of the irrigation water are available for crop production. This practice reduces losses from deep percolation if DU is high but allows a cumulative SMD to develop in the bottom portion of the root zone. Depletion of deep moisture augments the limited irrigation supply. Frequent checks of the SMD are essential for obtaining the maximum benefit from this practice and to avoid the danger of running out of deep moisture reserves and stressing a crop at a critical period, such as maize at tasseling. The area of land irrigated should not exceed what can be irrigated economically with the limited supply of irrigation water plus the available reserve of deep soil moisture [16, 17]. Another means for maximizing the efficiency of water use and reducing required system capacity without reducing yields are to irrigate only part of the area at any one time. This method is effective in orchard or vineyard irrigation by furrows, emitters, or orchard sprinklers because trees and vines have extensive root systems. The full soil profile throughout the area should be wet annually from rain or early season irrigation. During the period of deficient water supply, irrigation should be restricted to applying the SMD to a reduced area near each plant. This substantially reduces the loss of water by surface evaporation and thereby increases the percentage of irrigation water transpired by the crop. High AMD in the area wetted stresses the crop slowly as it draws moisture from the areas of the unirrigated areas and the lower root zone. Location of the area watered is relatively unimportant because root systems in a mature orchard of vineyard are extensive. This technique of limited irrigation utilizes the available supply of water very efficiently. Certain cultural practices such as harvesting and propping trees suggest modification in planning and managing irrigation; this may result in using limited irrigation. For example, depth of the pre-harvest irrigation can be reduced by spreading the limited amount of available water wider and shallower. This permits the large mass of roots near the surface to function normally and thus reduces crop stress and improves crop quality. Sometimes the area is reduced since furrows cannot be plowed close to trees because of low branches or props. Often sprinklers have to be placed only in the tree row to reduce foliar interception.

A common practice in young orchards under basin, furrow, sprinkler, or trickle irrigation is to irrigate only the area immediately adjacent to the trees until their root systems become extensive. Even in mature orchards, much of the surface area is left dry to improve trafficability. In fact, ability to do this is a prime advantage of trickle and furrow irrigation, which is never intended to wet the total soil area of an orchard. The planned reduction of the area to be wetted is compensated by more frequent irrigation in inverse proportion to the wetted area [5, 6]. For example, if only half area is to be wetted, it is wetted at twice the normal frequency; this is a prime example of limited irrigation. However, great caution should be exercised if one plans to design a system to irrigate less than one-third of the volume of potential root soil. An excellent variation of limited irrigation is the use of alternate side irrigation. In this practice, all or part of the area on one side of the plant is wetted at a time, i.e., the full SMD is replaced on half the field. At the next irrigation, the SMD is replaced on the other side of the plant. At each irrigation, only half the usual application is applied but at half the usual frequency.

Stress irrigation applies to any of a number of practices which result in under-irrigation to conserve water at the expense of some reduction in potential yields. Irrigation procedures that are likely to stress a crop can be combined with alternate side irrigation to reduce the maximum stress.

Maximizing crop production from a limited amount of water is important either when the water supply is inadequate or when the value of water is measured by crop production per unit of water. In such areas, operating at a high MAD extends the interval between irrigations. This practice of stress irrigation may reduce yields per unit area but may produce total crop per unit of water on an enlarged area and thereby produce a greater net return.

Except for some of the special variations mentioned below, intentional under-irrigation puts a premium on having high values of DU and AELQ to reduce losses of water and results in a higher percentage of the irrigation water being transpired by the crop.

Reducing system capacities as discussed above, and/or accepting a lower DU enables the reduction of capital investment. When a system that achieves only low DU is used, the SMD may not be fully replaced in portions of the field even when the water supply is adequate. In such areas, management simply plans to accept a reduced yield from the dry portions of the field. Such systems require careful management, logical design, checks of SMD, and periodic evaluations of the success of the operation.

The above design logic anticipates moderate to low values of DU and AELQ as a trade-off for reducing the costs of system development. Wide spacing of sprinklers and operation at low pressures may reduce costs, but they may also cause deficiencies of soil moisture to cumulate in the drier spots. The dry spots may produce fewer crops, but profits may be increased because of the reduced cost of capital more than offset the crop losses. To eliminate the dry spots, abnormally large quantities of water must be applied which may be uneconomical or cause drainage problems.

For furrows and border strips, reduced land grading or use of longer-than-normal lengths of run is possible means for decreasing costs for capital and labor. However, these practices should be used only where resultant reductions in cost substantially exceed the losses resulting from reduced production at the under the irrigated end of the furrow or strip. Furthermore, salt accumulated in dry areas, which are not leached by occasional rainfall, may become a hazard.

Before using any of these forms of stress irrigation, a manager should determine that the resulting savings in capital, labor, water, and management will more than offset the value of the estimated decrease in crop yield per unit area.

6 High-Frequency Irrigation

Movable and permanent solid set (or full coverage) sprinklers, center pivot, and trickle (or drip) systems are normally managed to apply light frequent irrigations. High-frequency irrigation is used to achieve any or all of three major objectives: (1) to maintain a continuous low-stress high level of soil moisture to produce high yields or better quality of crops; (2) to avoid the runoff that often accompanies high rates of application; and (3) to control temperature, humidity, and/or wind erosion. Under some conditions, high-frequency irrigation may be conducive to diseases or excessive vegetative growth.

Under high-frequency irrigation, depth of each application is usually less than 3 cm unless an area is being intentionally under-irrigated; the SMD would also be less than 3 cm. It is practically impossible to estimate the SMD precisely enough for it to be useful in determining whether the soil is dry enough to require irrigation when the MAD is so low.

Estimates of the rate of a crop's use of water give a reasonable basis for scheduling high-frequency irrigation. A crop's use of water can be estimated from weather data, taken from measurements from evaporation pans, or can be based on experience. Except where under-irrigation is intended, ideal system management would exactly replace the water consumed in the areas that receive the minimum application.

It is impractical to attempt to estimate exactly the volume of water consumed between irrigation. Since over-irrigation is difficult to measure, it is good management to under-irrigate slightly when using systems other than trickle irrigation. The SMD can be checked periodically to spot areas where deficits of soil moisture have been cumulative. For such areas, scheduling of irrigation can be corrected accordingly. This practice of under-irrigation should not be risked if only a small portion of the root mass is irrigated as in trickle irrigation.

High-frequency irrigation is particularly well suited for use in conjunction with limited irrigation where the deep soil moisture is being gradually depleted over a whole area, as sometimes happens under center pivot and other automatic sprinkler irrigation systems. Light frequent watering of the topsoil plus the gradual withdrawal of moisture from the subsoil can produce optimum crop yield when the irrigation system capacity is limited. However, where subsoil moisture is inadequate, light frequent irrigation, causing heavy moisture losses from evaporation, may be an inefficient use of a limited supply of water and also increase salinity. Therefore, less frequent deeper irrigations may produce better crops [18].

While using supplemental irrigation in areas that receive high rainfall, it is good practice to apply shallow irrigation frequently while maintaining an SMD between 3 cm and 6 cm in the lower part of the root zone. Thus, the soil always has some storage capacity for rain but also has plenty of water for the crop.

7 Uniformity, Efficiency, and Economics

The efficiency of any operation, including irrigation, is a measure of how well its performance compares with some ideal level of performance. The following evaluation procedures usually imply that full irrigation with high DU and AELQ is the desired ideal. The concept of full irrigations in the areas receiving the average low quarter depth of application is useful for standardizing evaluation procedures in the field. However, this concept may provide a poor basis for evaluating and managing a system to optimize profit or any other value such as production per unit of land, production from a given quantity of water, or production per unit of energy input [12, 13].

Intentional under-irrigation of areas that are receiving the average low quarter depth of application may provide the optimum profitability. Rather than replenishing the water in almost all of the area, as is implied by PELQ, it may be more economical to leave a substantial area under-watered. This would be especially true for deep-rooted crops, low-value crops, and for crops growing in humid regions.

8 Conclusions

Basic concepts and terms of indicators of the technical and agronomic performance of smallholder irrigation schemes that used is very significant for monitoring and evaluating. Essential deficit irrigation between limited and stress irrigation and high-frequency irrigation depends on soil moisture depletion (SMD), management allowed depletion (MAD), irrigation methods, efficiency, and uniformity of irrigation.

9 Recommendations

For an irrigation scheme evaluation, the values of performance indicators should be applied in order to evaluate the performance level concerning the irrigation scheme. For their calculation, performance indicators call upon a certain number of parameters that have to be measured in the areas of irrigation schemes for the progressive elaboration.

A detailed study is needed to optimize profit which would be beyond the scope of the following evaluation procedures described here. In addition to evaluation of system performance in the field, which indicates the location and magnitude of water losses, such a study would require a thorough knowledge of system costs, plus the relation between water and crop production in the area studied.

References

1. Casley DJ, Kumar K (1990) Project monitoring and evaluation in agriculture. London
2. FAO (2002) Monitoring the technical and financial performance of an irrigation scheme. Irrigation Manual 14, Harare, Zimbabwe
3. IIMI (1996) Méthodologie d'évaluation des performances et de diagnostic des systèmes irrigués. Projet Management de l'irrigation au Burkina Faso, Novembre, p 1996
4. Sally H (1995) Performance assessment of rice irrigation in the Sahel: major indicators and preliminary results from Burkina Faso and Niger. Paper presented at Workshop on irrigated rice in the Sahel: prospects for sustainable development, 27–31 Mar 1995
5. Merriam JL (1978) Farm irrigation system evaluation: a guide for management
6. Merriam JL (1966) A management control concept for determining the economical depth and frequency of irrigation. Trans ASAE 9(4):492–498
7. Fao SAFR (2000) Socio-economic impact of smallholder irrigation in Zimbabwe: case studies of ten irrigation schemes. Harare, Zimbabwe
8. Abd El-Hafez SA, El-Bably AZ (2006) Irrigation improvement project (IIP), on-farm water management (OFWM), Kafr El-Sheikh and El-Behira Directorates, Final report
9. FAO (1992) CROPWAT: A computer programme for irrigation planning and management. FAO Irrigation and Drainage Paper No. 46. Prepared by: Smith M Rome, Italy
10. SCS National Engineering Handbook (1976) Planning Farm Irrigation Systems, Chapter 3, Section 15, USDA, Washington, DC, July 1967
11. Smerdon ET, Glass LI (1965) Surface irrigation water distribution efficiency related to soil infiltration. Trans ASAE 8(1):1965
12. Griddle WD, Sterling D, Claude HP, Dell GS (1956) Methods for evaluating irrigation systems. Agricultural handbook No. 82, SCS, USDA, Washington
13. Merriam JL (1968) Irrigation system evaluation and improvement. Blake Printery, San Luis Obispo, California
14. Willardson LS, Bishop AA (1976) Analysis of surface irrigation application efficiency. J Irrig Drain Div ASCE 93(No. IR2), June 1967
15. Beaudoux E, de Crombrugghe G, Douxchamps F, Gueneau M, Nieuwkerk M (1992) Supporting development action from identification to evaluation. Macmillan Education Ltd
16. FAO (1989) Guidelines for designing and evaluating surface irrigation systems. FAO Irrigation and Drainage Paper No 45. Developed by: Walker, W.R. Rome, Italy
17. Casley DJ, Lury DA (1981) A handbook on monitoring and evaluation of agriculture and rural development projects. World Bank
18. United Nations ACC Task Force on Rural Development Panel on Monitoring and Evaluation (1984) Guiding principles for the design and use of monitoring and evaluation in rural development projects and programmes. Rome

Evolution of Irrigation System, Tools and Technologies

Elmesery

1 Introduction

Irrigation has played a major role in the development of ancient civilization. Despite the immense advance of civilization in the last few thousand years, irrigation practice has remained virtually at a standstill. It is astonishing that many current irrigation practices are almost identical to those used in ancient times. Egypt has large arid and semi-arid areas; ancient Egypt has some knowledge of irrigation practices. Today, Egypt has high techniques in modern irrigation methods.

The main objective of this chapter is to give vision around history of growth irrigation in Egypt, water lift techniques in ancient Egypt and some knowledge of a Nile River and its constructions.

2 Stages of Development

The use of Nile water for irrigation purposes is old in Egypt, where life has developed. Irrigation arts themselves have developed with the passage of ages and the wide circle of science and knowledge. They have used scientific methods to control, tame and restrain the Nile River and exploit its maximum resources to the limit [1].

The history of irrigation in Egypt can be divided into several stages. These stages are as follows [2]:

1. Pre-history.
2. Post-history in its extension until the early nineteenth century.

Elmesery (✉)
Water and Irrigation Systems, Faculty of Agricultural Engineering, Al-Azhar University, Nasr City, Cairo, Egypt
e-mail: alaa.elmesery@azhar.edu.eg

E.-S. E. Omran and A. M. Negm (eds.), *Technological and Modern Irrigation Environment in Egypt*, Springer Water,
https://doi.org/10.1007/978-3-030-30375-4_4

3. Nineteenth-century stage.
4. The first half of the twentieth century.
5. The continuous storage of Nile water.

2.1 Pre-history

We can imagine the land of Egypt in the eras of history. It was in the form of desert barren, surrounded by a Nile River toward the outlet in the sea (Delta land). The Nile River was carry with flooding huge amounts of sand and silt, sinking to Upper Egypt until the desert hills on both sides. The flooding dumped the remaining sand and silt at its outlet in the sea, a component of Delta with the centuries. It is likely that the first population of Egypt had settled in Upper Egypt before the Delta, because Upper Egypt is characterized by the adjacent hills and the population were fleeing from the edge of the desert during the flood and then descended into the plains after its became dry and plants were appeared.

In the Delta regions, the Nile River was ravaged by a swarm of water, and it had many branches, wide wagons and lagoon (lakes). The Nile River swept most of the Delta lands in flood every year, leaving behind vast stretches of marshes, which are not as populated as it was in Upper Egypt.

The Egyptians knew how to cultivate the land. They threw their seeds into the wet soil after exposure, harvested the crops after their maturity and then ran to the edge of the desert before being overwhelmed by the flood.

This system of origin and instinct was the origin of the idea that was later inspired by the pharaohs when they laid the rules of basin irrigated irrigation on a sound engineering basis [2].

2.2 Post-history Stage

The kings of the pharaohs learned that the agricultural lands that extend in Upper Egypt descend slightly from the south to the north. Another slope descends between the banks of the Nile River and the desert on both sides. King Mina established the first of its kings a bridge on the left bank of the Nile River along its length, to prevent water from tyranny on this bank, and leave the right bank without modification.

Mina then proceeded to divide the canals through the high beaches to deliver the floodwaters to the lands far from the Nile River's bed. He also created vertical bridges on the Nile River's shore that extended up to the Libyan hills to hold the water in the basin and prevented it from flowing north. The ponds are natural and industrial and have enough time to sink the silt they carry.

When the kings of the Twelfth Dynasty ascended the throne of Egypt, they completed the right bridge, but they alerted to the imprisonment of the Nile River between two high bridges might be destroy by the water of river during high floods. So that, they connected the Nile course to Faiyum depression, which was known as "Morris Lake" to descend the excess flood water in the Faiyum depression then return again to the river when the level water of the river was drop. This was the first human knowledge of storage projects.

The steps were followed by the consolidation of the basins irrigation that prevailed throughout the country until the early nineteenth century. The land was divided into a basin bordered by the Nile River on one side and the desert on the other. Also, the vertical bridges were built on the Nile to separate the basin from the other, and separated the high lands from the low bridges which are parallel to the Nile River. Then, the basins had divided into chains and the various canals which had different lengths, widths and levels and erected bridges and estuaries [1, 2].

2.3 Nineteenth-Century Stage

The nineteenth century was a major change in the history of Egyptian irrigation, which can be attributed to two main factors:

- Start the application of the permanent irrigation system;
- discovering the sources, determining the features of its basin and starting a hydrological study of the Nile River.

The system of permanent irrigation was the result of the desire to cultivate cotton, cane, fruit and some other summer crops, which are grown in the season of need.

The beginning of this system has not been easy without devising a method whereby these crops can be irrigated, especially during the months of the fall of the Nile River (February–July), and the decline of the species is significantly lower than that of cultivated land.

The application of this system began first with expensive means, which combined the two systems in the Delta provinces and based on the completion of the Nile River bridges in the Delta. The basis of the construction of multiple bridges on these canals is to raise the water in front of them to the extent that the expenses of raising the pellets to the cultivated land were low. The Delta lands under this system were considered to grow grain and alfalfa after the drainage of water that had been flooded during the flood [2].

In August, the canals were cut down to irrigate the lower parts of the soil. The highlands continued to be irrigated, and the corn was grown. Seventy days later, all the canals were completely filled with water on the drying land.

The depth of the excavation was no less than five meters in this period, and these strenuous processes were followed by unpaid workers called "Anfar al-Awna". The method taken in the early 1900s to circumvent the permanent irrigation system by combining it with the basin irrigation system was complex and overbearing.

Finally, thinking of building a large bridge (barrage) at the head of the Delta helps to raise the level of the Nile River as it descends, so that its water flows into the main wind canals without having to be deepened, and the water flows high in the branch canal fed by these main canals [2].

It was started in 1843 and named after Al-Qanatir Al-Khairiya Society (Fig. 1). It was accompanied by the construction of three main canal branches (Monofy, Tawfiqy and Behery) that take water in front of it.

Fig. 1 Al-Qanatir Al-Khairiya in Qalyubia Governorate, Egypt. It is the location of the Delta barrages, the first modern irrigation structure across the Nile, located at the apex of the Nile Delta (https://en.wikipedia.org/wiki/El_Qanater_El_Khayreya)

Al-Qanatir Al-Khairiya was built in 1861, but its weak structure did not enable the officials to raise the water level to the required degree until it was strengthened at the end of the last century.

In 1873 the excavation of the Ibrahimiya canal, which takes water from the Nile River at Assiut and branched at Dirout after 61 km, was started in four branches: the Sea of Yusef, Dirutia, Ibrahimeya and Sahli.

The permanent irrigation system in Upper Egypt and in Middle Egypt was firmly established at the beginning of the twentieth century when it was supported by the major irrigation projects that have been implemented since then.

The past century has witnessed a great deal of effort by the pioneers and explorers until the features of the Nile River and its upper tributaries have become clear. The study of the Nile River and its basin has begun the study of the water sources, the monitoring of its movements and changes and the thinking of the projects of exploiting the Nile River's water [1].

2.4 *The First Half of the Twentieth-Century Stage*

At the end of the nineteenth century, when the need to expand the cultivation of the cotton crop by increasing the area under the permanent irrigation system, two major problems were risen as:

- The scarcity of water in the season of need (February–July) and the inability to meet the needs of summer crops.
- Lower levels of the Nile River in this season than the level of the cultivated land.

The scarcity of water in the Nile River's fall season has been overcome by storing part of the floodwater each year in low-capacity reservoirs and gradually releasing stored water in the season of the need to boost the Nile River's revenue to help meet the agricultural demands of summer crops.

The Aswan Dam was built in 1902 and its height was raised twice, the first in 1912 and the second in 1933. Jabal al-Awliya Dam was structured on the White Nile in 1937, bringing the total amount of water stored annually to about 7.5 billion cubic meters or about half the average revenue of the Nile River in the season of need.

It was also possible to overcome the problem of low levels of Nile in the season of the need to proceed in the structure of a group of major bridges on the Nile River at Esna and Nag Hammadi, Assiut, Zifta and Edvina, as well as the structure of the Delta bridge, which replaced Al-Qanatir Al-Khairiya.

In addition to these major projects, this period of Egyptian irrigation history is characterized by scientific research and extensive studies in irrigation affairs, especially concerning water measurement, fair distribution, river hydrology and the full utilization of its water. Egyptian irrigation has a good place and reputation around the world [2].

2.5 Stage of Continuous Storage of Nile Water

When the revolution in July 1952 and the government reviewed the water situation to increase the agricultural land for the development of national production, it found that the storage of the Aswan and Jabal al-Awliya tanks had been exhausted by the previous projects. Therefore, efforts were directed to study the High Dam project to store the floodwater, once in 1946. Once the validity of the project has been established, it has embarked on its implementation to achieve its enormous advantages.

However, the government did not want to stand idly by until the High Dam reaches its desired fruit, prompting the implementation of a range of urgent projects with the aim of raising some additional water resources to push the reclamation.

At the head of these projects was the use of drainage water for irrigation purposes after mixing it with the water of the canals and the depletion of underground water in the Egyptian soil inside the valley and in the desert.

As well as the development of a new policy for the implementation of public drainage projects and field drainage covered to achieve the expansion of agriculture vertically increases the yield of cultivated land and maintains the fertility of the land.

The implementation of the High Dam project is considered as the first step in the continuous storage of Nile water. Similar projects are being studied for continuous storage of the Nile River's water in the tropical lakes and Lake Tana [2].

2.6 Basins Irrigation System in Egypt

The basins system of irrigation was the prevailing system in Egypt until the beginning of the nineteenth century. Gradually, this system was reduced to the transformation of the system of permanent irrigation. The effects of the High Dam project have ended on the final regime that has existed for centuries since the dawn of history [3].

The cultivated land in Egypt until the early nineteenth century was entirely dependent on the Nile River's income in the flood season. If the flood is high and its water is high, it controls the Nile River in the ponds and submerges it, and then, the farmers start to sow their seeds in the land. After a few months, they harvested their winter crops. Under this system, the land used to produce only one crop every year except the basins whose owners supplied them with artesian wells to irrigate summer crops during the river's low water season [2, 3].

If the income is low, the country will be affected by disasters and drought, although this situation was dealt with in the twentieth century by erecting the Nile River bridges to raise the flood levels to the extent that helps feed the basins. In 1933, the Ministry of Works (currently the Ministry of Irrigation) developed a water policy that sets agricultural expansion programs within 20 years, not later than the water stored in the Aswan reservoir and the Jabal al-Awliya reservoir. This water policy included the conversion of about 524,500 feddan in Upper Egypt to permanent irrigation. However, the country was surprised by two dangerous floods in 1934 and 1938, which reminded the need to turn away from the idea of conversion, considering the pelvic ground security valve for high flooding, reducing the climax of the pressure on the Nile River bridges in central Egypt.

It was possible to reconcile the high flood fears and the demands of agricultural development with the introduction of a new system of double irrigation by supplying the pelagic waters with the storage water to irrigate the summer crops and then prepare them to receive the flooding when the flood comes. This system was applied in an area of 114,200 feddan (1 feddan = 4200 m^2) [2, 3].

When the Second World War began in 1939, the need for grain was increased, and a similar system called cereal irrigation system was applied in an area of 257,500 feddan leaving about 590,800 feddan in Upper Egypt, except for the 11,000 feddan.

Since it is expected that the reservoir of the High Dam will only be able to meet the actual needs of the crops, the flood phenomenon on which the irrigation system is dependent will disappear, and therefore, all the migration should be transformed into a permanent irrigation system, which was implemented since 1959/1960 [2].

3 Ancient Egyptian Irrigation Tools

3.1 Shadoof

A *shadoof* or spelled *shaduf* is an irrigation tool. It is also called a counterpoise lift, well pole and well sweep. It is a hand-operated device for lifting water invented in ancient times and still used in Egypt, India and some other [3]. The *shadoof* was an early tool used in irrigation by the ancient Egyptians who lived around the Nile River. The *shadoof* was used to lift water from a river or lake onto land or into another river or lake. It is a long arm or pole with a bucket attached to the end of it [4–6].

3.1.1 Construction

The *shadoof* consists of a long, horizontal pole mounted like a seesaw. A bucket is hung on a rope from the long end, and a counterweight (a large rock) is hung on the short end [3] (Fig. 2).

The *shadoof* consists of a long, horizontal pole mounted like a seesaw. A bucket is hung on a rope from the long end, and a counterweight (a large rock) is hung on the short end. The operator pulls down the rope attached to the long end to fill the bucket with water and allows the counterweight to raise the bucket. To raise water to higher levels, a series of shadoofs are sometimes mounted one above the other [3]. At the end of each movement, the water is emptied out into runnels which convey the water along irrigation ditches in the required direction.

Fig. 2 Egyptian *Shadoof*, an early tool used in irrigation by the ancient Egyptians who lived along the Nile River. http://www.waterhistory.org/histories/nile/shaduf.jpg

3.2 *Tanbour*

The Archimede's screw also called the Archimedean screw or screw pump is "a machine historically used for transferring water from a low-lying body of water into irrigation ditches. Water is pumped by turning a screw-shaped surface inside a pipe" (Fig. 3) [7], http://windermeresun.com/2017/02/22/archimedes-screw-at-disney-springs-fl/.

3.2.1 Construction

It consists of a piece of wood in the form of a screw surrounded by a nestled disk. The lower part of the *tanbour* is placed in the water and rotated, causing the water to rise to higher levels, and today farmers still use it in times of low water levels [3].

The introduction of the *Saqia* was apparently during Ptolemaic times in the second century BCE, but the actual origin of *Saqia* is debated, with many saying that it was a Persian invention. This device was originally built from wood. The device consists of a large cogged wheel placed horizontally on a vertical shaft to which one or two of the animals are yoked. A series of clay jars (*qadus*) is attached to a vertical wooden wheel. Such jars were found extensively in the mains of Kulubnarti. The oxen then rotate the vertical beam, sending each of the clay jars into the water to be filled. The water is then emptied into a trough that carries the water by furrows into the cultivated field [8].

The functions of water wheels were water lifting for irrigation purposes and as a power source. When used for water lifting, power can be supplied by human or animal force or by the water current itself.

3.3 *Water Wheel (*Saqia*)*

3.3.1 Hydraulic Noria (*Naura*/*Saqia*)

A hydraulic noria (*Naura*) is a machine activated by water power and used for lifting water into a small aqueduct, either for irrigation or for the use in towns and villages.

According to John Peter Oleson, both the compartmented wheel and the hydraulic noria (Fig. 4) appeared in ancient Egypt by the fourth century BCE, with the *Saqia* being invented there a century later.

Also, there were other types of water wheels in Egypt. One of these types is called a *Nubian wheel*. It is a vertical wheel bearing a number of pottery vessels. When the wheel spins in the water, the pottery vessels are filled with water and then poured into the basin to be irrigated (Fig. 5). This wheel is connected to another wheel parallel to it and has wooden gears clipped in the wheels of another horizontal wheel. Connect

Fig. 3 Archimedean screw (screw pump) in ancient Egypt. It is a machine used for transferring water from a low-lying body of water into irrigation ditches. http://hesed.info/blog/simple-water-pump-diagram.abp

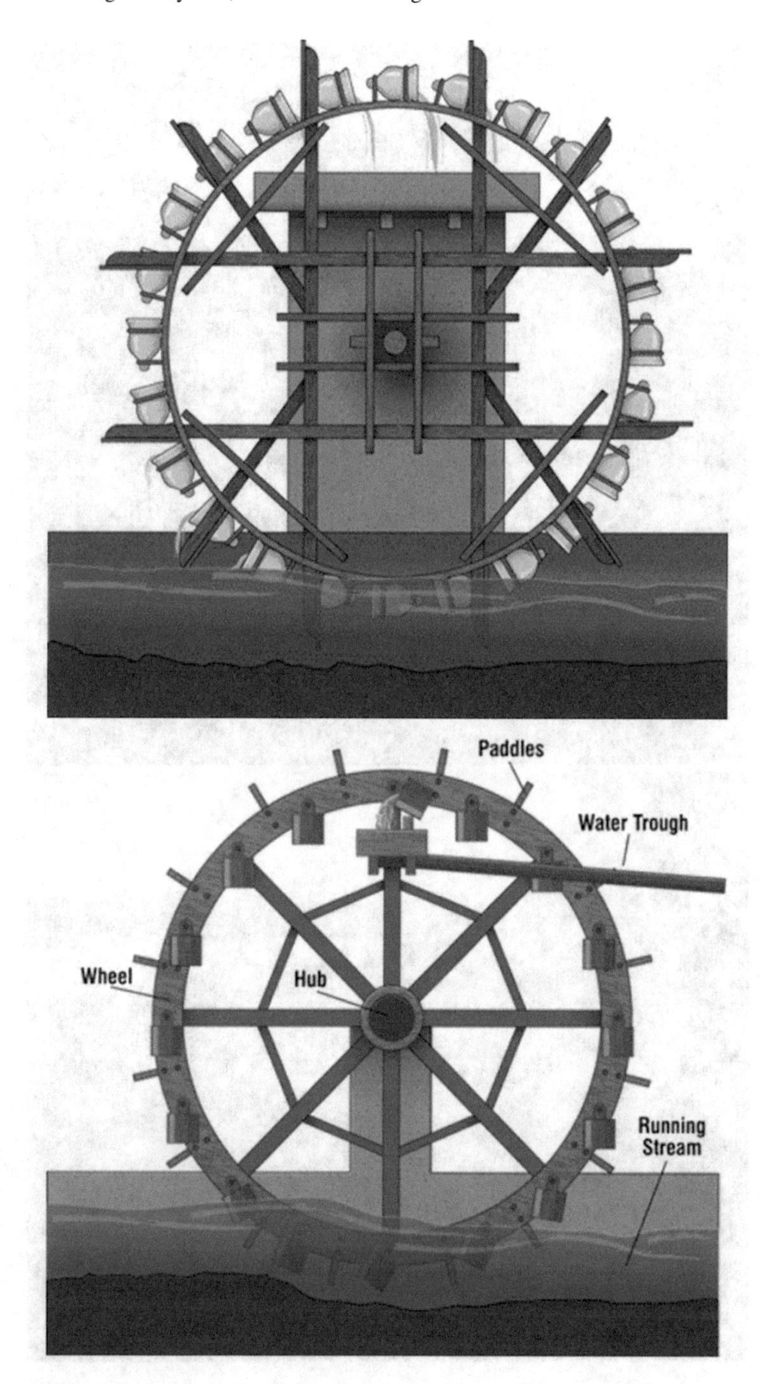

Fig. 4 Hydraulic noria (*Naura*) in ancient Egypt. http://asterion.almadark.com/wp-content/uploads/2008/12/noria2.thumbnail.gif

Fig. 5 Nubian wheel in ancient Egypt

the cow or buffalo or the camel in a connected arm, and when the animal revolves around the last axis of the wheel, the stem rotates accordingly.

In Faiyum, where water is increasing in some locations, drivers are driving with the water push, called the drivers of the roar. Some drivers are still in the heart of Faiyum (Fig. 6).

4 Irrigation in Egypt Today

In Egypt, food production is fully dependent on High Dam in Aswan that stores water for irrigation of the Nile valley and some of the adjoining desert and guarantees food supply to the population [9]. After the construction of the High Dam in Aswan, irrigation systems in Egypt have developed.

In Egypt, based on the method of applying water to the land, there are four categories of irrigation systems: (1) surface irrigation systems, (2) sprinkler irrigation systems, (3) localized (or trickle) irrigation systems and (4) sub-surface irrigation systems.

The sprinkler systems back to date 1800s in USA, major advances in sprinkler irrigation systems came with the introduction of wheel lines and center pivot. In 1960s, the semi-portable sprinkler irrigation systems were expanded in Egypt as a large project in reclamation lands. Also, set-move and hand-move were expanded. Today, the center pivot expansion on large-scale is at special form.

To improve the use of water in agriculture, farmers and irrigation professionals thought of two main points: (1) the water application directly to the root zone; (2) reducing evaporation from the soil surface by using the sub-surface water application as clay gar (or clay pot). These technologies were used by the ancient Egypt farmers [10]. In 1975, the first experiments on modern trickle irrigation by El-Awady were made, and after this time, the trickle irrigation with its different types was applied and expanded in reclamation lands [11].

Fig. 6 Al-Hader wheels (Sawaqi Al-Hader) in Faiyum Governorate, Egypt

References

1. Badr AA (1990) Irrigation engineering and drainage. Cairo University, Giza, Egypt (In Arabic)
2. Al-Sherbini HA, Yaseen AM (1970) Irrigation and drainage engineering, 1st edn. Dar El-Maaref, Cairo, Egypt (In Arabic)
3. Noaman MN, El Quosy D (2017) Hydrology of the Nile and ancient agriculture. In: Irrigated agriculture in Egypt. Springer, Cham, pp 9–28
4. Potts DT (2012) A companion to the archaeology of the ancient Near East. Wiley, Hoboken, NJ
5. Knight EH (1877) Knight's American mechanical dictionary, vol 3. Hurd and Houghton, Sutton
6. Roberts JM, Westad OA (2013) The penguin history of the world. 6th edn. Penguin, UK. https://books.google.com.eg/
7. Wikander O (2000) Handbook of ancient water technology. Brill, Leiden, 741p. ISBN: 414176647
8. Oleson JP (2008) The Oxford handbook of engineering and technology in the classical world. Oxford University Press, Oxford
9. Sne M (2011) Sprinkler irrigation technology and application. CINADCO: 1–3. https://www.scribd.com/doc/20157961/Micro-Irrigation-Technology-and-Applications
10. Sne M (2005) Drip irrigation. 2nd edn. Shirley Oren, Pub Coord, CINADCO:1–39. https://www.scribd.com/document/19942977/Drip-Irrigation-05
11. Awady MN, Amerhom GW, Zaki MS (1975) Trickle irrigation trial on pea in conditions typical of Qalubia. Ann Agric Sci Moshtohor 4:235–244

Smart Irrigation Technology

Smart Sensing System for Precision Agriculture

El-Sayed E. Omran and Abdelazim M. Negm

1 Introduction

The strategic question for sustainable land management is where and when a specific activity can carry out without degrading our natural resources and fulfilling the food requirements within social boundary conditions [1]. Note the spatiotemporal aspect of these questions where a specific crop can grow and when we can do so. With these basic questions, sustainable agricultural management at different (small, medium, large) scales comes in. Each scale used needs different data from several sources. Separate processes act on diverse hierarchical levels in which each level has its data needs. A main obstacle to describe these processes and to apply dynamic models in different application is the requirement for high-quality, high-frequency data. Much of the information has a spatial component and is held as digital spatial data.

In 1998, the digital Earth (individual computing) concept which converted the real Earth into a virtual one on the Internet was proposed [2]. Digital Earth, which incorporates geographical information system (GIS) and virtual reality innovation with spatial data, made it conceivable to study the Earth and the environment online and enables to reach wherever you want. However, geospatial information service data are developing quickly with increasing data acquisition volumes, handling, and updating. A key challenge facing the geospatial community specifically has been dealing with the large extents of spatial data being managed and accessed

E.-S. E. Omran (✉)
Soil and Water Department, Faculty of Agriculture, Suez Canal University, Ismailia 41522, Egypt
e-mail: ee.omran@gmail.com

Institute of African Research and Studies and Nile Basin Countries, Aswan University, Aswan, Egypt

A. M. Negm
Water and Water Structures Engineering Department, Faculty of Engineering, Zagazig University, Zagazig 44519, Egypt
e-mail: amnegm85@yahoo.com; amnegm@zu.edu.eg

E.-S. E. Omran and A. M. Negm (eds.), *Technological and Modern Irrigation Environment in Egypt*, Springer Water,
https://doi.org/10.1007/978-3-030-30375-4_5

particularly remote sensing (RS) data. In 1999, the "Internet of Things" (IoT) concept was introduced [3]. IoT is the system in which present reality items are associated with an embedded system including electronics and sensors through which the data can be transferred reliably.

In 2005, Omran [1] proposed spatial data infrastructure to support land evaluation applications (SDILEA) and to enhance data sharing as a facility to access and search spatial data through Geo-portal and clearinghouse. This facility includes services to help find out the data. In 2009, the accessibility of high-capacity networks, low-cost computers and storage devices as well as the extensive adoption of service-oriented architecture leads to the "smart Earth" (cloud computing) concept [4]. Cloud computing offers sharing of resources and offers Infrastructure as a Service (IaaS), Software as a Service (SaaS), and Platform as a Service (PaaS) with cheap cost. Data storage has gone from printed copy, file copy, network share to service-oriented cloud computing. Cloud computing has been utilized for agricultural data storage [5, 6]. Computing and storage devices have also improved from a personal computer, laptop to smart devices. Smartphones became common in the late 2000s, which produced from 2012 onward have high-speed broadband, numerous built-in sensors (e.g., positioning, motion, and cameras), which improves the smartphones' capability to assist users in achieving various tasks.

Many workers have adopted these new technologies (e.g., smartphones) to aid their work [7, 8]. The fascinating application in agriculture fields having an increasing need for a decision support system is precision agriculture (PA). The out-of-date technique of agricultural system is not appropriate, because farming productivities depend mainly on the environmental conditions (e.g., weather and water), global warming issues (droughts and floods), and plant disease outbreaks, which are disruptive of farming productivity. The smartphones application for agricultural sensing offers significant benefits to the PA. Agricultural IoT can be viewed as a network of sensors, cameras, and devices, which will work toward a common goal of helping a farmer, who does his job in an intelligent manner. Through sensor networks, agriculture can be connected to the IoT, which permits us to make connections among agronomists, farmers, and crops regardless of their geographical differences. Spatial data integrated with IoT and cloud computing will bring our agriculture world online [9]. The main goal was to make fully utilized new technologies in PA fields, which can diminish the farmer problems. The need to provide the decision-makers with fast, reliable, and up-to-date information, which will help farmers make the right decisions has become a requirement for PA users and decision making. Therefore, how emerging technology can provide real-time data to support the PA process is the most imperative issue that needs investigation.

Therefore, the goal of the current study is to review the potential of smart sensors application for sustainable agriculture management within a specific system for PA. To be specific, it is of essential to understand: (1) what is the status of smart sensors, cloud computing, and IoT development to support PA and (2) propose a system architecture and key technical elements of new technology for sustainable agricultural management.

2 Precision Agriculture from Remote Sensing to Proximal Sensing

After years of diminishing investment and intellectual efforts, soil science (e.g., pedology) which is the core of PA is thriving again [10]. The observational and interpretative techniques of soil profile description have not changed much in the past decades despite the technological revolution that is taken place in many subdisciplines. Although soils can be studied without digging a pit or taking a soil core, the soil profile is at the heart of many soil studies. The soil profile can be dug by workers or by a tractor, which can be expensive (Fig. 1). When we go to traditional study, pedologists would look at the soil profile and utilize their senses (seeing, feeling, smelling, hearing, and tasting), which represented basic activities in the early phases of soil survey, and they still are highly relevant to infer properties about the soil that is being observed. Experienced soil researchers can draw on their experience to mentally visualize soil profiles based on written descriptions, which require considerable practice and experience. Different soil researchers will write similar, but not necessarily identical, descriptions of the same soil profile. When going to RS, there will be limitations because of atmospheric conditions, spatial and spectral resolution, and field conditions. In addition, the reflectance data come only from the soil surface. Moreover, spectral bands are fixed, which may be unsuitable for a given application. Some extra limitations, which make RS not suitable for PA are too coarse spatial resolutions for within-field analysis, inadequate repeat coverage for intensive agricultural precision, and long periods between image acquisition and distribution to users. Aircraft-based sensors avoid these limitations but are sometimes difficult to calibrate and hard to register to map coordinates for large scale [11]. The point is as follows: Are these approaches sufficient to determine the profile characteristics

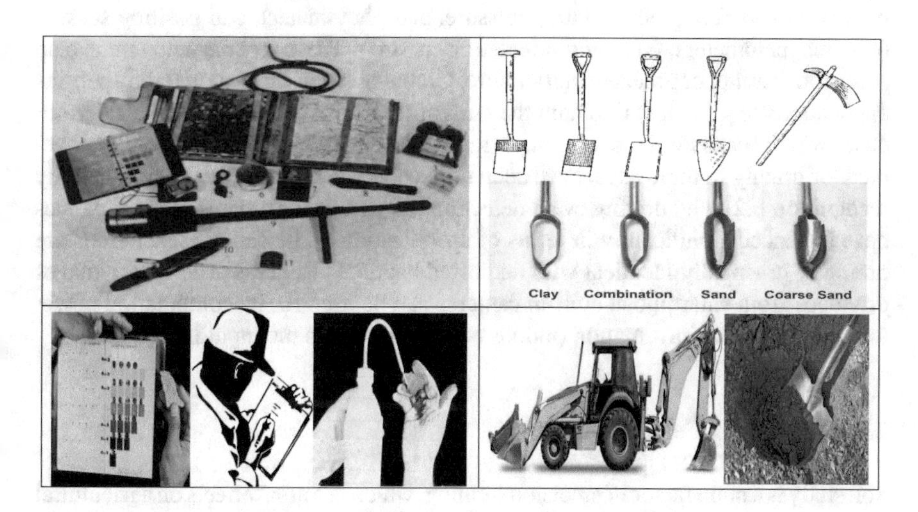

Fig. 1 Traditional tools to open and describe soil profiles (labor, machine, auger, and others)

accurately? Can we obtain new information more powerfully and accurately using sensing tool technologies? There is a hope that such limitations will be overcome by proximal sensing. In situ proximal sensing has been utilized to acquire the desirable and accurate information about the soil, which compliments the pedologists's field kit to observe soil profiles. It closes an obvious gap in the pedological toolkit because so far profile descriptions were just that "descriptions." A proximal soil sensing method has been verified on soil profiles and permits to measure many soil properties beyond the visible range of the spectrum [12]. Examples include X-ray fluorescence [13], visible–near-infrared (VNIR) spectroscopy [14–16], imaging [17] spectroscopy, digital camera [18], a chromameter [19], digital photography [20], and digital soil morphometric [12]. While invasive methods may temporarily fill the gap, the holy grail of soil description would be through a noninvasive sensor. In reality, no single sensor or technique has the capability to predict all attributes of interest. The great power comes from putting the data gained from multiple sensors together (data fusion combined with soil inference systems) that will provide the most useful information.

2.1 *Smartphone-Based Precision Agriculture*

Low-cost smartphones arranged with different sensors are opening the way for farmers to up-to-date agricultural information and assistance from agricultural experts and government extension workers. At a large-scale area, farmers can utilize smartphone-based sensors to collect massive amounts of data using simpler, faster, cheaper, and less laborious techniques than traditional methods. Sensors installed in smartphones can be classified into motion sensors (e.g., accelerometers, gravity sensors), environmental sensors (e.g., temperature, pressure, and photometer), and position sensors (e.g., magnetometers, GPS, and orientation sensors). These sensors help to make up precision farming or precision agriculture. Global positioning system (GPS) permits applications (e.g., maps) to obtain the current location of interest [21]. Accelerometers, which measure the acceleration force whether caused by the phone's movement or gravity in three axes, have been utilized as sensors in detection [8], activity recognition [22], and driving event detection [23]. Built-in cameras of smartphones have enhanced significantly in terms of specifications. The camera resolutions are presently practically identical with dedicated handheld cameras. Therefore, images or videos from smartphone built-in cameras can be used for its computation power to supply valuable information (indoor positioning) from the input images [24].

2.1.1 Soil (e.g., Color) Sensing

Soil study is a main factor in precision farming, which has great effects on agricultural success. Farmers who have soil data obtain a benefit in farming including in PA. The most significant soil property is soil color, which is a diagnostic criterion for anthraquic horizon, cambic, fragic, fulvic, melanic, plaggic, umbric horizons, gleyic, and albic material [25]. Soil parameters that have been assessed utilizing color include organic matter, OM [18, 19], texture [26], water table depth [27], iron oxide [28], and others. Traditionally, soil researchers have decided the soil sample color by matching a soil aggregate against color patches (produced by Albert H. Munsell in the early twentieth century). Munsell color has far-reaching implications for the description, and classification of soils [25, 29], and in addition to examining soil genesis and evaluation. However, there is some subjectivity in the Munsell soil color assessment. The color determination is problematic even for experts due to the quality and age of Munsell charts. The result of soil color assessment on-site is affected by personal experience. Color readings depend on the moisture status of the soil profile and light quality. In addition to differences in light under which color estimations are being made, it is also well established that physiological differences in the eye mean that not everyone distinguishes color in the same manner and that color perception drifts with aging due to the gradual yellowing of the eye's lens [30]. The visual judgment of soil color between individuals utilizing the same Munsell chart under intensely controlled conditions was variable. Sánchez-Marañón et al. [31] investigated color charts and found that production differences and non-uniform fading can significantly affect color determination.

Some smartphone applications in the literature utilized smartphone sensors in examining the soil for PA. Improvements to digital cameras in smartphones have brought about high-quality and consistent imaging. The pixels number in a smartphone camera is presently more than expected just to decide the soil color. Spectral accuracy of cameras is an issue as they just give color information across broad spectral ranges. This restricts application for spectroscopic analysis. Soil color characteristics have been measured on various distinctive ways, ranging from naked eye in comparison with Munsell color charts [18] to electronic measurement (proximal sensing) of soil with digital cameras. Digital cameras have already been proposed as soil color sensors [20, 32, 33] due to the likelihood of gaining reliable color information [hue H, value V (lightness), and chroma C (intensity)] from RGB images. Smartphone with built-in cameras was utilized as imaging devices to obtain and transmit digital data to an off-site laboratory or external computer, which handled the information and returned the analysis results to the phone. The few works that have measured soil color with digital cameras (e.g., [32, 34, 35] required external software for calculations. Gomez-Robledo et al. [36] explored the utilization of smartphone camera as a soil color sensor, utilizing it to determine Munsell color of soil samples. Levin et al. [37] utilized color indices from digital photography to estimate iron oxide content and textural parameters in sandy soils, whereas Gregory et al. [38] assessed soil OM utilizing a digital camera with VNIR wavelength capacity. The use of extra functionality of mobile phone cameras enhances data interpretation and transmission abilities. Aitkenhead et al. [18] demonstrated a smartphone application linking camera, image analysis, and server-side processing for the soil carbon estimation.

GPS circuitry installed in most of smartphone and tablet gives them the same functionality as a standard GPS device. Essential GPS location information is given in latitude/longitude rather than in individual national grid reference systems and so may be converted to match spatial data sets. GPS positioning permits the user's location to be captured at the time of making other sensor readings. This positional data are then inserted into the photograph's header, which is uploaded and can be used to decide the parameter values of the spatial extent of the user's location. The accuracy of smartphone GPS is less than standard GPS, because of the built-in antenna limitation. With a mid-range smartphone, the location accuracy is usually within 35–40 m more than 95% of the locations [39].

2.1.2 Soil (Thin Section) Micromorphology

Of course, quantitative analyses have been a successful part of micromorphological analyses over the years, but the preparation of thin sections is time-consuming and expensive. Even though micromorphology has made major impacts on soil genesis studies, it has never become a standard tool in soil survey interpretations. Although not analyzed directly in the field, thin sections were taken, and image analyzers were used to study pore distributions and how they work [40, 41]. Mesomorphological analysis was put in place to bridge the gap between the micromorphological methods and field description, which was largely qualitative [42]. Instead of measuring discrete points, a continuous description of properties such as pore size and porosity could be initiated, which allow understanding the relationship between structure and redoximorphic. Hyperspectral cameras [17] and laser scanners [71] can scan in two dimensions and capture this vertical and lateral variation.

Saturated soil hydraulic conductivity (Ksat) is one of the most important soil parameters in hydrological modeling because it characterizes water movement through soils [43]. Unfortunately, Ksat is also the most difficult properties to accurately evaluate due to its variability over short-range distances and over time. Most Ksat field methods and techniques are expensive and time-consuming and require many Ksat measurements over large areas and extended periods to capture spatial and temporal variability. Researchers have developed various approaches to up scaling, but mostly rely on pedotransfer functions [43]. Pores of the thin section were analyzed via Environment for Visualizing Images (ENVI) software after performing image enhancement. The enhanced image was brought into ENVI as RGB and transformed it to hue, saturation, and value (HSV) formats. The soil pore classification was assessed in ArcMap using an unsupervised classification.

2.1.3 Soil Monolith in the Laboratory

Soil profile description in the field restricts the equipment choice due to the lack of control over ambient conditions (e.g., lightning and soil moisture) and the time available for controlled measurements. Soil monoliths are extracted from whole profiles and subsequent preparation, examination, and measurement under controlled conditions in the laboratory. Traditional box-fitting methods [44, 45] are time-consuming to apply, and modern hollow auguring techniques require heavy machinery and capture a small profile width [46]. A soil monolith extractor has been established in the house and facilitates the implementation of digital techniques. Digital images have been gathered on these soil monoliths and calibrated using a set of reference color chips. The spectral resolution of these images has been enhanced by combining the spatial resolution of the images with the spectral resolution of VNIR spectroscopy. Image processing methods (e.g., principal component analysis) and image segmentation have been enhanced to support the delineation of soil horizons and collect the soil structure information. From these enhanced images, spectral and morphological metrics were derived and classified.

2.1.4 Lab-on-a-Smartphone for Water Quality

The smartphone application is dedicated to encouraging users to submit information about water quality conditions, which affect farming and agriculture, that is, water clarity, salinity, algae cover, temperature, water level, and accompanying photographs [39]. Hussain et al. [47] exhibited the operation of a smartphone-based platform salinity sensor for exact and reliable monitoring of the salinity level in the water environment. Two freely accessible Android applications have been utilized for identification and examination of salinity level. The smartphone sensor can measure salinity level variation as low as 0.1 parts per thousand (ppt) with high accuracy and repeatability. Levin et al. [48] present a field deployable colorimetric analyzer based on an inexpensive smartphone for taking a photograph of the colored solution. A software was prepared with the phone for recording and analyzing the RGB color picture. Water samples were tested using a smartphone for fluoride concentration, which displayed a significant positive correlation coefficient between 0.9952 and 1.000. Gunda et al. [49] detect total coliform and *Escherichia coli* bacteria in contaminated drinking water samples. The test method, called Mobile Water Kit (MWK), includes a set of custom chemical reagents, syringe filter units, and a smartphone platform that would serve as the detection/analysis system. With MWK, the total coliform and *Escherichia coli* bacteria were detected in water samples within less than 30 min, depending on the bacteria concentration. For one of the field samples, the MWK was able to detect total coliform within 35 s, which is faster than other rapid test methods available. Garcıa et al. [50] exported a smartphone platform for detecting ion concentrations through image characteristic of H (hue) value. This allows an unprofessional user to analyze potassium in water with a wide linear range from 3.1×10^{-5} to 0.1 mol/L and high reproducibility of $<1.6\%$.

Moonrungsee et al. [51] determined iron in zeolites by a smartphone for quantifying red, green, and blue light intensities. The blue color was suitable for creating an equation relationship with iron from 0 to 1.2% (w/w) iron and kept in the database for analysis. This procedure was rapid, simple, inexpensive, and produced little chemical waste. Lopez-Ruiz et al. [52] offered an Android application for nitrite and pH determination. Under controlled conditions of light, using the flash of the smartphone as a light source, the image captured by the built-in camera is processed using a customized algorithm for the colored areas detection. Image processing allows decreasing the light source influence in the picture. Then, the *H* (hue) and *S* (saturation) coordinates of the HSV color space are taken-out and correlated to pH and nitrite concentration. The results validate significant utilization of a mobile phone as an analytical instrument. For the pH, the resolution accuracy obtained is 0.04 units, however, in nitrite, 0.51% at 4.0 mg L^{-1} of resolution and 0.52 mg L^{-1} as the detection limit were achieved.

2.1.5 Smartphone for Smart Farming

The following are some examples of smartphone applications for precision farming. First, smartphones are devoted to plant pest and disease detection/diagnosis in farms, which help diseases detection because of their computing power, high-resolution displays, and extensive built-in sets of accessories. Prasad et al. [53] propose a mobile vision system, which helps in plant disease identification process. The system worked by capturing images of plant leaves for diseases and then by preprocessing those images. The preprocessing image was vital for saving the transmission cost of sending diseased leaf images to plant pathologists in remote laboratories. A clustering algorithm segmented three areas of leaf images: background, non-diseased portion, and diseased portion(s) of the leaf. Leaf pictures were cropped to the diseased patch only and transmitted over the available network to remote laboratories (laboratory experts) for further disease identification and suggesting cure and prevention for the diseases. The application is constantly accessible and enables the farmer to recognize plant diseases without expert knowledge. Rafoss et al. [54] used GPS-enabled smartphones as a way to fight fire blight. As mobile phones associated with the Internet give an approach to get data and report disease outbreaks rapidly, stakeholders (e.g., farmers, policy makers, and field workers) can take right actions to mitigate the damage. The application was actualized for mobile devices to be able to show disease outbreak reports on a map and users can report or edit disease outbreaks. GPS is utilized to get to the mobile phone location to recover the adjoining map and the information. Cameras can be utilized to take pictures or videos, which are stored on the cloud for future reference or further inspection [39, 55, 56], or to be image-processed further [36, 39, 57, 58]. Thermal sensors have good potential for early detection of plant disease, especially when the disease directly affects transpiration rate [59].

Second, real-time monitoring of crop nutritional status is important to track crop growth and advance dynamics over time. In doing so, fertilizer and pesticide application can be adjusted based on the application rate to accommodate the growth requirement of crops for enhanced agricultural productivity [60]. Crop nutrient demand is typically dynamic across different growth stages. A shortage of any nutrients (e.g., NPK) can result in restricted shoots and roots growth, early defoliation of older leaves, and decreased biomass yield [61]. Moreover, excessive fertilization adversely influences water, soil and air quality, and ecosystem biodiversity. Fertilizer application is an essential farming activity with a possibility to incredibly influence farm productivity. Farmers are responsible for decisions on which fertilizer to apply and their crop-specific applicable to reduce environmental concerns. Traditional methods for fertilizer management were based on either soil or plant analyses, which are expensive, slow, or labor-intensive [62]. Consequently, scientific and political communities are calling for exploring technological tools to reduce contamination [62]. Smartphone sensing, as a timely and nondestructive tool, could be an alternative to traditional plant testing for diagnosis of crop nutrient status. Sumriddetchkajorn [63] utilized a smartphone-based color estimator dedicated to rice leaves' chlorophyll evaluation. The application evaluated the leaves color level and recommended requiring amounts of fertilizer for applying to the field [63, 64].

Third, irrigation water management requires timely application of the right amount of water. For better irrigation design, data about the soil moisture patterns are important. Managing irrigation water needs to combine an easy and cheap method of measuring soil-wetting front, which is of great importance for precision agriculture. Traditional procedures to monitor distribution patterns of wetting front patterns are time-consuming, laborious, and expensive. Omran et al. [65] propose an approach using the camera sensor to assess and map wetting front (area and depth) by image analysis for smart precision irrigation farming. Imaging analysis acquires both spectral and spatial information to identify and quantify solute infiltration into the soil. Using imaging techniques with image processing algorithms may open a new avenue for inferred useful information from soil characteristics and soil quality for sustainability.

Finally, an innovative application using smartphone-based sensors was developed to determine fruits ripeness [63]. A small infrared sensor to control fruits and vegetables ripeness has been set up. Fruits pictures under white and ultraviolet light sources were taken to detect ripeness levels for green fruits. Farmers could integrate the system by separating fruits of different ripeness levels into piles before sending them to the markets. With the help of computer vision techniques, this process could be done in bulk rather than farmers manually inspecting each fruit. Aroca et al. [66] proposed a mobile sensing platform that integrates different sensors, for instance, touch pressure, imaging, inertial measurements, and a radio frequency identification (RFID) reader, which is suitable for several applications (e.g., a fruit classification and grading), which help workers during manual harvesting.

2.2 *Portable X-Ray Fluorescence (PXRF) for Soil Analysis*

The soil classification system has criteria that depend on field observations complemented by laboratory analyses, which are biased, time, and cost demanding and performed on definite profile location. Most soil description (e.g., texture, color, structure) is dependent upon the senses and the expert experience, hence is subjective. The field decision on several soil diagnostic units and taxonomy has to be decided after the laboratory results are available. To provide interpretations, that are more objective and enhance existing field soil descriptions, there has become a demand to digitally collect and quantify soil characteristics in situ [12]. There has been increasing concern in using proximal soil sensors to assess soil properties. Portable XRF (pXRF) spectrometer has shown potential as a field diagnostic device as it can provide much data promptly. X-ray fluorescence (XRF) spectrometry is a quick, proximal scanning innovation, which allows for total metal quantification in soils within approximately 2 min. pXRF has been used in several soil applications from pedology to environmental quality assessment [13]. To determining soils heavy metals content in the field, XRF gives a multi-element analytical method for the routine nondestructive analysis of soils with minimal sample preparation. XRF has a widespread concentration range for many elements in a sample [67]. Portable XRF tools can be employed in soil investigation for faster and efficient metals analysis in soil profiles. As pXRF can be applied in the field, sample preparation is not necessary.

Soil minerals directly or indirectly influence almost all soil properties. As such, soil mineral composition has a large impact on soil behavior and to gain insight into soil function. Routine soil mineral analysis includes laboratory-based X-ray diffraction (XRD) of random powder samples, and oriented clay with suitable pretreatment (e.g., Mg/K saturation, ethylene glycol, heating to 550 °C) is a time-consuming task. Nondestructive, in situ XRD devices commonly have a limited angle range greater than 20° 2θ [68]. This is problematic as most phyllosilicates have primary peaks at lower angles (<10° 2θ), and thus, these devices are less utilized to estimate these important soil constituents. Expert interpretation is still essential, and minerals estimation remains semi-quantitative. However, field pXRF has been useful for lithological investigation [69] and archeological investigations (e.g., [70]). pXRF has been used in a wide range from sulfur (S) to uranium (U), plus light element analysis (magnesium (Mg), aluminum (Al), silicon (Si), and phosphorus (P). These devices detect a wide concentration range of contaminated soil, on-site, in seconds.

2.3 *Multistripe Laser Triangulation (MLT) Scanning for Soil Physical Analysis*

Soil structure is a differentiation criterion in the WRB for mollic and umbric horizons anthraquic, nitic, vertic, petrocalcic, and calcic. The structure determination is critical in the case of natric—columnar, prismatic (or blocky) structure required—because it determines the Solonetz reference soil group. However, no device is available that can measure the distinct aspects of the structure in situ [12]. MLT scanning method is accurate, simple, and cheap (e.g., [71]), which opens up the possibility to measure both soil surfaces and 3D soil specimens (e.g., individual peds), and therefore, nondestructively. Figure 2 shows MLT scanner in the field and in the laboratory. An individual soil clod was weighed and scanned in the laboratory. A 3D object is digitized from various positions and merged into a continuous surface [72] to quantify soil structure and the pores distribution (i.e., soil architecture) of the soil.

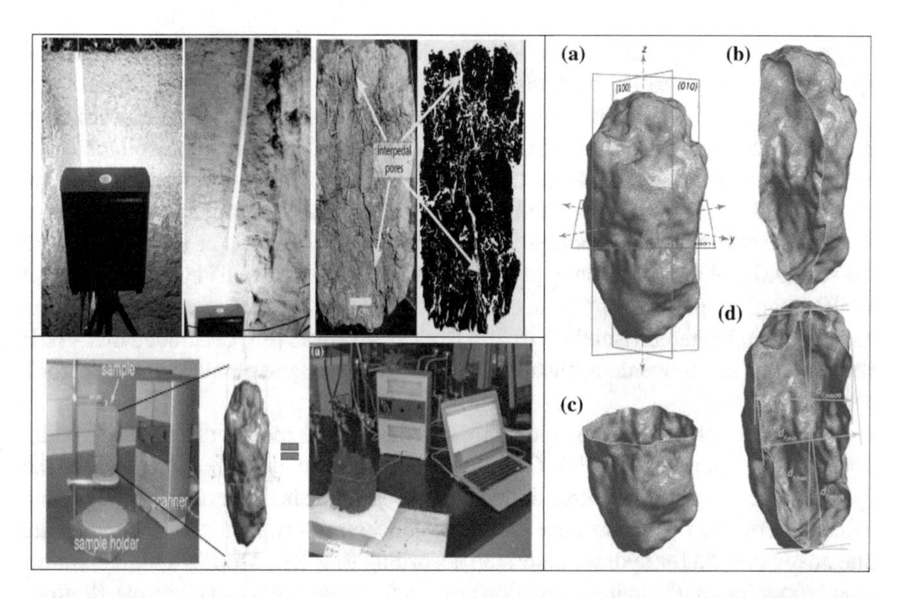

Fig. 2 Multistripe laser triangulation (MLT) scanner in the field (upper-left) and in the laboratory scanning a large clod (lower-left). A scanned prismatic ped (right) showing cross sections through the mid of the aggregate and parallel to the **b** {100}-plane, **c** {010}-plane, and **d** {001}-plane [73]

Bulk density determination from clod method was benefits from a noninvasive (MLT scanning) technique [74] for obtaining clod volume. The accurate bulk density determination on very small soil aggregates samples is measured by the accuracy of the volume determination [75]. Some methods have been established to measure these small samples, but they need sealing the aggregates by filling up the pores either with an organic liquid and submerging the saturated aggregate in the same liquid or in a liquid that is immiscible with it [75]. MLT scanning offers the opportunity to enhance soil architecture analyses by allowing repeated bulk density measurements (i.e., at each soil water potential) on the same clod on which water retention is determined. This may play an important role in the volumetric determination of water content for water retention in swelling and shrinking soils (e.g., Vertisols). Finally, in the examination of soil surface deformation from shrinkage, laser triangulation scanning has helped to accurately assess processes such as curling (e.g., [76] and crack formation dynamics [77].

2.4 *Reflectance Spectroscopy for Soil–Plant Properties Measurement*

There has been interest in using proximal soil sensors to acquire information on soil properties. VNIR spectroscopy has shown potential as a field diagnostic device as they can provide huge data promptly and predict many soil attributes [78]. Reflectance spectroscopy was verified by Waiser et al. [79] for in situ clay content quantification of soils. In situ spectroscopic measurements coupled with chemometrics were effectively applied by Viscarra Rossel et al. [80] to measure soil color, soil minerals, and clay content. Lagacherie et al. [81] used reflectance spectrometry to measure clay and calcium carbonate in the laboratory. VNIR spectroscopy has been applied to measure soil organic carbon and clay content [36]. Steffens et al. [82] measure the soil OM and composition applying imaging spectroscopy. They concluded that VNIR imaging spectroscopy is an effective tool for OM mapping even if the layers are not distinguishable visually. Viscarra Rossel and Hicks [83] concluded that VNIR spectroscopy is a convenient, cheap technique to observe and monitor organic carbon composition.

Mottles, which are colored spots in a soil matrix, are mostly the result of oxidation and reduction of iron (Fe). Concreted mottles of oxides are diagnostic for the hydragric, ferric, plinthic, petroplinthic, and pisoplinthic horizons and for the stagnic color pattern. Iron or manganese coatings, concentrations, or redox depletions are diagnostic criteria for hydragric horizon according to WRB. Mottles and redoximorphic properties are the main distinction criteria for Stagnosols and Gleysols. Steffens and Buddenbaum [17] determined that laboratory spectroscopy assists the spatially correct soil classification including the soil mottling quantification. To distinguishing soil horizons, VNIR reflectance spectroscopy coupled with principal component variables can be effectively used as variables describing the spectral properties along with the soil profile.

The effectiveness of VNIR spectroscopy for prediction of some soil properties (e.g., clay, iron oxides, salinity, calcium carbonate, organic matter, and heavy metals) from Bahr El-Baqar region, Egypt, was evaluated using soil samples scanned in the 350–2500 nm region. Figure 3 displays the main spectral signatures and corresponding soil attributes. Some samples have high clay minerals, and other samples have calcite and iron oxides. The predictive capacity of reflectance spectroscopy and partial least square regression (PLSR) was high for the soil salinity and clay content. These results can be explained by the sharp spectral activity of organic carbon and clay in the VNIR-SWIR region. The wavelengths utilized in the clay minerals prediction were between 890 and 2430 nm, and all relate to the soil's mineralogy. Absorptions between 400 and 1000 nm are caused by the iron oxides present, mainly hematite and goethite, those between 1300 and 1500 nm are from absorption by hydroxyl groups in clay minerals and water. Absorptions between 1700 and 1800 nm are produced by carbon. The strong absorption near 1900 nm is due to water (hygroscopic water and water held within clay mineral structures). The clay minerals (kaolinite, illite, and smectite), carbonates, and organic compounds cause absorptions between 2100 and 2500 nm. The optimal estimation models of two types of clay mineral contents using specific wavelengths revealed that the recovery accuracy was acceptable. Therefore, the reflectance spectra of soil acquired from Bahr El-Baqar could be used for calcite, iron oxides, and clay minerals detection (illite and montmorillonite) in the soils.

Spectroscopic methods are also given significant potential for plant mineral analysis. Imaging allows to monitor crop health and detect nutrient deficiency (i.e., NPK).

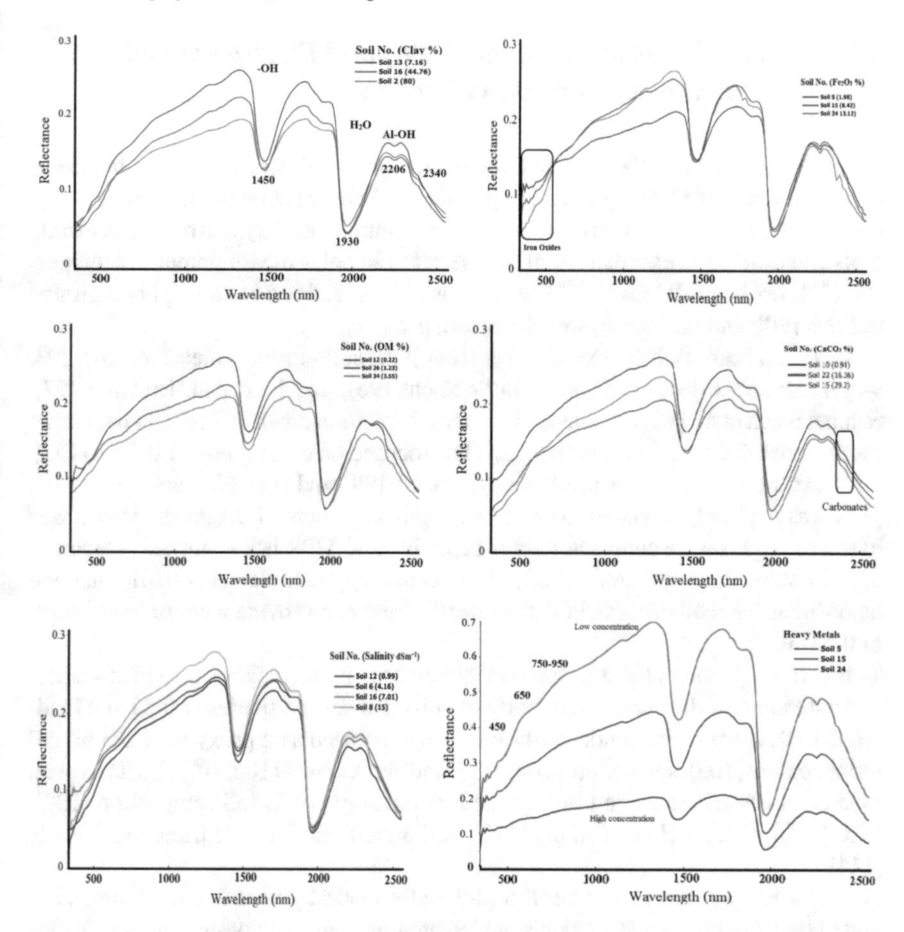

Fig. 3 Reflectance spectra of Bahr El-Baqar soil indicating the spectral features of the most important constitutes (iron oxides, clay minerals, salinity, carbonate minerals, organic matter, and heavy metals)

For N status assessment of crops, portable devices with a direct link to fertilizing equipment are being used increasingly [84–86], and Mn deficiency can be diagnosed directly in situ utilizing a handheld instrument [87]. Mostly, fast spectroscopic techniques require no or only little sample preparation and are therefore often nondestructive.

2.5 Ground-Penetrating Radar (GPR) and Electromagnetic Induction for Underground Sensing

Quantifying soil subsurface using outdated methods is destructive, labor-intensive, and point-based [88]. Geophysical (e.g., GPR and electrical resistance tomography) methods are fast, nondestructive subsurface measurements [89], and on-site detecting tools that provide an excellent compromise between point-measurement and regional RS. GPR has been effectively utilized in many investigations, including underground utilities [90], and measuring soil characteristics [91].

GPR has been used in tree root detection [92], soil moisture measurements [93, 94], water table depth [95], soil clay content [96], and hard pan detection [97]. Schmelzbach et al. [98] obtained soil water information down to 7 m with decimeter resolution. GPR for plant root detection has focused on coarse roots (>0.2 cm) [92]. GPR can be utilized to measure root diameter [99] and root biomass [100, 101]. GPR was explicitly intended to chart soil depth and extent of diagnostic subsurface horizons. To measure compacted layer depth in soils, GPR is a noninvasive technology for subsurface soil study [102]. This technology complements traditional and labor-intensive method of field data collection and can provide a continuous image of the soil.

Electromagnetic induction (EMI) is widely used by soil scientists to gain a better understanding of the spatial variability of soils and soil properties at field and landscape scales. Apparent conductivity has also been used as a proxy measure of soil water content [103], soil texture [104, 105], and clay content [106, 107]. EMI has been used to assess difference in lithology and mineralogy [108], soil compaction [109], CEC [110], $CaCO_3$ [111], soil pH [112], soil organic carbon [113], and available N [114].

Soil salinity mapping has been traditionally made by visual observations supported by laboratory measurements, which are time and cost consuming. It provides a limited number of point measurements that may or may not be representative of the field or soil landscape. EMI produces a large number of georeferenced, quantitative measurements that can be associated with the spatial variability of salinity and sodicity at field and landscape scales. ECa maps with EMI have the potential to provide higher levels of resolution and greater distinction of soil types than soil maps prepared with traditional methods [115]. However, a major challenge in using EMI to map soil salinity has been the conversion of apparent conductivity (ECa) into the conductivity of the saturated paste extract (ECe). A number of models were developed to predict ECe from ECa (e.g., [116]). Unfortunately, models are imperfect and are often both time-dependent and site-specific [117]. As a consequence, calibration equations and modeled results usually cannot be extrapolated to other sites [118, 119]. Another challenge to the use of EMI to map salinity occurs at high conductivity values, which are more than 100 mSm^{-1} [120] when the quadrature component of the received electromagnetic field is not linearly proportional to soil conductivity (breakdown of low induction number approximation).

2.6 Thermal Infrared and Acoustic Sensor for Disease Detection

Two cases use the thermal and acoustic sensor to early predict plant and animal's diseases. First, disease damage assessment in plant is being done traditionally by a visual approach. Early and in situ disease detection methods that can control large areas within a short period during the season with limited labor and funds do not exist in practice. Omran [121] assesses the potential of thermal infrared (TIR) image of early sensing of peanut leaf spot disease. Heat sensing camera is depicting the thermal behavior of the healthy and diseased canopy. Figure 4 (left) demonstrates the on-site thermal measurements for the peanut canopy. The discrimination between infected and healthy leaves through the temperature difference is essential before the appearance of visible necrosis on leaves.

Second, date palm is one of the most important fruit trees cultivated in the Middle East and North Africa [122]. One of the major threats to date palm over the world is the red palm weevil (RPW), Rhynchophorus ferrugineus (Olivier), (Coleoptera: Curculionidae) [123]. RPW is an economically "important invasive tissue borer that has a broad host range restricted to palm trees, mostly young trees less than 20 years old" (https://www.coursehero.com/file/p3d6jik/The-sustainability-of-oil-palm-cultivati). Visual detection [124] is the traditional method of RPW detection. Since direct visual detection of the RPW is quite difficult, [125] in situ detection to detect the RPW presence with its larva by sensing its (sound and heat) activity in offshoots is important to avoid the extensive damage to palm trees. This helps farmers to avoid heavy sprays of pesticides and take the necessary actions to restrict dangerous infestations [126]. Two sensors were used to early detect the RPW presence in situ (Fig. 4, right). On the one hand, acoustic sensor was used to detect infected trees (on a small scale) with RPW. The sound probe is established of three components: the microphone, the probe, and the signal conditioning stage. The probe delivers the signal caught from the palm tree with the highest possible quality. The audio probe is responsible for securing of sounds from the RPW, conditioning and legitimately enhancing the caught sound signal. A low-power processor will have the capacity to run and process the sound taken by the audio probe and decide the RPW presence. A wireless microphone, which contains a radio transmitter, is ready to convey messages reporting the results of RPW activity. It transmits the audio as a small FM radio rather than via a cable to a nearby receiver connected to the sound system. On the other hand, the thermal sensor was utilized to detect infected trees (on a large scale).

The hypothesis was that the tunneling of RPW destroys the vascular system of the palm and creates local conditions of water stress. Mozib and El-Shafie [127] confirm that the infested date palm average temperature was significantly higher (32.60 °C) than the average temperature recorded at the same time inside the healthy trees (29.53 °C) and in the ambient atmosphere (29.35 °C). If the tree is healthy, a uniform surface temperature distribution exists and the TI displays a uniform coloring, but if the color is not uniform, then deterioration/cavities may be present. Four to five

Fig. 4 Thermal behavior of the healthy and diseased canopy for the peanut plant (left) and red palm weevil detection (right)

thermal images are usually sufficient to recognize the situations of an entire tree. A faster assessment usually needs 2–3 min, while an in-depth analysis requires less than 10 min. This technique wants less time compared to other technologically advanced investigation systems. The system has been utilized in all weather conditions, by night and day, in the summer and winter, with temperatures ranging from +2 to +35 °C.

3 An Internet of Things and a Cloud-Based for Smart Precision Agriculture

In the near future, the projected seven trillion sensing devices will be collecting data for users every minute or possibly even every second [128]. Most of these data require real-time processing and analysis in order to provide immediate information and services. Due to the huge number of sensing devices, and the individual demands for services, service providers must provide a large enough pool of resources and sufficient mechanisms for flexibility.

3.1 On-Line Smart Farming Concept

Spatial data infrastructure is a concept, architecture, and set of standards for how spatial data can be remotely searched and accessed [1]. The geospatial sector, through organizations such as the Open Geospatial Consortium (OGC) and ISO, has defined standards for interoperable spatial data and architecture. Widespread adoption of OGC standards and various SDI initiatives has prepared the geospatial industry to embrace cloud computing for hosting spatial data and spatial processing services.

Internet of Things (IoT) is a concept that incorporates all the objects around us as a component of the Internet. Coverage IoT is very large and includes different smart devices such as smartphones, digital cameras, and smart sensors. When these devices are interconnected, they provide much more intelligent processes and services that can be used in PA. Devices and sensors connected to the Internet provide services and produce a huge amount of data. Cloud computing is a model for on-demand access to the repository of configurable resources (networks, servers, storage, applications, services, software, etc.), which can easily provide such infrastructure, applications, and software. In addition, the data acquired are uploaded to the web services based on Simple Object Access Protocol (SOAP) and Representational State Transfer (REST), using messaging mechanisms such as emails, SMS, and blogs. Cloud acts as a front end to access the IoT. Wireless sensors contributing to the system that integrates the IoT paradigm into planting management processes can be beneficial to address expected solutions. Various objects interact with each other in IoT to achieve an intelligent PA.

There are many IoT characteristics in PA, various data sources, real-time processing, and spatial–temporal attributes. For example, ArcGIS Online is a mapping platform designed by Esri to achieve tasks in the cloud. These functions can be grouped into "mapping," "analysis," and "applications." The ArcGIS Online map gives the user the ability to include data from different sources. These incorporate "any web service discoverable from ArcGIS Online or an entity's own on-premise ArcGIS Server implementation, in addition to data uploaded from a desktop. There is access to a number of Esri-sourced base maps. Users can add their own base map to the ArcGIS Online map if desired. Once a web map is authored and shared, it can be utilized to create a web application" (https://files.nc.gov/ncdit/documents/files/Assessing-Geospatial-Cloud-Solutions). This application can reside within ArcGIS Online or the user's own server. Google and Microsoft have offered different cloud computing services. Although current services focused on online storage and utility computing, there are few cloud platforms dedicated to IoT environment in PA. General cloud computing has some features potential to meet some IoT requirements. However, there are necessities of IoT such as real-time service delivery and autonomy in subsystems, which are beyond the capability of general clouds. Although researchers have proposed few models in the agriculture domain utilizing one or more of the technologies mentioned, the dynamic model that provides an integrated approach is needed. Therefore, a cloud model for IoT in PA was proposed.

3.2 Overall Framework of the Proposed System for Precision Agriculture

The system framework definition is critical for facilitating the PA process. The five core components of the system should be based on IoT, sensors, cloud, and spatial data infrastructure components along with models and software. The interconnected nature of the five components means that modifying one component will require modifications to be made for other components. Figure 5 shows the overall augmented components for smart precision agriculture. These components are (1) Spatial data and Metadata. The metadata required for PA data sets documentation is metadata, identification, data quality, reference system, and citation and responsible party. (2) Shareholders (PA users and providers). (3) Technology (access network and access service). (4) Standards and protocols. Metadata standard, data transfer, data storage format, data schema, and web mapping standard. (5) Institutional framework. Policy issues: legal aspect (copyright, liability, privacy); culture aspect; pricing aspect; organizational arrangements; social consideration; political support; and educational issues.

The proposed system will connect the PA world in both a sensory and intelligent manner through combining technological developments in item identification ("tagging things"), wireless sensor network (WSN) ("feeling things"), embedded system ("thinking things"), and nanotechnology ("shrinking things"). The proposed

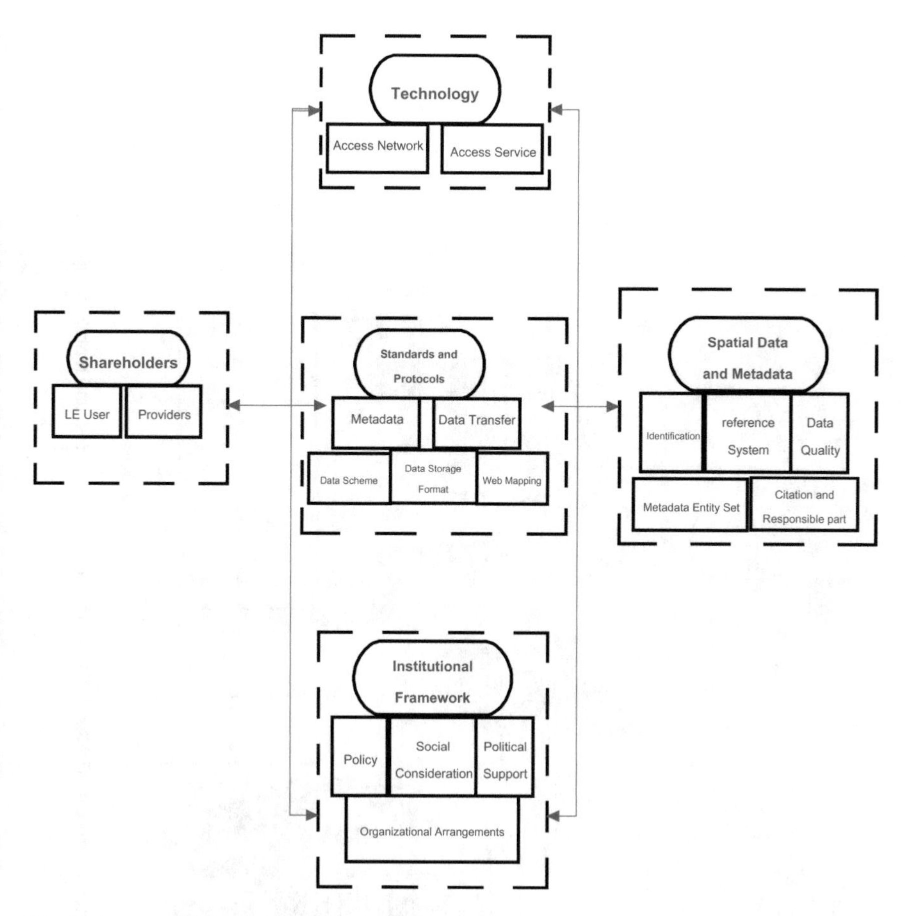

Fig. 5 Augmented system for smart precision agriculture

model consists of hardware and software components, which have their potential to be helpful in the PA process. Limited access to data, information, and services providers leads to data duplicity by many users. To promote access to this system, it may be more desirable to share data than to develop or install duplicate data. In order to overcome this type of problem, the role of standards is important. To share PA data, one should follow appropriate standards for data production. A common data standard for metadata, data transfer, and data storage format accepted by all potential users will reduce data duplicity and data redundancies. IoT is the fastest and most efficient way to distribute, find, and get the data. Here, the technology role comes. The requirements of IoT in PA were organized into common services, employ service-oriented architecture (SOA) and modular design method.

The proposed system for PA revolves around the data collection for use by the farmers and other stakeholders. The system architecture is divided into four layers (Fig. 6): data collection layer, network layer, cloud services layer, and the application

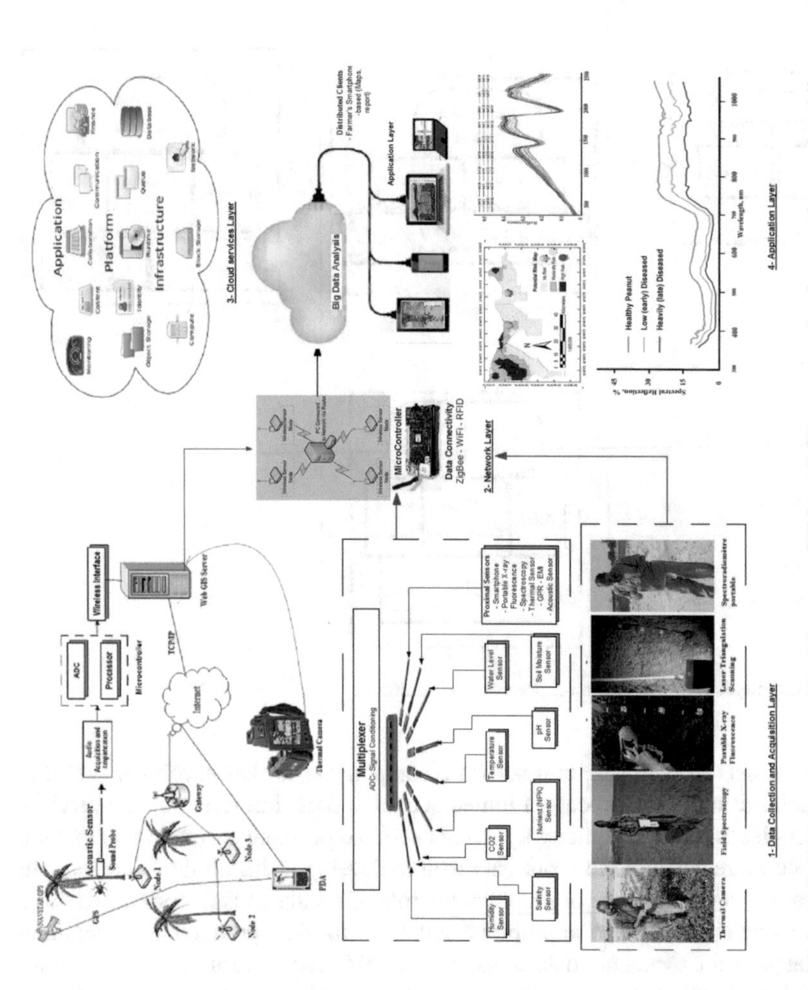

Fig. 6 Proposed model for smart agricultural system combined Internet of Things (IoT), sensors, cloud computing, and mobile computing

layer, in which their functions are as follows:

1. *Data Collection and Acquisition Layer*: The central job of acquisition layer is to realize automatic and actual gathering and transformation of the physical figures of real agricultural production through diverse sensors into digital data that can be managed in the virtual world through various means. One of the most important IoT elements is WSNs. WSNs consist of smart sensing nodes with embedded CPUs, low-power radios and sensors, which are used to monitor the agricultural field. The information categories of that IoT for agriculture collects are temperature, humidity, pressure, salinity, nutrient, and gas concentrations, etc.
2. *Network Layer*: The key task of the network layer is to gather the agricultural data acquired through the sensor layer for processing. Transport layer is the nerve center of the Internet of Things for agriculture, transmitting, and processing data. The network layer comprises the combination of the Internet network and telecommunication, network management center, information center, and intelligent processing centers. The network will aid to transfer the data generated by the devices as mentioned earlier. Different network technologies like GSM, LTE, WiFi, 3G, etc., may be used depending upon the availability and requirements.
3. *Cloud services Layer*: The main task of this layer is data saved and computing technology like cloud services. The cloud servers can be made accessible "independent of the locations and hence most suitable for IoT type of systems. The data can be stored and computed upon such servers. Cloud services can be taken on a pay-per-use policy as they are becoming popular for this reason" (https://www.ijrter.com/papers/volume-2/issue-4/the-agriculture-internet-of-thing).
4. *Application Layer*: The key task of the application layer is to analyze and process the data collected to develop digital awareness of the real agricultural field. Big data analytics tools can work on the vast amount of data generated and stored on the cloud servers, to excavate important patterns and trends in the data.

3.3 Strengths and Weaknesses of the System

The incorporation of IoT and cloud computing is of great significance and benefits for precision agriculture. Cloud computing's power storage, processing, and service ability, combined with the IoT's ability of information collection, compose a real network between people and items and the items themselves. Cloud computing is a flexible, powerful, and cost-effective framework for providing users with real-time data at any time with high quality. Regardless of the application, there are strengths to use a system in PA. The following are the system application benefits in agriculture:

- The system connects huge heterogeneous sensors, which constantly gather the data of real PA and produce a large volume of information, which can directly benefit from IoT and the vast storage capacities of cloud computing.

- Access can be via any Internet connection, anytime, anywhere as very easy, giving the operator a simple-to-understand user experience/interface.
- If you have many remote users, then the cloud makes the data distribution and analysis very simple. No need to send data using DVD or to download large data sets to update a local server.
- The cloud infrastructure can be scaled to meet the needs. This is a benefit if the consumption of the maps, data, and web services fluctuates.
- Applications that interact with physical devices need high computing power to enable real-time processing of mass data. Distributed cloud computing can provide powerful utility to perform complex computing.
- Having a cloud allows data collected in real or near real time to post directly on your system.
- Use efficiency improvement for inputs (soil, water, fertilizers, pesticides, etc.) and reduced cost of production. There are not any maintenance costs, which perhaps the largest cost savings. There is no requirement to procure the hardware infrastructure required to run the application.

Although there are numerous advantages to using a system, one should also be aware of the potential disadvantages and limitations of the system. The following downsides are identified.

- Spatial data have always been big, taking up gigabytes on your own server. If you move to a system, there is both the time to migrate the data to the server and for end users to access and download.
- Security is an imperative aspect of the system, and you will need to make yourself sure that the system has good security, so users data are not accessed your data.
- After all, they are a business to make a profit, so they will be making efforts to provide you with a valuable service, so they do not want things to go wrong.
- Another downside of the system is the absence of control you may have over the way your data are used, displayed, manipulated, and analyzed.

4 Conclusions

Two objectives are identified in the introduction. The first is to assess the status of smart sensors, cloud computing, and IoT development to support PA. Some conclusions from the systematic review can be drawn. Proximal soil sensing allows measuring many of soil–plant properties in situ. Examples include portable X-ray, visible–near-infrared spectroscopy, digital camera, smartphone, multistripe laser triangulation scanning, ground-penetrating radar, and electromagnetic induction sensor. There is an availability of a number of smartphone applications for the target farmers. Most of these applications are easily accessible if target users have access to smartphones. Direct estimation of soil properties in the field is possible and can be measured with accuracy levels suitable for soil and plant monitoring requirements.

Second, this chapter proposed a system that integrates the Internet of Things (IoT) with cloud computing and sensors, which are crucial to building smart precision agriculture. The cloud is "a collection of platforms and infrastructures on which data are stored and processed, allowing farmers to retrieve and upload their data for a specific application" (https://rd.springer.com/chapter/10.1007/698_2017_63), at any site with available Internet access. Hence, the cloud is a pool of resources accessible via the Internet. Combining the cloud, IoT, and sensors is crucial for the system. Four layers are proposed in the system, which are sensor layer, transmission layer, cloud services layer, and application layer. Benefits and possible limitations of the proposed system are identified.

Finally, researchers and farmers reading this chapter may have an impression of a research gap (missing technology) regarding agricultural applications to develop utilized phone-based sensors. Current techniques need to be tested and utilized; new technologies need to be adapted as they arise and overlooked technologies resurrected. Large sections of the electromagnetic spectrum are being used, as well as ultrasonic, electrical resistivity, and physical measurements but others, such as magnetic susceptibility, seem to be underexploited, which might be useful for future investigation.

Acknowledgements Abdelazim Negm acknowledges the partial support of the Science and Technology Development Fund (STDF) of Egypt in the framework of the grant no. 30771 for the project titled "A Novel Standalone Solar-Driven Agriculture Greenhouse—Desalination System: That Grows Its Energy And Irrigation Water" via the Newton-Mosharafa funding scheme.

References

1. Omran ESE (2005) Spatial data infrastructure to support land evaluation applications in Egypt. MSc Thesis GIRS-2005–016, Centre for Geo-Information, Wageningen University, The Netherlands
2. Gore A (1999) The digital Earth: understanding our planet in the 21st century. Photogr Eng Remote Sens 65:528
3. Bian F, Xie T, Cui X, Zeng Y (2013) Geo-informatics in resource management and sustainable ecosystem. In: (eds) International symposium, GRMSE 2013, Proceedings, Part 2. Wuhan, China, 8–10
4. The Economist (2009) Cloud computing: clash of the clouds. http://www.economist.com/node/14637206. Retrieved 09 Oct 2016
5. Prasad S, Peddoju S, Ghosh D (2013) AgroMobile: a cloud-based framework for agriculturists on mobile platform. Int J Adv Sci Technol 59:41–52
6. Channe H, Kothari S, Kadam D (2015) Multidisciplinary model for smart agriculture using internet-of-things (IoT), sensors, cloud-computing, mobile-computing & big-data analysis. Int J Comput Technol Appl 6(3):374–382
7. Mosa ASM, Yoo I, Sheets L (2012) A systematic review of healthcare applications for smartphones. BMC Med Inf Decis Mak 12(1):67
8. Habib MA, Mohktar MS, Kamaruzzaman SB, Lim KS, Pin TM, Ibrahim F (2014) Smartphone-based solutions for fall detection and prevention: challenges and open issues. Sensors 14(4):7181–7208

9. Duan, YE (2011) Design of intelligent agriculture management information system based on IOT.In: International conference on intelligent computation technology and automation (ICICTA), vol 1, pp 1045–1049. 28–29 Mar 2011
10. Omran ESE (2008) Is soil science dead and buried? Future image in the world of 10 billion people. CATRINA 3(2):59–68
11. Moran MS, Inoue Y, Barnes EM (1997) Opportunities and limitations for image-based remote sensing in precision crop management. Remote Sens Environ 61:319–346
12. Hartemink AE, Minasny B (2016) Digital soil morphometrics. In: (eds) Progress in soil science
13. Weindorf D, Zhu Y, Chakraborty S, Bakr N, Huang B (2012) Use of portable X-ray fluorescence spectrometry for environmental quality assessment of peri-urban agriculture. Environ Monit Assess 184:217–227
14. Ben-Dor E, Taylor RG, Hill J, Demattê JAM, Whiting ML, Chabrillat S, Sommer S (2008) Imaging spectrometry for soil applications. In: Sparks DL (ed) Advances in agronomy, Academic Press, Elsevier 97:321–392
15. Roudier P, Hedley C, Ross C (2015) Prediction of volumetric soil organic carbon from field-moist intact soil cores. Eur J Soil Sci 66(4):651–660
16. Omran ESE (2016) Inference model to predict heavy metals of Bahr El Baqar soils, Egypt using spectroscopy and chemometrics technique. Model Earth Syst Environ 3:2: 200
17. Steffens M, Buddenbaum H (2013) Laboratory imaging spectroscopy of a stagnic luvisol profile—high resolution soil characterisation, classification and mapping of elemental concentrations. Geoderma 195:122–132
18. Aitkenhead MJ, Coull M, Towers W, Hudson G, Black H I J (2013) Prediction of soil characteristics and colour using data from the national soils inventory of Scotland. Geoderma 200:99–107
19. Liles GC, Beaudette D E, O'Geen A T, Horwath W R (2013) Developing predictive soil C models for soils using quantitative color measurements. Soil Sci Soc Am J 77(6):2173–2181
20. O'Donnell TK, Goyne K W, Miles R J, Baffaut C, Anderson S H, Sudduth K A (2011) Determination of representative elementary areas for soil redoximorphic features identified by digital image processing. Geoderma 161:138–146
21. Gong H, Chen C, Bialostozky E, Lawson C T (2012) A GPS/GIS method for travel mode detection in New York City. Comput Environ Urban Syst 36(2):131–139
22. Anjum A, Ilyas MU (2013) Activity recognition using smartphone sensors. In: Proceedings of the IEEE 10th consumer communications and networking conference (CCNC'13), pp 914–919
23. Chaovalit P, Saiprasert C, Pholprasit T (2014) A method for driving event detection using sax with resource usage exploration on smartphone platform. EURASIP J Wirel Commun Netw 2014(135)
24. Werner M, Kessel M, Marouane C (2011) Indoor positioning using smartphone camera. In: Proceedings of the international conference on indoor positioning and indoor navigation (IPIN'11), 6(1)
25. IUSS Working Group WRB, World reference base for soil resources World Soil Resources Reports, 2006. No. 103. FAO, Rome
26. Ibanez-Asensio S, Marques-Mateu A, Moreno-Ramon H, Balasch S (2013) Statistical relationships between soil colour and soil attributes in semiarid areas. Biosys Eng 116(2):120–129
27. Humphrey C, O'Driscoll M (2011) Evaluation of soil colors as indicators of the seasonal high water table in coastal North Carolina. Int J Soil Sci 6(2):103–113
28. Gunal H, Ersahin S, Yetgin B, Kutlu T (2008) Use of chromameter-measured color parameters in estimating color-related soil variables. Commun Soil Sci Plant Anal 39(5–6):726–740
29. Soil Survey Staff (2014) Keys to soil taxonomy, 12th edn. USDA-Natural Resources Conservation Service, Washington, DC
30. Billmeyer F, Saltzman M (1981) Principles of color technology. Wiley, New York, NY
31. Sánchez-Marañón M, Huertas R, Melgosa M (2005) Colour variation in standard soil-colour charts. Soil Res 43(7):827–837

32. Viscarra Rossel RA (2008) The soil spectroscopy group and the development of a global spectral library. In: 3rd global workshop on digital soil mapping. Utah State University, Logan, Utah, USA, 30 Sept–3 Oct 2008
33. Aydemir S, Keskin S, Drees LR (2004) Quantification of soil features using digital image processing (DIP) techniques. Geoderma 119(1–2):1–8
34. Pongnumkul S, Chaovalit P, Surasvadi N (2015) Applications of smartphone-based sensors in agriculture: a systematic review of research. J Sens 2015:18 (ID 195308)
35. Han P, Dong D, Zhao X, Jiao L, Lang Y (2016) A smartphone-based soil color sensor: for soil type classification. Comput Electron Agric 123:232–241
36. Gomez-Robledo L, Lopez-Ruiz N, Melgosa M, Palma A, Capitan-Vallvey L, Sanchez-Maranon M (2013) Using the mobile phone as Munsell soil-colour sensor: an experiment under controlled illumination conditions. Comput Electron Agric 99:200–208
37. Levin N, Ben-Dor E, Singer A (2005) A digital camera as a tool to measure colour indices and related properties of sandy soils in semi-arid environments. Int J Remote Sens 26(24):5475–5492
38. Gregory S, Lauzon J, O'Halloran I, Heck R (2006) Predicting soil organic matter content in southwestern Ontario fields using imagery from high-resolution digital cameras. Can J Soil Sci 86(3):573–584
39. Aitkenhead M, Donnelly D, Coull M, Black H (2013) E-smart: environmental sensing for monitoring and advising in real-time. IFIP Adv Inf Commun Technol 413:129–142
40. Murphy CP, Bullock P, Turner RH (1977) The measurement and characterisation of voids in soil thin sections by image analysis. Part I. Principles and techniques. Eur J Soil Sci 28(3): 498–508
41. Bouma J, Jongerius A, Boersma O, Jager A, Schoonderbeek D (1977) The function of different types of macropores during saturated flow through four swelling soil horizons. Soil Sci Soc Am J 41:945–950
42. Koppi A, McBratney A (1991) A basis for soil mesomorphological analysis. J Soil Sci 42(1):139–146
43. Guber A, Pachepsky Y, van Genuchten M, Rawls W, Simunek J, Jacques D, Nicholson T, Cady R (2006) Field-scale water flow simulations using ensembles of pedotransfer functions for soil water retention. Vadose Zone J 5:234–247
44. Berger K, Muckenhirn R (1945) Soil profiles of natural appearance mounted with vinylite resin. Proc Soil Sci Soc Am 10:368–370
45. Brown L (1963) Lacquer cement method of making soil monoliths. University of California, Division of Agricultural Sciences, California Agricultural Experiment Station
46. Haddad N, Lawrie R, Eldridge S (2009) Improved method of making soil monoliths using an acrylic bonding agent and proline auger. Geoderma 151:395–400
47. Hussain I, Das M, Ahamad K, Nath P (2017) Water salinity detection using a smartphone. Sens Actuators B: Chem 239:1042–1050
48. Levin S, Krishnan S, Rajkumar S, Halery N, Balkunde P (2016) Monitoring of fluoride in water samples using a smartphone. Sci Total Environ 551–552:101–107
49. Gunda N, Naicker S, Shinde S, Kimbahune S, Shrivastava S, Mitra S (2014) Mobile water kit (MWK): a smartphone compatible low-cost water monitoring system for rapid detection of total coliform and E. coli. Anal Methods 6(16 21):6139–6590
50. Garcıa A, Erenas M, Marinetto E (2011) Mobile phone platform as portable chemical analyze. Sens Actuators B Chem 156:350–359
51. Moonrungsee N, Pencharee S, Peamaroon N (2016) Determination of iron in zeolite catalysts by a smartphone camera-based colorimetric analyzer. Instrum Sci Technol 44(4)
52. Lopez-Ruiz N, Curto V, Erenas M, Benito-Lopez F, Diamond D, Palma A, Capitan-Vallvey L (2014) Smartphone-based simultaneous pH and nitrite colorimetric determination for paper microfluidic devices anal. Chem 86(19):9554–9562
53. Prasad S, Peddoju SK, Ghosh D (2014) Energy efficient mobile vision system for plant leaf disease identification. In: Proceedings of the IEEE wireless communications and networking conference (WCNC'14), pp 3314–3319

54. Rafoss T, Sælid K, Sletten A, Gyland L F, Engravslia L (2010) Open geospatial technology standards and their potential in plant pest risk management-GPS-enabled mobile phones utilising open geospatial technology standards web feature service transactions support the fighting of fire blight in norway. Comput Electron Agric 74(2):336–340
55. Saha B, Ali K, Basak P, Chaudhuri A (2012) Developmentof m-sahayak-the innovative android based application for real-time assistance in Indian agriculture and health sectors. In: Proceedings of the 6th international conference on mobile ubiquitous computing, systems, services and technologies (UBICOMM'12), pp 133–137
56. Mesas-Carrascosa FJ, Castillejo-Gonz´alez I L, de la Orden M S, Garc´ıa-Ferrer A (2012) Real-time mobile phone application to support land policy. Comput Electron Agric 85:109–111
57. Confalonieri R, Foi M, Casa R, and et al (2013) Development of an app for estimating leaf area index using a smartphone. Trueness and precision determination and comparison with other indirect methods. Comput Electron Agric 96:67–74
58. Frommberger L, Schmid F, Cai C (2013) Micro-mapping with smartphones for monitoring agricultural development. In: Proceedings of the 3rd ACM symposium on computing for development (DEV'13)
59. Raza S-e-A, Prince G, Clarkson J, Rajpoot N (2015) Automatic detection of diseased tomato plants using thermal and stereo visible light images. PLoS ONE 10(4):e0123262
60. Duveiller G, Baret F, Defourny P (2012) Remotely sensed green area index for winter wheat crop monitoring: 10-year assessment at regional scale over a fragmented landscape. Agric Meteorol 166–167:156–168
61. Gianquinto G, Orsini F, Fecondini M, Mezzetti M, Sambo P, Bona S (2011) A methodological approach for defining spectral indices for assessing tomato nitrogen status and yield. Eur J Agron 35:135–143
62. Bagheri N, Ahmad H, Alavipanah S K, Omid M (2013) Multispectral remote sensing for site-specific nitrogen fertilizer management. Pesqui Agropecuária Bras 48(10)
63. Sumriddetchkajorn S (2013) How optics and photonics is simply applied in agriculture? In: International conference on photonics solutions of Proceedings of SPIE, vol 8883
64. Intaravanne Y, Sumriddetchkajorn S (2012) Baikhao (rice leaf) app: a mobile device-based application in analyzing the color level of the rice leaf for nitrogen estimation. In: Opto-electronic imaging and multimedia technology II, Proceedings of SPIE, vol 8558. The International Society for Optical Engineering, Washington
65. Omran E, El-Masry G, Rashad A (2012) A new approach to assess wetting front map by image analysis technique for precision irrigation farming. In: International conference of agricultural engineering CIGR-AgEng2012, Papers Book, Valencia 8–12 July 2012. ISBN: 10-84-615-9928-4
66. Aroca RV, Gomes R B, Dantas R R, Calbo A G (2013) A wearable mobile sensor platform to assist fruit grading. Sens (Basel) 13(5):6109–6140
67. Hettipathirana T (2004) Simultaneous determination of parts-per-million level Cr, As, Cd and Pb, and major elements in low level contaminated soils using borate fusion and energy dispersive X-ray fluorescence spectrometry with polarized excitation. Spectrochim Acta Part B 59:223–229
68. Gianoncelli A, Castaing J, Ortega L, Dooryhee E, Salomon J, Walter P, Hodeau J, Bordet P (2008) A portable instrument for in situ determination of the chemical and phase compositions of cultural heritage objects. X-Ray Spectrom 37(4):418–423
69. Downs R (2015) Determining mineralogy on mars with the CheMin X-ray diffractometer. Elements 11(1):45–50
70. Cannon K, Mustard J, Salvatore M (2015) Alteration of immature sedimentary rocks on Earth and Mars: recording aqueous and surface–atmosphere processes. Earth Planet Sci Lett 417:78–86
71. Eck D, Hirmas D, Giménez D (2013) Quantifying soil structure from field excavation walls using multistripe laser triangulation scanning. Soil Sci Soc Am J 77:1319–1328

72. Usamentiaga R, Molleda J, Garcia D, Bulnes F (2014) Removing vibrations in 3D reconstruction using multiple laser stripes. Opt Lasers Eng 53:51–59
73. Hirmas D et al (2016) Quantifying soil structure and porosity using three-dimensional laser scanning. In: Hartemink AE, Minasny B (eds) Digital soil morphometrics. Springer, Dordrecht
74. Rossi A, Hirmas D, Graham R, Sternberg P (2008) Bulk density determination by automated three-dimensional laser scanning. Soil Sci Soc Am J 72:1591–1593
75. Subroy V, Giménez D, Hirmas D, Takhistov P (2012) On determining soil aggregate bulk density by displacement in two immiscible liquids. Soil Sci Soc Am J 76:1212–1216
76. Zielinski M, Sánchez M, Romero E, Atique A (2014) Precise observation of soil surface curling. Geoderma 226–227:85–93
77. Sanchez M, Atique A, Kim S, Romero E, Zielinski M (2013) Exploring desiccation cracks in soils using a 2D profile laser device. Acta Geotech 8:583–596
78. Viscarra Rossel R, Webster R (2011) Discrimination of Australian soil horizons and classes from their visible-near infrared spectra. Eur J Soil Sci 62(4):637–647
79. Waiser T, Morgan C, Brown D, Hallmark C (2007) In situ characterization of soil clay content with visible near-infrared diffuse reflectance spectroscopy. Soil Sci Soc Am J 71(2):389–396
80. Viscarra Rossel RA, Cattle S R, Ortega A, Fouad Y (2009) In situ measurements of soil colour, mineral composition and clay content by vis–NIR spectroscopy. Geoderma 150:253–266
81. Lagacherie P, Baret F, Feret J, Madeira Netto J, Robbez-Masson J (2008) Estimation of soil clay and calcium carbonate using laboratory, field and airborne hyperspectral measurements. Remote Sens Environ 112:825–835
82. Steffens M, Kohlpaintner M, Buddenbaum H (2014) Fine spatial resolution mapping of soil organic matter quality in a histosol profile. Eur J Soil Sci 65:827–839
83. Viscarra Rossel R, Hicks W (2015) Estimates of soil organic carbon and its fractions with small uncertainty using visible–near infrared transfer functions. Eur J Soil Sci 66:438–450
84. Van Maarschalkerweerd M, Husted S (2015) Recent developments in fast spectroscopy for plant mineral analysis. Front Plant Sci 6:169
85. Tremblay N, Wang Z J, Ma B L, Belec C, Vigneault P (2009) A comparison of crop data measured by two commercial sensors for variable-rate nitrogen application. Precis Agric 10:145–161
86. Samborski SM, Tremblay N, Fallon E (2009) Strategies to make use of plant sensors-based diagnostic information for nitrogen recommendations. Agron J 101:800–816
87. Schmidt SB, Pedas P, Laursen K H, Schjoerring J K, Husted S (2013) Latent manganese deficiency in barley can be diagnosed and remediated on the basis of chlorophyll a fluorescence measurements. Plant Soil Environ 372:417–429
88. Castro ACM, Meixedo J P, Santos J M, Góis J, Bento- Gonçalves A, Vieira A, Lourenço L (2015) On sampling collection procedure effectiveness for forest soil characterization. Flamma 6:98–100
89. Liu X, Xuejun D, Daniel IL (2016) Ground penetrating radar for underground sensing in agriculture: a review. Int Agrophys 30:533–543
90. Cheng N, Conrad Tang H, Chan C (2013) Identification and positioning of underground utilities using ground penetrating radar (GPR). Sustain Environ Res 23(2):141–152
91. Doolittle J, Butnor J (2008) Chapter 6: Soils, peatlands, and biomonitoring. In: Jol HM (ed) Ground penetrating radar: theory and applications. Elsevier, Amsterdam, The Netherlands, pp 179–202
92. Guo L, Chen J, Cui X, Fan B, Lin H (2013) Application of ground penetrating radar for coarse root detection and quantification: a review. Plant Soil 362:1–23
93. Qin Y, Chen X, Zhou K, Klenk P, Roth K, Sun L (2013) Ground-penetrating radar for monitoring the distribution of near-surface soil water content in the Gurbantünggüt Desert. Environ Earth Sci 70:2883–2893
94. Van Dam RL (2014) Calibration functions for estimating soil moisture from GPR dielectric constant measurements. Comm Soil Sci Plant Anal 45:392–413
95. Mahmoudzadeh M, Francés A, Lubczynski M, Lambot S (2012) Using ground penetrating radar to investigate the water table depth in weathered granites-Sardon case study. Spain J Appl Geophys 79:17–26

96. Tosti F, Patriarca C, Slob E, Benedetto A, Lambot S (2013) Clay content evaluation in soils through GPR signal processing. J Appl Geophys 97:69–80
97. Raper RL, Asmussen L, Powell JB (1990) Sensing hard pan depth with ground-penetrating radar. Trans ASAE 33:41–46
98. Schmelzbach C, Tronicke J, Dietrich P (2012) Highresolution water content estimation from surface-based ground-penetrating radar reflection data by impedance inversion. Wat Resour Res 48:W08505
99. Barton CV, Montagu KD (2004) Detection of tree roots and determination of root diameters by ground penetrating radar under optimal conditions. Tree Physiol 24:1323–1331
100. Guo L, Lin H, Fan B, Cui X, Chen J (2013) Impact of root water content on root biomass estimation using ground penetrating radar: evidence from forward simulations and field controlled experiments. Plant Soil 371:503–520
101. Zhu S, Huang C, Su Y, Sato M (2014) 3D ground penetrating radar to detect tree roots and estimate root biomass in the field. Remote Sens 6:5754–5773
102. De Benedetto D, Castrignano A, Rinaldi M, Ruggieri S, Santoro F, Figorito B, Gualano S, Diacono M, Tamborrino R (2013) An approach for delineating homogeneous zones by using multi-sensor data. Geoderma 199:117–127
103. Tromp-van Meerveld HJ, McDonnell JJ (2009) Assessment of multi-frequency electromagnetic induction for determining soil moisture patterns at the hillslope scale. J Hydrol 368:56–67
104. Heil K, Schmidhalter U (2012) Characterisation of soil texture variability using apparent electrical conductivity at a highly variable site. Comput Geosci 39:98–110
105. White ML, Michele L, Shaw JN, Raper R L, Rodekohr D, Wood C (2012) A multivariate approach for high-resolution soil survey development. Soil Sci Aoc Am J 177(5):345–354
106. Cockx L, Van Meirvenne M, Vitharana U W A, Verbeke L P C, Simpson D, Saey T, Van Coille F M B (2009) Extracting topsoil information from EM38DD sensor data using neural network approach. Soil Sci Soc Am J 73(6):1–8
107. Harvey OR, Morgan CLS (2009) Predicting regional-scale soil variability using single calibrated apparent soil electrical conductivity model. Soil Sci Soc Am J 73:164–169
108. Doolittle J, Chibirka J, Muniz E, Shaw R (2013) Using EMI and P-XRF to characterize the magnetic properties and the concentration of metals in soils formed over different lithologies. Soil Horiz 54(3):1–10
109. Al-Gaadi K (2012) Employing electromagnetic induction techniques for the assessment of soil compaction. Am J Agric Biol Sci 4:425–434
110. Triantafilis J, Lesch S M, La Lau K, Buchanan S M (2009) Field level digital mapping of cation exchange capacity using electromagnetic induction and a hierarchical spatial regression model. Aust J Soil Res 47:651–663
111. Vitharana UWA, Van Meirvenne M, Simpson D, Cockx L, De Baerdemaeker J (2008) Key soil and topographic properties to delineate potential management classes for precision agriculture in the European loess area. Geoderma 143:206–215
112. Van Meirvenne M, Islam M M, De Smedt P, Meerschman E, Van De Vijver E, Saey T (2013) Key variables for the identification of soil management classes in the aeolian landscapes of North–West Europe. Geoderma 199:99–105
113. Martinez G, Vanderlinden K, Ordóñez R, Muriel J L (2009) Can apparent electrical conductivity improve the spatial characterization of soil organic carbon? Vadose Zone J 8(3):586–593
114. Wienhold BJ, Doran JW (2008) Apparent electrical conductivity for delineating spatial variability in soil properties. In: Allred BJ, Daniels JJ, Ehsani MR (eds) Handbook of agricultural geophysics. CRC Press, Taylor and Francis Group, Boca Raton, Florida, pp 211–215
115. Shaner DL, Kosla R, Brodahl M K, Buchleiter G W, Farahani H J (2008) How well do zone sampling based soil electrical conductivity maps represent soil variability? Agron J 100(5):1472–1480
116. Johnston MA, Savage M J, Moolman J H, du Plessis H M (1997) Evaluation of calibration methods for interpreting soil salinity from electromagnetic induction measurements. Soil Sci Soc Am J 61:1627–1633

117. Lesch SM, Herrero J, Rhoades JD (1998) Monitoring for temporal changes in soil salinity using electromagnetic induction techniques. Soil Sci Soc Am J 62:232–242
118. Doolittle J, Brevik EC (2014) The use of electromagnetic induction techniques in soils studies. Publications from USDA-ARS/ UNL Faculty. Paper 1462. http://digitalcommons.unl.edu/usdaarsfacpub/1462
119. Cassel F, Goorahoo D, Zoldoske D, Adhikari D (2009) Mapping soil salinity using ground-based electromagnetic induction. In: Metternicht G, Zinck JA (eds) Remote sensing of soil salinization. CRC Press, Taylor and Francis Group, Boca Raton, Florida, pp 199–233
120. Morris ER (2009) Height-above-ground effects on penetration depth and response of electromagnetic induction soil conductivity meters. Comput Electron Agric 68:150–156
121. Omran ESE (2016) Early sensing of peanut leaf spot using spectroscopy and thermal imaging. Arch Agron Soil Sci 1–14
122. Sawaya WN (2000) Proposal for the establishment of a regional network for date-palm in the near East and North Africa. A Draft Discuss FAO/RNE
123. Dembilio Ó, Jacas JA, Llácer E (2009) Are the palms Washingtonia filifera and chamaerops humilis suitable hosts for the red palm weevil, Rhynchophorus ferrugineus (Col. Curculionidae). J Appl Entomol 33:565–567
124. Mahmud AI, João F, Eleonore RAV (2015) Red palm weevil (Rhynchophorus ferrugineus Olivier, 1790): Threat of Palms. J Biol Sci 15(2):56–67
125. Faleiro JR (2005) Insight into the management of red palm weevil Rhynchophorus ferrugineus Olivier: based on experiences on coconut in India and date palm in Saudi Arabia, Fundación Agroalimed. Jorn Int Sobre El Picudo Rojo Las Palmeras 27–29:35–57
126. Yones MS, Arafat SM, Abou Hadid A F, Abd Elrahman H A, Dahi H F (2012) Determination of the best timing for control application against cotton leaf worm using remote sensing and geographical information techniques. Egypt J Remote Sens Space Sci 15:151–160
127. Mozib ME, El-Shafie HA (2013) Effect of red palm weevil, Rhynchophorus ferrugineus (Olivier) infestation on temperature profiles of date palm tree. J Entomol Nematol 5(6):77–83
128. Li D, Yao Y, Shao Z, Wang L (2014) From digital Earth to smart Earth. Chin Sci Bull 59(8):722–733

Development of Recent Information and Data on Irrigation Technology and Management

S. A. Abd El-Hafez, M. A. Mahmoud and A. Z. El-Bably

1 Introduction

In Egypt, expert systems in the agriculture field apply the application of agro-climate in agriculture that was mainly utilizing data in a set of models, including water management, planting dates of major field crops, etc., there was another technique depend on information-based, instead of a database. This led to the emergence of agriculture expert systems (www.claes.sci.eg). This required an intelligent assessment and through an understanding of all the climatic and hydrological as well as agricultural foundation and community conditions where such updated information is purposed to exercise in irrigation so that it could be widely applied [1]. Today, agriculture based on irrigation consumes the largest part of the water resources of the nation. Water consumption occurs via transpiration, evaporation, incorporation into products or crops, consumed by humans or livestock, or otherwise extracted from the water environment. Development of recent information on on-farm irrigation technique and management is very significant to improve irrigation technologies and advanced farm management practices that offer an opportunity for agriculture to utilize irrigation water more efficiently. Ongoing updated information for farmers is vital to adopt advanced water management practices to rationalize water without sacrificing crop yields [2]. Vanino et al. [3] revealed the suitability of the MultiSpectral Instrument

S. A. Abd El-Hafez · M. A. Mahmoud · A. Z. El-Bably (✉)
Water Requirements and Field Irrigation Research Department, Soil, Water and Environment Research Institute (SWERI), Agricultural Research Center (ARC), 9th Cairo Univ. Street, Giza, Egypt
e-mail: elbabli@gmail.com

S. A. Abd El-Hafez
e-mail: s.a.abdelhafez@gmail.com

M. A. Mahmoud
e-mail: mahmoud_abdalla96@yahoo.com

E.-S. E. Omran and A. M. Negm (eds.), *Technological and Modern Irrigation Environment in Egypt*, Springer Water,
https://doi.org/10.1007/978-3-030-30375-4_6

sensor on-board Sentinel-2A to estimate water requirements of tomatoes is providing useful information for optimizing the irrigation over extended farmland, at field level. The hypothesized positive links between Information and Communications Technologies (ICTs) and development [4–6].

Irrigated agriculture surely faces an increased water scarcity and drought. Irrigation technology improvement and related developments in water management depending on offer an opportunity for agriculture to mitigate water shortage more efficiently for the allocated water. Increasing water supplies from surface or groundwater sources are not the solution for meeting water shortage that they were in the past. Today, decreasing chances to develop new surface supplies, deficits in the Government budget and increasing concern for tourism, industry and environment make large-scale water projects, i.e., reservoirs and dams unlikely. Concurrently, users of groundwater in many areas are facing dwindling groundwater, low yields, and higher energy costs.

A question of considerable importance to policymakers concerns how farmers will adjust production to rationalize irrigation water in response to less information and data. Updating information to farmers will assist them to adopt more efficient irrigation techniques, convert irrigated land to dry-land farming, shift to crops that use less water, and apply less water per acre for a given technology and crop [7]. Irrigation improvement schemes require data collection and evaluation, planning and design of the scheme, and implementation. Collection data and its evaluation involve collecting information from soil surveys, water analyses, and topographic surveys [8]. The scale of the topographic map depends upon the type of irrigation improvement schemes that are proposed. For example, a surface irrigation scheme requires a more detailed scale due to the precision required for preparing the land. When the surveys and analyses concerning acceptable water quality and suitable soil characteristics are completed, the team can begin the planning and design stage. The cost of the surface irrigation scheme will largely depend on the results of the surveys. Based on the detailed data collected from the surveys, the engineers can begin to plan and design the pumping stations and the layout of the irrigation scheme. Upon completion of the detailed design, the team will begin the implementation of the irrigation scheme [9].

The aim of this chapter is to develop and manage the recent information and data on irrigation technology to promote intensive and sustainable irrigated agriculture and improve food security in Egypt.

2 Data Records

2.1 Data and Records of Irrigator

Taking a sample of irrigators' agricultural and financial records (or a census if the population is small), will allow for conclusions about scheme-level performance.

When comparing the enterprise records kept by the individual irrigators, conclusions about seed rates, fertilizer levels per acre for various crops (i.e., adoption rates) can be executed about irrigators and so can conclusions be made about yields and incomes. Analysis of the scheme level can be done by the agricultural extension worker (AEW) for the scheme and/or by an external evaluation team [10].

2.2 *Data and Records of Irrigation Management Advisor (IMA)*

In a scheme where irrigators have the responsibility of managing scheme affairs, the IMA will keep records of scheme-level costs and responsibilities. Scheme-level costs kept by the IMA include:

- Energy bills
- Costs of servicing and repair
- Security guard cost
- Replacement costs
- Levies (e.g., subscription fees to irrigation associations, etc.).

These scheme records (fixed costs) will allow for the budgeting of scheme costs per acre. Individual farmers may keep a record of their contributions to these costs in addition repairs and other costs relating to their plots. This will allow for irrigation profit estimating per plot.

The IMA should also keep other records, like the frequency of breakdown of equipment. Monitoring the breakdown frequency, for example, engine breakdown will allow decisions to be made on whether, for example, a new engine should be bought.

The IMA also needs to keep records on the consumption rate of energy. A sudden rise in consumption of the energy rate may indicate that the equipment needs servicing. Servicing equipment in time may prevent costly repairs following equipment breakdown. Records of power failure, where national electricity supply is available, should also be kept.

The IMA may also keep other non-financial records such as, for example:

- List of plot holders
- Gender disaggregation of plot holders
- List of IMA post bearers
- Gender disaggregation of IMA post bearers. These data can be updated as and when the need arises.

They will be useful in monitoring issues, i.e., advancement of women, for example, whether, over time, women access irrigation plots in their own right or hold leadership positions.

2.3 Data and Records of Agricultural Extension Worker (AEW)

The AEW assigned to the scheme may keep the same data as the IMA. This is because the AEW is well placed to communicate the data (should the need arise) to various stakeholders such as the scheme engineer, planners, and other researchers in the scheme [9].

Over and above these data, the AEW will keep records of:

- Crops cultivated in the scheme
- Total area per crop
- Cropping program
- Recommended agronomic practices
- Irrigation infrastructure condition
- Irrigation scheduling
- Details of courses run for committees and farmers and attendance
- Courses attended by the AEW
- Problems encountered, such as disease outbreaks and conflicts.

2.4 Data and Records of Experts

Some data and records of expert, which may have financial implications, must be collected by experts since they are impossible or difficult for farmers to collect. Such data include technical performance such as discharge rate, the assessment of equipment condition, and irrigation infrastructure.

The discharge rate can be taken at the scheme working. Monitoring the discharge rate on a regular basis, for example, yearly, will allow decisions to be made on whether equipment should be kept as it is, or whether it should be serviced or replaced altogether. Estimating the water used volumes can be the means for estimating the scheme efficiency.

Some of the environmental performance data, i.e., soil pH and water pollution, will also need regularly measuring, for example, yearly, throughout the life of the project in order to detect whether there are any changes in quality of water and soil as a result of the project [11].

2.5 Data and Records of the External Evaluators

A team of outsiders working closely with the stakeholders in the scheme can survey irrigators just before they start irrigating their new plots (baseline study).

After that, mid-term evaluation, ex-post evaluation (at the project completion), and impact evaluation (some years after completion) missions should be undertaken.

The missions do not need to be executed by the same team. For example, a mid-term and ex-post evaluation may be commissioned by a donor agency using its staff [9, 10].

Issues to be known in a formal survey include:

- Asset ownership by irrigators
- Family nutritional status
- Ability to pay school fees
- Employment creation
- Food security status
- Disease incidence among irrigating households and the surrounding community.

Farmer organization and management ability, and also the external evaluating team may have informal discussions with various stakeholders and make observations to cover issues such as:

- Advancement of women
- Different linkages with the scheme food security situation of the area
- Appropriateness of the technology, for example, treadle pumps to women
- Erosion
- Waterlogging/drainage problems.

The external evaluating team may also use agricultural and financial data from a sample of farmers and scheme-level data from IMA to cover issues such as:

- Financial viability of the scheme
- Change in irrigators' yields and/or incomes.

3 Data and Information Required for Irrigation

3.1 Crop Evapotranspiration (ET_c)

The main data for crop evapotranspiration which sometimes called crop water consumption or crop water requirement are climatic factors, crop characteristics, management practices, and environmental aspects. The main climatic factors that affect evapotranspiration or water consumptive use are wind speed, solar radiation, air temperature, and air humidity. There are many factors affect crop water consumptive use such as the type of crop, variety and development stages. Moreover, crop resistance differences to transpiration, the height of crop, canopy and characteristics of the crop root system lead to different evapotranspiration in different crops types under normal environmental circumstances.

There are some factors such as fertility lack, the salinity of soil, limited use of fertilizers and chemicals, lack of disease and pest control, poor soil management and finite water availability at the root zone may limit the crop development and reduce evapotranspiration or water consumption. Other factors that affect evapotranspiration or water used are crop population and groundcover. Agricultural practices and the irrigation system type used can change the microclimate, affect the crop attributes or affect the moistening of the soil and crop surface [11, 12]. All these affect crop water consumption or crop evapotranspiration.

3.2 Crop Coefficients

Crop coefficients values over the growing period are represented on the curve of crop coefficient. Only three values of crop coefficients (K_c) are desired to construct and describe the K_c curve: those during the initial stage (K_{cini}), the mid-season stage (K_{cmid}), and at the season stage end ($K_{c\ \mathrm{end}}$).

The values of K_{cini}, K_{cmid}, and K_{cend} for different crops were listed in [13]. The coefficients are organized by group type similar to the way it was done for the growth stages length. There is a close similarity in coefficients among crops that belong to the same crop group because they have similar crop characteristics, i.e., plant height, etc., and water management is normally similar [14].

3.3 Crop Types

The large variation in crop coefficients values between prime groups of crops is due to the lower transpiration of different crops due to close stomata during the day in pineapple crop and waxy leaves in citrus. Also, the differences in ground cover, the height of crop, the roughness of the crop, and produce various crop coefficients values [13].

3.4 Climate

Wind and humidity factors affect crop coefficients where the wind speed changes the crop aerodynamic resistance and their crop coefficients, in particular, for those crops that are taller compared to the grass reference crop. Crop aerodynamic resistance is changing with climate, in particular, relative humidity. Crop coefficient increases when wind speed rises and relative humidity decreases. Higher crop coefficient values produce from more arid conditions and climates that have greater wind speed. However, lower crop coefficient values produce from more humid conditions and climates that have lower wind speed.

Average climatic data were used in estimating the $\mathrm{ET_o}$. Since the weather condition differs from year to year, $\mathrm{ET_c}$ will change from year to another and from period to period. Monthly ET_c values can vary from one year to another year by 50% or more. For irrigation projects establishment, the variations with time become very important. When sufficient climatic data are available for 10 years or less, ET_c could be calculated for each year, and a probability analysis could be done. The value of ET_c then selected for design is commonly depend on a probability of 75–80%, which could be similar to the probability of water availability.

Changes in the microclimatic environment should be considered because of the cultivated area. Climatic data are gathered before irrigation progress has taken place and normally the agro-meteorological stations, from which data are taken, are located where there is irrigation development. Irrigation fields will produce a different microclimate and ET_c may not accurate to the predicted values, depend on agro-meteorological data that used. This is more clear for large the cultivated area in arid windy climates [14, 15].

Agro-meteorological elements determining crop water used or evapotranspiration are weather parameters, which give energy for evaporation. The main weather factors to be considered are wind speed, solar radiation, air humidity, and air temperature. Agro-meteorological data are registered at various weather stations. Agro-meteorological stations are sited in cropped areas where instruments are exposed to atmospheric conditions similar to those for the surrounding crops. In these stations, air temperature, sunshine duration, wind speed, and air relative humidity at each site were measured at the height of two meters above the ground surface. The gathered data at stations other than agro-meteorological stations require careful analysis of their validity before use.

In most regions in Egypt, a national agro-meteorological society commonly lists processed climatic data from the different stations and issues agro-meteorological bulletins as a service for the stakeholders. This service should be extended for information on the collected climatic data at various weather stations in Egypt to be used as input data in the FAO Penman-Monteith equation.

The databases can be checked in order to confirm the matching of the existing database. They should be used for preliminary studies as they contain, in general, mean monthly data only and some stations have incomplete data. The information found from the databases should never replace actual long-term data.

3.5 Data and Records of Water and Soil

Water used by crop or evapotranspiration of the crop (ET_c) is a combination of transpiration by the crop and evaporation from the soil surface. Crop coefficient is integrated within the variation of soil evaporation and crop transpiration among field crops.

According to FAO [16], if plants are sufficiently established and there are favorable growing conditions, i.e., available nutrients and water, soil aeration, etc., the ET_c is not affected, even when rooting depth is severely restricted. However, the following conditions must be considered:

Table 1 General soil data of heavy texture (clayey soil)

Parameters	
Total available soil moisture (FC-WP)	200.0 mm/m
Maximum rain infiltration rate	40 mm/day
Initial soil moisture depletion	0%
Initial available soil moisture	200.0 mm/m

Available soil water: The soil water content effect on crop water used or evapotranspiration varies with the crop and is conditioned primarily by the soils type and characteristics of water retention, characteristics of crop root and the agro-meteorological factors determining the transpiration. When evaporative conditions are lower, the crop may transpire at the predicted evapotranspiration rate even though available soil water depletion is greater. Evapotranspiration of crop or water used by the crop will be reduced if the rate of water supply to the roots zone is unable to commensurate with transpiration losses. This is more clear in heavy textured compared with light textured soils.
Groundwater: Because crop growth is influenced by a shallow water table, the ET_c is also affected.
Salinity: evapotranspiration of the crop or water consumption is affected by soil salinity, as the absorption of the crop to the added irrigation water into the soil salinity is reduced due to the higher osmotic potential of saline soil water.
Water and crop yield: Various crops have different critical periods for soil water stress. Therefore, the timing and duration of the shortage are important and associated with the yield.

General soil data of heavy texture such as total available soil moisture (field capacity (FC), welting point (WP), maximum rain infiltration rate, initial soil moisture depletion, and initially available soil moisture are presented in Table 1.

3.6 Stages of Crop Growth

The crop coefficient for a given crop changes during the growing period as the groundcover, crop height, and leaf area changes. Four growth stages are recognized for the selection of crop coefficient: an initial stage, crop development stage, mid-season stage, and stage of the late season [10].
Crop coefficient and the calculation of ET_c:

- Crop growth stages identification, determination of their lengths, and selection of the corresponding K_c values
- Adjustment of the selected K_c values for moistening frequency or climatic conditions during each stage

- Crop coefficient curve creation that allows one to determine K_c values for any period during the growing period
- Calculation of ET_c from reference evapotranspiration and crop coefficient.

4 Evapotranspiration Concepts

4.1 Crop Evapotranspiration and Irrigation Requirements

Evapotranspiration of crop, which called crop water requirement (CWR) or water consumption of crop, describes the gross water used by crop or evapotranspiration [14]. Defined crop water requirements as 'the depth of water required to meet the water loss resulted from evapotranspiration of crop or water consumptive use, being disease-free, growing in the fields under normal conditions, comprising soil fertility, soil water content, and achieving full production potential under the growing environment. The use of computer software for the estimation of ET_c or CWR will be explained later.

Irrigation requirements (IR) refer to the water that must be supplied through the irrigation system to guarantee that full evapotranspiration of the crop or water consumption of the crop was received. If some other water sources are used by crops such as rainfall, lateral and underground seepage, and water stored in the soil, and then the irrigation requirement can be considerably less than the evapotranspiration of the crop or water consumption of the crop.

4.2 Irrigation Scheduling

When evapotranspiration of crop and irrigation requirements have been calculated, the coming process is the preparation of field irrigation schedules. There are three parameters have to be taken into consideration in preparing an irrigation schedule:

- The daily evapotranspiration of the crop or water consumption of the crop.
- The soil, in particular, total available moisture content or water-holding capacity.
- Depth of the efficacious root zone. Response crop to irrigation is affected by the soil fertility, soil biological status, and chemical and physical condition of soil. Soil conditions having depth, organic matter, texture, structure, bulk density, salinity, sodicity, drainage, topography, fertility, and chemical characteristics all affect the extent to which a crop root system penetrates into the soil and uses available moisture and nutrients. These factors affect the water movement, the ability of soil to retain water, and the ability of the crops to use the water. The irrigation system should be commensurate with most of these conditions.

The estimated values for available soil water content and intake have wide ranges. The soil database values need to be constantly refined to appropriate the field conditions. In the field, the actual value may differ from location to another, season-to-season, and even within the season. Within the season, it varies based on the farm type and equipment tillage type, tillage operations a number, management of the residue, and crop type and water quality. Adequate surface drainage is necessary for the irrigated soil. Internal drainage in the crop root zone can be natural or from an installed mole drainage system.

4.3 Reference Evapotranspiration (ET_o)

After all the parameters of the FAO Penman-Monteith equation have been determined, it is now possible to calculate ET_0. For Sakha Agricultural Research Station, these calculated values are shown in Table 2 in Chap. 8. From this table, it can be seen that the peak ET_0 at north delta area is 5.69 mm per day and it occurs in June, if calculated manually through an equation of the FAO Penman-Monteith [17, 18].

The manual calculation of ET_0 is a long and tedious procedure, and the risk of making arithmetical errors is fairly high. A computer program has been developed to accelerate the calculations and make them less tedious to perform. One such software is the FAO CROPWAT computer program for estimating reference evapotranspiration (ET_0) and evapotranspiration of the crop. At this stage, it is only important to compare the monthly reference evapotranspiration values obtained through manual calculations with those obtained using CROPWAT. From the climatic data in Table 2 for Sakha Agricultural Research Station, ET_0 was estimated with CROPWAT Version 8.0.

5 Calculation of Crop Water and Irrigation Requirements Using Computer Software

5.1 Model of the FAO CROPWAT

CROPWAT is a program that can calculate evapotranspiration of crop or water consumption of crop and irrigation requirement from data of crop and climatic. The program is interactive. Besides, the model allows the irrigation schedules development for various management conditions and the project water supply estimation for varying cropping patterns.

Model of the CROPWAT is depending on a model of water balance where the soil moisture level is determined on a daily basis from calculated evapotranspiration and inputs of rainfall and irrigation. Methodologies for water consumption of crop

Table 2 Monthly ET_o Penman-Monteith data

Country: Egypt				Station: North Delta (SAKHA)			
Altitude: 20 m				Latitude: 31.11° N Longitude: 30.95			
Months	Min Temp (°C)	Max Temp (°C)	Humidity (%)	Wind (m/s)	Sun (h)	Rad (MJ/m^2/day)	ET_o (mm/day)
January	6.0	19.3	84	1.30	6.2	11.4	1.54
February	6.2	20.5	82	1.40	6.9	14.2	2.03
March	7.8	23.0	73	1.70	7.8	18.1	3.04
April	10.3	27.0	62	1.51	8.7	21.6	4.15
May	14.1	31.1	54	1.51	9.6	24.1	5.23
June	17.0	32.0	58	1.51	10.8	26.1	5.69
July	19.0	34.0	63	1.30	10.5	25.5	5.66
August	18.3	33.5	67	1.30	10.2	24.2	5.27
September	17.6	32.0	71	1.10	9.5	21.0	4.31
October	15.5	29.8	73	1.00	8.5	16.9	3.24
November	12.5	25.8	77	1.10	7.3	12.9	2.25
December	8.2	21.5	84	1.10	5.9	10.4	1.54

or evapotranspiration of crop and yield response to irrigation water are used, while the actual evapotranspiration is determined from the available soil water status.

Monthly climatic data use in the program. Data of monthly climatic variables are temperature, wind speed, sunshine hours, relative humidity, and rainfall for the calculation of reference evapotranspiration. It also has four different methods to calculate effective rainfall but to be capable of doing this it requires dependable rainfall as input.

Through the input of crop data, i.e., growth stages, K_c factors, depth of root zone, and admissible soil moisture deficit factor, the program calculates the crop evapotranspiration on a decade basis (10 days).

The application of CROPWAT in calculating crop evapotranspiration and irrigation requirement is best illustrated by using an example of irrigated areas for smallholder in Egypt, as is shown later. In typical smallholder irrigation schemes in Egypt, each farmer is allocated on average a plot of between 0.5 and 1.5 acres. Smallholder farmers normally prefer to grow from two to four crops per season to obtain crops variety for home consumption, to allow agronomic considerations (rotations) and to spread their risk when it concerned with marketing.

5.2 *Program of Crops Sequence or Rotation*

This program is the initial step in calculating crop evapotranspiration, depend on the system capacity or the area to be completed by irrigation is determined. With the full participation of farmers, a selection of what crops to grow in summer and winter, respectively, is made. Factors to be taken in our consideration in crop selection include farmers' wishes and aspirations, financial considerations, climate and soils, water availability, labor requirements, marketing aspects, availability of inputs, rotational considerations, and susceptibility to diseases. These factors are normally sited specific.

Once the crops are selected, a cropping program showing the seasonal cropping patterns and indicating the place and the occupying area for each crop is made. Of importance are the sowing or transplanting dates, the growing season length and the needed time for crop harvest and soil preparation for the coming crop. It must be noted that the needed time for harvest and preparation of soil ought not to be included when estimating the evapotranspiration of the crop. It is, therefore, useful to indicate on the cropping program diagram the needed time to harvest.

In order to decrease the risk of pests and diseases and to avoid elimination of certain nutrients through plant uptake, the cropping program should allow the sequence of the crops among the subplots. Vegetables such as cabbages, carrots, onion, and field crops like wheat, maize, groundnuts, cotton, and beans could safely be planted on the same subplot every two years.

Cropping programs are not fixed, and they belong to the farmers. This should be considered when planning the system of irrigation. For design purposes, a cropping pattern should be made in such a condition that the water requirements for other crops that the farmer intends to grow could be satisfied. This includes a careful study of all points mentioned above and detailed discussions with the farmers.

5.3 *The Efficacious Rainfall and the Reference Crop Evapotranspiration (ET_o) Calculation*

The next stage is to input data on climatic variables, which are wind speed, temperature, relative humidity, sunshine hours, and rainfall into CROPWAT, to calculate the effective rainfall and the reference crop evapotranspiration. The required data concerning climatic variables for input into CROPWAT are normally contained in climatic handbooks issued by a national meteorological institution in Egypt. Alternatively, different climatic data files on disk saved after earlier sessions or from the CLIMWAT database [18] can be applied for calculating ET_o and effective rainfall. The input of relevant climatic data and dependable rainfall for Sakha climatic station result in computer printouts such as presented in Table 3. Method of the USDA Soil Conservation is used for the estimating of the efficacious rainfall from the gross rainfall.

Table 3 Effective rainfall method: USDA S.C. method

Months	Rainfall (mm)	Effective rainfall (mm)
January	17.0	16.5
February	13.0	12.7
March	5.0	5.0
April	5.0	5.0
May	2.0	2.0
June	0.0	0.0
July	0.0	0.0
August	0.0	0.0
September	0.0	0.0
October	3.0	3.0
November	8.0	7.9
December	12.0	11.8
Total	65.0	63.8

Not all the dependable rainfall is effective because it may be lost through evaporation, deep percolation or surface runoff. Only a portion of the rainfall can be effectively used by the crop, depending on the depth of root zone and the capacity of soil storage. Different methods exist to estimate the effective rainfall and the reader is referred to [16] for details.

One of the most commonly used methods is the method of the USDA Soil Conservation Service. The relationship between mean monthly effective rainfall and mean monthly rainfall is shown for different average monthly ET_c.

Depending on the cropping program adopted, the coming step is to enter the crop data into CROPWAT to enable the program to estimate crop water consumption or evapotranspiration of crop for the various crops. The crop data required is the crop planting dates, values of the crop coefficient at the various growth stages, growth stages length, depth of the crop roots at the different growth stages are given in Table 4, the allowable soil moisture deficit or depletion levels and the yield response factors (K_y). K_y is a factor to estimate yield reductions because of water stress condition [19].

This information should be based on local data, obtained through surveys or recommendations of agriculture research stations and extension service in each site. The agro-methodologies for estimating the above crop data should be accomplished. CROPWAT also contains data files for 30 different crops, based on global values, which can be recovered and adjusted for local conditions.

After the input of the crop data, CROPWAT proceeds to calculate the evapotranspiration of crops and irrigation requirement of the given cropping pattern, using the entered crop data and the ET_o and values of effective rainfall calculated earlier. The calculation of evapotranspiration of crop or consumption is done on a decade (10 days

Table 4 Maize crop data

Stage	Initial	Develop	Mid	Late	Total
Length (days)	15	35	40	30	120
Kc Values	0.30	→	0.20	0.35	
Rooting depth (m)	0.30	→	1.00	1.00	
Critical depletion	0.55	→	0.55	0.80	
Yield response f.	0.40	0.40	1.30	0.50	1.25
Crop height (m)			2.00		
Planting date: 06/06					
Harvest: 30/10					

period) basis. For reasons of simplicity, all months are taken to have 30 days, subdivided into three decades of ten-day each. The mistake caused by this assumption is negligible.

After setting up the crop coefficient curve, the next process is the calculation of evapotranspiration of crop or consumption. The crop coefficient value can be determined for any period during the growing period of the crop coefficient curve. Once the crop coefficient values have been defined, the evapotranspiration of the crop can be calculated by multiplying the crop coefficient values for the corresponding reference evapotranspiration values are shown in Table 5

Table 5 Crop water requirements

ETo station: SAKHA (North Delta)				Crop: Maize (Grain)			
Rain station: SAKHA				Planting date: 06/06			
Month	Decade	Stage	Kc coeff	ETc (mm/day)	ETc (mm/dec)	Eff rain (mm/dec)	Irr. req. mm/dec
Jun	1	Init	0.30	1.66	8.3	0.1	8.2
Jun	2	Init	0.30	1.71	17.1	0.0	17.1
Jun	3	Deve	0.44	2.50	25.0	0.0	25.0
Jul	1	Deve	0.69	3.93	39.3	0.0	39.3
Jul	2	Deve	0.95	5.36	53.6	0.0	53.6
Jul	3	Mid	1.17	6.45	70.9	0.0	70.9
Aug	1	Mid	1.19	6.42	64.2	0.0	64.2
Aug	2	Mid	1.19	6.27	62.7	0.0	62.7
Aug	3	Mid	1.19	5.89	64.8	0.0	64.8
Sep	1	Late	1.11	5.15	51.5	0.0	51.5
Sep	2	Late	0.84	3.62	36.2	0.0	36.2
Sep	3	Late	0.56	2.21	22.1	0.1	22.0
Oct	1	Late	0.38	1.36	4.1	0.2	3.8

Monthly, ten-day or weekly values for crop coefficient are necessary when ET_c calculations are prepared on a monthly, ten-day, or weekly time basis, respectively. A general procedure is to establish the K_c curve, overlay the curve with the length of the weeks, decades or months and to obtain graphically from the curve the K_c value for the time under consideration.

The CROPWAT computer outputs, showing the irrigation needs or requirements for the various crops in the cropping program, have to be combined to get the irrigation requirements of all crops together, and which are irrigated at the same time. In addition, the corrected water consumption of crop for the months of peak demand must be calculated in order to present the crop and irrigation needs or irrigation requirement in a comprehensive way and allow the correction for peak demand, a summary table should be composed showing on a monthly basis the ET_0, the effective rainfall, the corrected ET_c, the irrigation requirements, expressed in mm, as well as the total requirements, expressed in m^3.

A properly designed, constructed and operated irrigation system shall not have effect on ET_c, with the exception of localized irrigation. Hence, the variation in the amount of water that applied for irrigation under the one or the other method was not attributed to the effect of the method on ET_c, but to the corresponding efficiency being achieved under the one or the other method localized irrigation (drip, spray jet, etc.) only moisten part of the soil and since evapotranspiration includes plant transpiration and the evaporation from the soil, the overall ET_c should be expected to be less under localized irrigation systems. However, ET_c is not affected by the method when the crop is near or at full groundcover. For the period before 70% groundcover reduced ET_c should be expected, since evaporation is limited to the wet areas of the soil only.

As for cultural practices, the used fertilizers have only a slight effect on ET_c, as long as the nutrient requirements for most favorable growth and yield are provided. The crop density will affect ET_c in the same way as percentage groundcover. For low plant populations, when the soil in the area in-between the rows is kept dry, the evaporation will be less, and thus ET_c will be less to a higher crop population. Tillage produces little, if any, the effect on ET_c. Rough tillage will accelerate evaporation from the plow layer, and deep tillage increases water losses when the land is fallow or when the crop cover is sparse. As far as mulching is concerned, while polyethylene and asphalt mulches are effective in reducing ET_c, residues of the crop are often considered of little net benefit in reducing ET_c. Crop residues as a barrier to soil evaporation are ineffective in irrigated agriculture. According to FAO [13, 14], the lower temperature of the covered soil and the higher reflective capacity of the organic matter are easily outweighed by evaporation of the often re-wetted residue layer. Windbreaks, relying on the distance covered and the height of the windbreak can reduce ET_c by 5–30% in windy, warm, and dry climate because of their effect on wind velocity. Anti-transpirants have been used in research for the reduction of ET_c. Their use has so far been limited to research and pilot projects.

5.4 Net Irrigation and Irrigation Requirements

A properly designed, constructed and operated irrigation system will not have any effect on ET_c, with the exception of localized irrigation. Hence, the variation in the water amount that used for irrigation under the one or the other method was not attributed to the effect of the method on ET_c, but to the corresponding efficiency being achieved under the one or the other method localized irrigation such as drip, spray jet, etc., only wets part of the soil and since evapotranspiration includes plant transpiration and the evaporation from the soil, the overall ET_c should be expected to be less under localized irrigation systems. However, ET_c is not affected by the method when the crop is near or at full groundcover. For the period before 70% groundcover reduced ET_c should be expected, since evaporation is limited to the wet areas of the soil only. Predicted values, depend on agro-meteorological data. This is more pronounced for large cropped areas in arid and windy climates. As for water requirements versus irrigation requirements, it is important to make a distinction between crop water requirement (CWR) which called evapotranspiration of the crop or water consumption of crop and irrigation requirement (IR). Whereas CWR refers to the water used by crops for the construction of cells and transpiration, the irrigation requirement is the irrigation water that has to be applied through the irrigation system to ensure that the crop receives its full water requirement. If the source of water supply is sole irrigation for the crop, then the irrigation requirement will be at least equivalent to the crop water requirement and is generally greater to allow for inefficiencies in the irrigation system. If the crop receives other sources of water, i.e., rainfall, lateral and underground seepage, water stored in the soil, etc., and then the irrigation requirements can be less than the crop water requirement.

The net irrigation requirement (IR_n) does not have losses that are occurring during the irrigation process. IR_n and losses consist of the gross irrigation requirement (IR_g). It is significant to realize that the estimation of crop evapotranspiration or water consumption is the initial process in the estimation of irrigation requirements of a given cropping program. Importance of calculating irrigation requirements and the crop evapotranspiration or water consumption for a proposed cropping pattern is a necessary portion of an irrigation system establishment.

The irrigation requirement (IR) is one of the main parameters for the operation of irrigation and water resources systems. Detailed information of the irrigation requirement quantity and its temporal and spatial variability is essential for evaluating the adequacy of water resources, for assessing the need of reservoirs of storage, and for the determining the capacity of irrigation systems. It is of prime importance in the policy formulation for the optimal water resources allocation as well as in decision making concerning operation and management of irrigation systems.

Incorrect estimation of the irrigation requirement causes serious problems in the performance and to the waste of valuable water resources. It leads to disadvantageous control concerning the soil moisture deficit in the root zone; it causes water logging due to the shallow water table, salinity or leaching of nutrients from the soil. It may lead to the inappropriate ability of the irrigation network or reservoirs of storage, to reduce field water use efficiency or water productivity and to a reduction in the irrigated area. Overestimating irrigation requirement at peak demand may also lead to increasing development costs.

The net irrigation requirement is derived from the field balance equation:

$$\mathrm{IR}_n = \mathrm{ET_c} - (P_e + G_e + W_b) + \mathrm{LR}\,\mathrm{mm} \tag{1}$$

where:

IR_n Net irrigation requirement in mm
ET_c Crop evapotranspiration in mm
P_e Effective dependable rainfall in mm
G_e Groundwater contribution from water table in mm
W_b Water stored in the soil at the beginning of each period in mm
LR Leaching requirement in mm

In most situations encountered in the planning of smallholder irrigation schemes at Nubaria site in the summer season, the project sites are located in dry areas with no rainfall. Hence, for planning purposes, the contribution of water stored in the soil is not considered in such schemes.

Crop evapotranspiration or water consumption can be partially covered by rainfall in the winter season. However, while the rainfall contribution may be substantial in a few years, in other years, it may be limited. Therefore, in establishing irrigation projects, the average values of rainfall that used should be avoided if more than 10 years of annual rainfall data are available, as is the case for Sakha Agriculture Research Station as presented in Table 3. In such cases, by using these data, a probability analysis can carry out so that a dependable level of rainfall is selected. The reliable rainfall is the rain that can be accounted for with a certain statistical probability, determined from a historical rainfall records. Before one carries out statistical analysis, it is always important to check with the meteorological station.

In most situations encountered in the planning of small-scale irrigation projects in winter season at North Delta, the cropped area sites that are located in dry areas with very low rainfall. Hence, for planning purposes, the water stored contribution in the soil is considered negligible in such areas.

It should be realized that the calculation of evapotranspiration of the crop or water consumption for crop and irrigation requirement is a theoretical exercise, based on statistical analysis of climatic parameters. However, climatic variables are unsteady. Consequently, the calculation of irrigation water requirements at the planning level can only be an approximation and it is not appropriate or recommended to attempt detailed accuracy.

Crop irrigation schedule is tabulated in Table 6 and Fig. 1, where for each maize crop data input is presented together with the corresponding water consumption for a crop called evapotranspiration of crops and irrigation requirements as calculated by CROPWAT.

The result of the totals is shown in Table 7 where maize crop is shown together

Table 6 Crop irrigation schedule

Date	Day	Stage	Rain (mm)	K_s fraction	ET_a (%)	Depl. (%)	Net Irr. (mm)	Deficit (mm)	Loss (mm)	Gr. Irr (mm)	Flow (l/s/ha)
9 Jul	34	Dev	0.0	1.00	100	55	85.7	0.0	0.0	107.1	0.36
29 Jul	54	Mid	0.0	1.00	100	58	115.6	0.0	0.0	144.5	0.84
16 Aug	72	Mid	0.0	1.00	100	57	114.7	0.0	0.0	143.4	0.92
5 Sep	92	End	0.0	1.00	100	58	115.6	0.0	0.0	144.5	0.84
3 Oct	End	End	0.3	1.00	100	43					

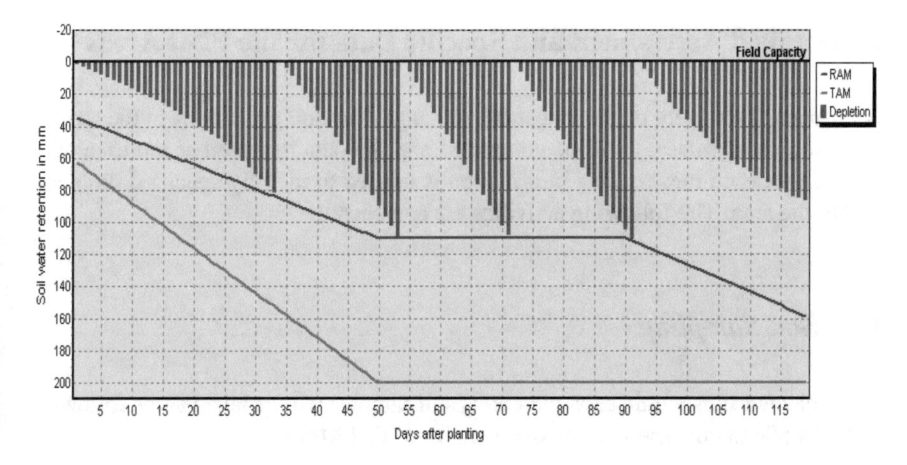

Fig. 1 Crop irrigation schedule

Table 7 Totals

Total gross irrigation	539.4 mm
Total rainfall	0.5 mm
Total net irrigation	431.6 mm
Effective rainfall	0.5 mm
Total irrigation losses	0.0 mm
Total rain loss	0.0 mm
Actual water use by crop	518.4 mm
Moist deficit at harvest	86.4 mm
Potential water use by crop	518.4 mm
Actual irrigation requirement	517.9 mm
Efficiency irrigation schedule	100.0%
Efficiency rain	100.0%
Deficiency irrigation schedule	0.0%

with the corresponding crop evapotranspiration or water consumption and irrigation requirements as calculated by CROPWAT 8.0.

The CROPWAT 8.0 model can be used for scheduling a single crop or multiple crops in the same scheme, thereby estimating water requirements for the same irrigation management. The combination of climatic analysis, soils types, and water use modeling provides an excellent opportunity for matching water needs to appropriate crops in different regions.

6 Detailed Assessment and Specific Data for the Pilot Areas

A detailed assessment of the irrigation, drainage, and soil conditions in the tertiary irrigation canal (mesqa) will be completed and the data analyzed at representative sites within each mesqa. This information is needed to identify specified problems within the area. The following activities are considered.

6.1 Soil Sampling

- At selected sites within each pilot area, soil samples will be taken at the following depths; 0–15 cm; 15–30 cm; 30–60 cm; and 60–120 cm.

6.2 Soil Analysis

Each soil sample will be analyzed for the following:

- Physical properties
- Electrical conductivity (EC)
- SAR (calculated)
- pH
- Nitrate-Nitrogen (NO_3)
- Phosphorus (*P*) (0–15 cm layer only)
- Organic matter (0–15 cm layer only)
- Trace elements (Ni, Pb, Cd, Cr, and Co) (0–15 cm layer only), and
- Heavy metals (Zn, Mn, Fe, Hg) (0–15 cm, layer only).

6.3 Water Balance Calculation

- Measure the amount of irrigation water pumped from the mesqa to the pilot area and selected farms;
- Measure the amount of surface drainage water from the pilot area and the selected farms;
- Measure the subsurface drainage discharge from the pilot area and the selected farms; and
- Calculate the current water balance for the pilot area.

6.4 Groundwater Monitoring

- Water table observation wells will be installed at selected sites, and located mid-way between two subsurface drainage laterals. These observation wells are needed to measure groundwater fluctuations in each field. Each well will be 2.0 m in length, with 0.25 m above ground level. The bottom of each well will be plugged, and a tight-fitting cap installed on the top. Each well will be slotted from the bottom to within 0.25 m of the ground surface. Each well will be sealed at ground surface to stop surface water from flowing down the side of the well.

6.5 Irrigation and Groundwater Analysis

The irrigation water at the inlet to the pilot area plus groundwater samples from the observation wells will be analyzed for the following:

- EC
- SAR (calculated)
- pH
- Nitrate-Nitrogen (NO_3)
- Fecal Coliform bacteria
- Selected pesticides commonly applied in the region
- Trace elements (Ni, Pb, Cd, Cr, and Co), and
- Heavy metals (Zn, Mn, Fe, Hg).

6.6 Analysis of Surface and Subsurface Drainage

Surface drain water will be sampled at the site where the surface drainage system leaves the pilot area. Subsurface drainage water will be sampled at the collector drain manhole located closest to the downstream end of the mesqa. All water samples will be analyzed for the following:

- EC
- SAR (calculated)
- pH
- Nitrate-Nitrogen (NO_3)
- Fecal Conform bacteria
- Selected pesticides commonly used in the region
- Trace elements (Ni, Pb, Cd, Cr, and Co), and
- Heavy metals (Zn, Mn, Fe, Hg).

Based on an assessment of data from; the existing information; Environmental and Soil Baseline Survey and the detailed pilot area analysis; the specific irrigation; drainage; soil, environmental, and groundwater problems within each irrigation canal (mesqa) will be identified. This will be executed in close consultation with the farmers.

7 Conclusions

The required new knowledge, data and information for irrigation, evapotranspiration concepts, and computer software is vital to calculate evapotranspiration of crop and irrigation requirements. Furthermore, a detailed assessment and specific information for the experimental areas for irrigation schemes to update information and data on irrigation technology to promote intensive and sustainable irrigated agriculture and improve food security and support decision making in Egypt.

8 Recommendations

Development and recent data provide Egypt with a planning tool for rational exploitation of developing and manage recent information and data concerning soil, crops and water resources. This planning is intended to lead to an increase in crop production for local consumption, as well as promote the production of high-value crops. The planning tool will support decision making with direct regard to:

- Establishing agro-ecological zones (AEZ) maps
- Identifying water availability and selection of potential irrigation areas
- Identifying criteria for crop selection and estimated water requirements
- Identification of the most favorable regions to develop irrigation management practices
- Giving priority to irrigation water distribution
- Organization and control of irrigation supply management
- Developing irrigated agriculture in small, medium and large-scale projects on old and new lands. Further, decision making will be made with regard to:
  - Supporting the national policy options for irrigation water distribution.
  - Upgrading the Egyptian agricultural production.
  - Recommending options for water harvesting and storage in the coastal regions.
  - Suggesting solutions for mismanagement.
  - Identifying energy requirements for irrigation systems.

References

1. El-Bably A (2015) Advanced irrigation technology for enhancing field water use efficiency and precision irrigation for rice (a case study-Egypt). In: 26th ICID Euro mediterranean regional conference innovate to improve irrigation performance, at the occasion of the 66th international executive council of the international commission of irrigations and drainage (ICID) Montpellier, France, 11–16 Oct 2015
2. Abd El-Hafez SA, El-Bably AZ (2006) Irrigation improvement project (IIP). On-farm water management (ofwm), Kafr El-Sheikh and El-Behira Directorates, Final report
3. Vanino S, Nino P, De Michele C, Falanga Bolognesi S, D'Urso G, Di Bene C, Pennelli B, Vuolo F, Farina R, Pulighe G, Napoli R (2018) Capability of sentinel-2 data for estimating maximum evapotranspiration and irrigation requirements for tomato crop in central Italy. Remote Sens Environ 215:452–470. https://doi.org/10.1016/j.rse.2018.06.035
4. Njoh AJ (2018) The relationship between modern information and communications technologies (ICTs) and development in Africa. Util Policy 50:83–90. https://doi.org/10.1016/j.jup.2017.10.005
5. Car NJ (2018) USING decision models to enable better irrigation decision support systems. Comput Electron Agric 152:290–301. https://doi.org/10.1016/j.compag.2018.07.024
6. Knapp T, Huang Q (2017) Do climate factors matter for producers' irrigation practices decisions? J Hydrol 552:81–91. https://doi.org/10.1016/j.jhydrol.2017.06.037
7. Ministry of water resources and irrigation (MWRI) (2008) Monitoring and evaluation of an irrigation improvement project-report in W10 command area in Kafr El-Shiekh, Egypt
8. National Water Research Center (2013) Designing Local Framework for Integrated Water Resources Management Project, 1st Technical Report, pp 10–15
9. Fulazzaky MA, Akil H (2009) Development of data and information centre system to improve water resources management in Indonesia. Water Resour Manag 23:1055–1066
10. FAO (2002) Crop water requirements and irrigation scheduling. Irrigation Manual 4, Harare, Zimbabwe
11. FAO (2002) Monitoring the technical and financial performance of an irrigation scheme. Irrigation Manual 14, Harare, Zimbabwe
12. Jordans E (1998) Sector guide irrigation. Socioeconomic and gender analysis programme, FAO, SEAGA: Rome, Italy
13. Richard A, Pereira L, Raes D, Smith M (1998) Crop evapotranspiration: guidelines for computing crop water requirements. FAO Irrigation and Drainage Paper 56. FAO: Rome, Italy
14. Doorenbos J, Pruitt WO (1984) Crop water requirements. FAO Irrigation and Drainage Paper 24. FAO: Rome, Italy
15. Sharma K (2009) Application of the climate information and prediction in the water sector: capabilities. National Rainfed Area Authority, New Delhi, India
16. Smith M (1992) CROPWAT: a computer programme for irrigation planning and management. FAO Irrigation and Drainage Paper No. 46. FAO: Rome, Italy
17. El-Bably AZ (2007) Irrigation scheduling of some maize cultivars using class a pan evaporation in North Delta. Egypt. Bull Fac Agric, Cairo Univ 58:222–232
18. Smith M (1993) CLIMWAT for CROPWAT: a climatic database for irrigation planning and management. FAO Irrigation and Drainage Paper 49. FAO: Rome, Italy
19. Doorenbos J, Kassam AH (1986) Yield response to water. FAO Irrigation and Drainage Paper 33. FAO: Rome, Italy

Medicinal Plants in Hydroponic System Under Water-Deficit Conditions—A Way to Save Water

Eid M. Koriesh and Islam H. Abo El-Soud

1 Introduction

In the Holy Quran, God says "we made from the water everything live." In Egypt, the current water-deficit insufficiency is about 13.5 billion cubic meter per year (BCM/yr) and is conventional to increase continuously. The research was initiated with the objective of check distinct scenarios for 2025 using the Water Evaluation and Planning (WEAP) model and by implementing distinct water sufficiency measures. According to the National Water Resources Plan, the agricultural sector consumes 67% of the total withdrawal [1].

One of the main targets of the Egyptian government is to encourage the use of ecologically friendly agriculture as hydroponic systems [2].

FAO [3] presented in their publications through the United Nations team that water scarcity is one of the considerable ultimatums of the twenty-first century. Agriculture, envelop crops, livestock, fisheries, aquaculture, and forestry, are both a cause and a fatality of water scarcity. It responsible for a predicted 70% of global water withdrawals. Climate change also affects freshwater revenues adversely, in terms of both quantity and quality. More frequent and inconsiderate droughts impact agricultural production while increasing temperatures translate into increased water needs in agriculture sectors.

We need to find out policies, game plan, programs and field quantity for the adoption of agriculture to water scarcity, using context-specific passage and processes tailored to specific locality and needs, as an abutment for the formation of transformational projects.

E. M. Koriesh (✉) · I. H. Abo El-Soud
Department of Horticulture, Suez Canal University, Ismailia 41522, Egypt
e-mail: quriesheid@yahoo.com

I. H. Abo El-Soud
e-mail: islamhassan2010@hotmail.com

E.-S. E. Omran and A. M. Negm (eds.), *Technological and Modern Irrigation Environment in Egypt*, Springer Water,
https://doi.org/10.1007/978-3-030-30375-4_7

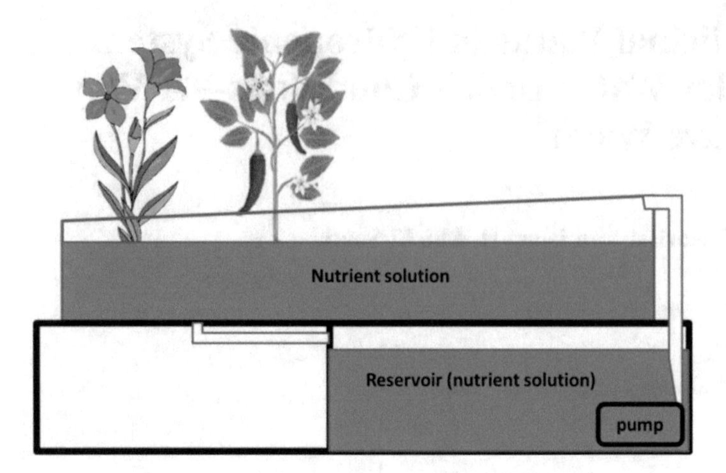

Fig. 1 Simple continuous flow hydroponic systems

Some of the adverse encounters of ordinary agriculture (geoponic) include the high and incompetent use of water, large land requirements, high application of nutrients utilization, and soil deterioration. The need for large amounts and high-grade plant products to meet the increasing demand of the world population justifies the improvement of technologies which harmonize the water and nutrient solution demand and source to greenhouse plants to attain crop yield optimization. The awareness about water and nutrient uptake by plants is urgent for developing control strategies which increase the opportunity to source the required amounts of water and nutrients for higher crop growth and development. Traditional agricultural systems use large amounts of irrigation freshwater and fertilizers, with relatively insignificant returns (Fig. 1).

The hydroponic systems have two categories, one soil-based and the other soilless. The present development is for growers to optimize root conditions that allow greater control of the vital root zone variables that is possible with soil. This led to use little or no soil other than, maybe, sand or gravel. Most hydroponic systems in commercial production use some sort of substrate to create a matrix that forms the root zone.

Irrigation managing in hydroponics is further sensitive than in soil-established systems. Hydroponic systems classically have a much smaller root zone volume, greater water-holding capacity, extra available water, lower water tension, better hydraulic conductivity, and a greater dissolved oxygen in the irrigation solution. Therefore, fertigation is possibly the most significant management factor through which planters can control plant growth and either yield quantity or quality in hydroponics [4]. Usually, hydroponics requires on-call irrigation. Water necessities can be determined either indirectly or directly. Indirect is the most common, where plant status is correlated with the environmental conditions including humidity, radiation and temperature. The irrigation systems and its control in hydroponics lessen over-watering and water stress of plants.

2 Irrigation in Hydroponics

Schroeder and Lieth [5] concluded that there are two points of vision to irrigation and water requirements in hydroponic systems that provide the basis for making hydroponic systems work efficiently. First of these realizes the root zone as a basin that must be refilled every time a definite level of reduction has occurred. Second is to treat the entire system as a duct of materials from a source to the root outward of the plants. They added that in optimizing irrigation, it is important to understand both of these two sides of the system [6]. When the irrigation pattern includes soluble fertilizers at various concentrations, this is called "fertigation." Also, the chemicals in the water may also include pesticides and oxygen; this is called "chemigation." In many soilless systems, fertigation is the favored approach to provide nutrients, especially where the substrate is not skilled of holding nutrient ions. One important plane in hydroponics that directly affects irrigation is a substrate individual known as "hydraulic conductivity."

2.1 Irrigation Control Methods in Medicinal Plants Grown in Hydroponic Systems [5]

For water regime strategy, decreasing the quantity of water that is used during the irrigation of plants is a top import for farmers worldwide. Hydroponics can reduce up to 70% of water. This practice will help to solve many of the globe's water challenges today. Intended for irrigation events and schemes, decisions must be made regarding when to irrigate and for how long. An irrigation event is a single application of water or nutrient solution. With soilless growing systems, an irrigation event takes a period of time, ranging from a few seconds to numerous minutes. Irrigation schemes are controlling strategies developed to achieve specific crop production goals utilizing many delivery and monitoring methods and mention the overall plan of managing the irrigation water for the length of the crop.

- Irrigation management and control of medicinal plant water status is one of the key environmental control elements. Irrigation management has many impacts on the final product of medicinal plant [7]. Numerous levels of irrigation control are probable, ranging from totally manual to fully automated operation. The type of control must be harmonized with the production system. If only manual hand irrigation is likely, then the substrate must be wisely optimized to allow a high water-holding capacity, first-rate aeration and high hydraulic conductivity. The irrigation control should confirm that the supply and uptake of nutrient solution and oxygen match the necessities of the plants. With fully automated hydroponic systems, it is probable to create best situations at nearly all times. It should be noted that the greatest disorder to efficient irrigation is a lack of homogeneity in the irrigation system. The main reason for this is that hydroponic systems are very investment-intensive and are used to produce high-value crops. This treasures that every plant in the crops is important in a marketable product. Thus, the emitter that brings the least amount of water runs the irrigation duration of the full circuit. It is not unusual to find irrigation systems where the slowest emitter is bringing water at a rate of only 30% of the fastest emitter. This means that a great amount of wastes must happen in such a system, anyway type of irrigation control that is used.

2.2 *Occasional Irrigation*

Conventional irrigation schemes can best be termed "occasional" irrigation. Many times in the horticulture literature, irrigation is described as being on an "as needed" basis, to enhance yield and water use efficiency [8]. This terminology states the use of occasional irrigation such that the root zone does not dry out to a degree that might result in "water stress." In this method, the objective is to supply water at intervals dictated by the plants' using of water, but applications frequently occur based on the timing of the observations of the irrigation director and secondarily on the plant's need. This irrigation system usually involves bringing the root zone to full capacity of water and then providing additional water which will result in a percentage of the applied water escape from the root zone.

2.3 *Pulse Irrigation*

'Pulse Irrigation' is a method that utilizes further frequent irrigations with smaller durations. The objective of "pulse" irrigation is to contest the application rate of the irrigation system with the absorption degrees of the substrate [9]. By applying the water in rotations of a short duration irrigation followed by a time for "rest," the substrate is allowed to absorb water that is pragmatic before additional water is added [10, 11]. Irrigations that are usually applied in a single quite long duration application are divided into numerous shorter intervals or "pulses." Pulsing may outcome in a shorter total irrigation period that may render into water savings. Dividing water applications over time increases the periods of high water availability. This generally increased water use efficiency [12].

2.4 *High-Frequency Irrigation*

In some types of soilless production systems, irrigation events may occur several times per day or hour and each (or at least some) of the irrigations results in bringing the root zone to its full water-holding capacity. In such situations, the system is dependent on the bulk flow of water, nutrients, and oxygen. This approach can only be used in systems where the substrate allows complete drainage of all excess water between irrigation events.

2.5 *Layout of Irrigation Systems in Hydroponics [13, 14]*

The importance of cultivation of medicinal plants has increased greatly in the last decade. The irregularity of weather events and quick climate change results in the high variability of active constituents of plants. However, hydroponics in protected horticulture can represent a sustainable solution for medicinal plants cultivation. Studies of Giurgiu [14] indicated that humidity and irrigation systems were manipulated gradually, until the parameters were considered stress factors for the plants. From lavender, St. John's wort and thyme, St. John's Wort showed the best results, with a significant growth and a shorter time, 30 days to the harvest peak, compared to plants cultivated in soil. They concluded that medicinal plants can be successfully produced with low water usage when cultivated in hydroponics.

Generally, hydroponic irrigation systems are grouped into open, semi-closed, and closed fertigation systems. Hydroponics without a substrate, like NFT, aeroponics, or ebb–flow systems, are closed systems [15–17].

2.5.1 Open Systems

Open system(s) of hydroponics are without recirculation of drainage water. The percent volume of drainage depends on the irrigation system used, the substrate and irrigation control methods [18]. Drainage can amount to between 0 and 50% of the supplied nutrient solution. Maintaining the EC in substrates or solution within boundaries requires drainage rates of 30% or more. However, environmental pollution due to the leaching of water and fertilizers is of concern. This system has a low water use efficiency [19].

2.5.2 Closed Systems

Closed systems were established as a consequence of environmental pollution since they support reuse of the drainage solution. Nevertheless, some unsolved problems still happen. Reuse of drainage water leads to an increase of nutrients, other ions, and hence changing the nutrient proportions [16]. In this system, the selected model flow rate of water will raise the water use efficiency for medicinal plants by optimization the contact time among the root and water flow rate [20, 21].

2.5.3 Semi-closed Systems

The system of semi-closed is exactly closed systems, which are opened on occasion to flush out the drainage nutrient solution, or to change the whole solution in the system. For present problems in closed systems, this is a phase to decrease water and nutrient solution leaching, but not to avoid. Irrigation control via tensiometer instruments allows overload of substrates without leaching, but a high capacity and uniformity of substrate is required [22]. This is achieved by placing tensiometers in descriptive pots or places in the system. So, this system results in increased soluble salts in the substrate [23].

2.6 Deficit Irrigation (DI)

Deficit irrigation (DI) has been widely explored as a valuable and sustainable production strategy in dry regions. By reducing water utilization to drought-sensitive growth phases, this process aims to maximize water production and to stabilize rather than maximize—yields [24]. Researches results confirm that DI is fruitful in rising water production for various plants without generating harsh yield minimizations. "Nevertheless, a certain minimum amount of seasonal moisture must be guaranteed" (https://csa.guide/csa/supplemental-irrigation-si-or-deficit-irrigation-di-of-rai). The DI desire precise knowledge of plant return to drought stress, as drought tolerance varies appreciably by genotype and phonological phase. In developing and optimizing DI approaches, field research should consequently be combined with plant water production modeling.

Water is one of the higher important components influencing plant growth, production, and quality and is the main component of plant cells and total fresh biomass content of plants. Typical greenhouse-grown herbaceous plants may contain 90–96% or even more water. For greenhouse plants, irrigation and its executive have special importance since natural rainfall is excluded, and if soilless culture systems (SCSSs) are used, often the under-groundwater sources are inapplicable. Irrigation executive includes all measures that guarantee an acceptable water source for plants. One could assume that for a specific plant species "only" the right water demand should be ensured. Irrigation systems have been under pressure to produce more with lower foodstuff of water. Various innovative processes can lead to an economic advantage [25].

Since plant water prerequisites and nutrient uptake models allowing a better alteration of the nutrient solutions are applicable, a decision abutment system (DSS) based on these nutrient uptake models can be advanced. This system affords a tool for better executive of irrigation systems with the aim to save water and nutrients and to weaken climate impact [26]. However, moisture deficits may limit some nutrients transport as well as uptake into the plant (potassium).

Controlled deficit irrigation is useful not only in reducing production costs but also in preserving water utilization and minimizing filtration of nutrients and pesticides into the under-groundwater. On the other hand, deficit water may increase the content of so-called health-promoting materials or active compounds [27]. Water possibility reciprocal to plant development may also signify the flavor of plants, and antioxidant content [28, 29].

Water availability is premeditated as an explanatory factor that affects plant growth. The economic medicinal value of a fragrant plant depends on the existence of specific active compounds that are changed under water deficit. It was reported that drought stress could be a dominant factor in increasing concentration of active metabolic products in some medicinal plants as *Ocimum* sp. [30]. He found that 75% of field water quantity produced the highest yield of herb and volatile oil for both species, and *Calendula* plants [31].

Moisture deficiency produces distinct plant anatomical changes which are harsh to respond to drought stress, ranging from morphological alteration and modification of root and shoot besides root/shoot ratio for desiccation prevention [32] as well as physiological and metabolic response (antioxidant buildup, stomatal stoppage and definition of stress-specific genes) [33].

Water availability is premeditated as a determinant factor that affects plant growth. The medicinal economic values of a fragrant plant rely on the presence of active constituents that are affected under water deficit [34]. In the context of improving water production, there is a growing interest of medicinal plant under the hydroponic systems as a system of deficit irrigation, an irrigation process whereby water source is diminished beneath higher levels, and mild stress is allowed with minimum effects on yield.

3 Hydroponics, Aquaponics, and Aeroponics

The production systems found in modern greenhouses can be branched into two divisions:

1. Soilless
2. Soil-based.

In general, hydroponic systems are soilless production systems. The public commonly think of hydroponic systems as water culture systems, without media; anyhow, advanced hydroponic systems in economic production use some sort of substrate to generate a matrix that forms the roots zone. The only fact that all hydroponic systems have in common is that no soil is used. Contrarily, all containerized soilless production systems (e.g., production of flowering plants grown in pots) in use today are hydroponic production systems in that they generate an artificial roots zone that encounters to raise water possibility for the roots. Such production systems consist of the roots zone, the aerial part of the plants, an irrigation system for the source of nutrient solutions to the roots zone, and a drainage system for dealing with runoff from the roots zone.

The nutrient solutions are the elementary function of the irrigation system to drop this solution to the plants in an excellent manner and appropriate dose.

Hydroponics, aquaponics, and aeroponics are modern agriculture systems that utilize nutrient-rich water more than soil for plant food. Because it does not require fertile land to be adequate, those modern agriculture systems need less water and space compared with the traditional agricultural systems; one more advantage of those technologies is the quantity to process the vertical agriculture management which increases the yield of the space unit. The benefits of the new state-of-the-art agriculture systems are higher yields and water competence.

Aeroponic cultivation of medicinal plants has shown excessive potential for producing root crops that are uniform, cleaner and earlier maturing [35]. Hydroponics can be a substitute horticultural system for crops susceptible to soil-borne diseases. The constant growing environment in a controlled greenhouse can produce medicinal plants with more reliable levels of active constituents, which is of concern to the phyto-pharmaceutical industry. There are some hydroponic systems suitable for rhizomatous medicinal plants such as ginger [36]. Most hydroponic systems are planned for crops that produce leaves or fruits which have fibrous roots. Rhizomatous producing medicinal plants have unusual requirements.

An experiment on ginger plants grown in aeroponics was achieved in a controlled greenhouse by Hayden et al. [37]. The unique aeroponic growing units incorporated a "rhizome compartment" separated and elevated above an aeroponic spray chamber. They used perlite or sphagnum moss or without any aggregate medium. Results indicated that plants grown in perlite matured faster.

3.1 *Type of Hydroponics*

Hydroponic technology [38] has started to permeate almost every fact and in agriculture, which include all of the action as plant monitor technology and mobile apps to aid farmers in "when," "where," "How," or "what" to plant and correctness agriculture.

Hydroponic uses about 5% of the water and a fraction of the land needed to produce an equivalent amount of production in traditional agriculture. In the San Joaquin Valley of California, the water use capability (WUE: kg tomato yield/ m^3 water applied) for tomato production was shown to be 10–12 kg m^{-3} for flood irrigation, 11–19 kg m^{-3} for sprinkler irrigation, and 19–25 kg m^{-3} for drip irrigation. Many researchers have expressed much higher WUE values for greenhouse tomato production. Open hydroponic irrigation systems (no recirculation of unused water by plants) in the Netherlands and France have been expressed as 45 and 39 kg m^{-3}, consideration. Closed irrigation systems (recirculation irrigation water) have been shown to carry out WUE values of 66 kg m^{-3} in the Netherlands.

Production of medicinal herb and roots plants in controlled climates (CEs) affords opportunities for improving the quality, purity, consistency, bioactivity, and biomass production of the raw material. Hydroponic systems in CE can produce high-quality herb and roots material free from accidental adulteration by weeds, soil or climate toxins such as heavy metals in soils. Hydroponic and aeroponics production of medicinal plants in controlled climates affords opportunities for improving quality, purity, consistency, bioactivity, saving water and biomass production on an economic scale. Ideally, the goal is to raise the climate and systems to maximize all characteristics [38]. Examples of plant production systems using perlite hydroponics, nutrient film technique (NFT), ebb and flow, and aeroponics were studied for various root, rhizome, and herb leaf plants. She added that biomass data comparing aeroponics vs. soilless culture or field has grown production of burdock roots (*Arctium lappa*), stinging nettles herb and rhizome (*Urtica dioica*), and roots and rhizome (*Anemopsis californica*) are presented, as well as smaller-scale projects observing ginger rhizome (*Zingiber officinale*) and skullcap herb (*Scutellaria lateriflora*).

3.1.1 Nutrient Film Technique (NFT)

This is the kind of hydroponic system higher people think of when they think about hydroponics. NFT systems have a constant flow of nutrient solutions so no need for a timer for the submersible pump. The nutrient solutions are "pumped into the growing tray (usually a tube) and flow over the roots of the plants, and then drains back into the container's (http://simplyhydro.com/system.htm). There is usually no growing medium used other than air, which saves the expense of replacing the growing medium after every plant. Normally, the plant is supported in a small plastic basket with the roots suspended into the nutrient solutions (http://simplyhydro.com/system.htm).

3.1.2 Ebb and Flow—(Flood and Drain)

The ebb and flow system work by temporarily flooding the grow tray with nutrient solutions and then draining the solution back into the container. This action is normally done with a submerged pump that is connected to a timer.

When the timer turns, the pump on nutrient solutions is pumped into the grow tray. When the timer shuts, the pump off the nutrient solutions flows back into the container. The timer is set to arise on some times a day, depending on the size and type of plants, temperature and humidity and the kind of growing medium used. The ebb and flow is an adaptable system that can be used with a variety of growing mediums. The perfect grow tray can be filled with grow rocks, gravel or granular rockwool.

The main disadvantage of this type of system is that with some types of growing medium (sand, gravel, perlite), there is an exposure to power outages in addition to pump and timer failures. The roots can dry out rapidly when the watering cycles are intermittent. This problem can be calmed somewhat by using growing media that holds more water (vermiculite, rockwool, coconut fiber or a good soilless mix like Pro-mix or Faffard's).

3.1.3 Water Culture

The water culture system is the simplest of all dynamic hydroponic systems. The stand that clamps the plants is typically made of Styrofoam and floats directly on the nutrient solutions. An air pump produces foodstuff air to the air stone that bubbles the nutrient solutions and foodstuff oxygen to the roots of the plants. Water culture is the system preferred for growing leaf lettuce, which is quick-growing water-loving plants, making them an ideal preferred for this type of hydroponic system. Very few plants other than lettuce will do well in this type of system.

This type of hydroponic system is good for the classroom and is beloved by teachers. A very cheap system can be made out of an old aquarium or other water compact container. We have free plans and instructions for a simple water culture system.

3.1.4 Aeroponic

The aeroponic system is apparently the higher high-tech type of hydroponic gardening. Like the NFT system, the growing medium is primarily air. The roots hang in the air and are misted with little amount of nutrient solutions. The misting is usually done every few minutes and stopped at night. Because the roots are revealed to the air like the NFT system, the roots will dry out rapidly if the misting cycles are interrupted.

Decreasing the amount of water that is used during the irrigation of plants is a top priority for farmers and politics around the world. Reducing 70% of water used will help to solve many of the globe's water ultimatums today. Hydroponics is a way of growing almost any plant that uses no soil, and higher importantly very little water. There are different systems. Hydroponic systems use a small pump to give water to the plant straight to the root. This design concedes any excess water to be arrested in an "overflow container" and re-used to irrigate other plants in the system. The soil is replaced with a nutrient-filled solution which means the plant does not waste time searching for nutrients and does not waste water.

The result is a plant with a smaller root system and a higher flowering probable with higher yield and a nutritious plant. The beauty of hydroponic systems is they can be grown in any weather conditions anywhere, whether it is indoors or outdoors. or in a greenhouse. Some systems have been planned to use up to 10 times fewer water than known growing methods. Due to soil not being used, plants can be located closer together which weakens the space needed to grow the same amount of plant.

Hydroponic systems are not new; they have been used for hundreds of years Anyhow this could be the time to carry out this technology. Also, in hydroponics, it can use solar panels to power the water pumps, consequently creating a fully sustainable growing system with minimum waste, and needing much less water. Increasingly, they are being used on a bigger scale, some large enough to switch full farms of production with much less space and water.

3.1.5 Water Stream Hydroponics

At Armenia, a low-cost modern system for soilless culture of medicinal and aromatic plants in soilless systems was developed. This system is called stream hydroponics [39]. It was established for the soilless production of different aromatic and medicinal plant species using polymeric film. The benefits of which are little cost, further automated system and beforehand instructed program. Its varieties are continuous and cylindrical hydroponics. They used *Bidens tripartita* L., *Leonurus quinquelobatus* Gilib., *Mentha piperita* L., *Ocimum basilicum* L., and *Salvia officinalis* L. [40]. Also, Daryadar [41] used *Mentha piperita* L. with the use of different hydroponics systems (cylindrical, continuous and classical). They found that it surpasses the soil culture with 1.5–2.7 more dry weight. On the other hand, essential oil synthesis was extra concentrated during the third cut (1.6–2.6 times) by cylindrical and classical hydroponics systems. It can be seen from evaluation of qualitative indices that high content (71–73%) of essential constituent menthol was noticed in cylindrical, classical and soil culture. Conversely, continuous alternates had a high content of izomenthon (11–15%).

3.1.6 Continuous Flow Hydroponic Systems

It is a cluster of hydroponic systems, where the nutrient solution is being regularly pumped over the plants' root systems. This system includes nutrient film technique (NFT), as is an elementary top feed of nutrient solution. In a continuous flow system, there is a continual flow of nutrient solution over the plants' root system. As a consequences of a continuous flow system the nutrient tank can be remote from the plant and its growing medium. This means that a large tank may be used to work a mass of plants. The larger a nutrient solution tank is, the easier it is to keep EC, pH and temperature levels. For providing oxygen to the root systems, this is done in either one of, or a blend of, two ways. First, the nutrient solution can be oxygenated in the tank by a powerful enough air pump. Second, in the NFT system, the roots are only somewhat suspended in the nutrient solution, whereas the ends of the roots are submerged and able to take in the nutrients; the chief mass of roots is exposed to the air where it can take on oxygen directly.

Continuous hydroponic system stimulated the great accumulation of extractive substances in the plants of *Leonurus quinquelobatus* Gilib., *Mentha piperita* L., *Ocimum basilicum* L., and *Salvia officinalis* L. [39].

The cylindrical hydroponics system promotes the increase of medicinal plants. This system promoted 1.3–2.9 more increase of yield of peppermint and basil and 1.3–3.3 more increase of the essential oils, total flavonoids and tannins. Sage's yield was increased by 1.2–5.4 times in NFT hydroponics [39, 40]. Cultivating the plants in cylindrical method and loading them beyond one another vertically achieved to increase the plant density by 47% when related to the rotary system (with the suitable lighting system) cylindrical type of hydroponics can be adopted for high-density crop production and permit leafy plants (such as basil, peppermint) to be grown indoors throughout the year without dependent on soil and climate. With a nearby loop, both growing systems aid water and minerals to be efficiently used with the elimination of runoff [42].

4 Aquaponics Can Serve as Saving Water [43, 44]

4.1 What Is Aquaponics?

Aquaponics has been about for centuries. It was conventionally a technique in tropical climates, using floating bamboo bundles with vegetation in freshwater ponds. This was merely the adaptation of agriculture to the tropics. The technique has developed into cutting edge over the last three decades. Aquaponics can be adapted to today's geographies and culture.

Aquaponics is a blending of aquaculture (the raising of aquatic animals) and hydroponics (growing plants in water without soil). In aquaponics, aquatic animals serve as the nutrition base for the plants. The good fact about aquaponics is that it is a closed system; it does not have to flow in one pipe and out of another.

4.2 Aquaponics System Does Not Require Machinery

Water possibility and water quality are the two very weaken components for growing plants around the world. Fortunately, since aquaponics powerfully deliver fewer water by recirculating it over the fish and plant system, it can benefit growing food in places that would then encounter. Locations with powerful water pollution or contamination should have acceptable filtration in place before build in an aquaponics system [45].

However, the fact that fewer than 10% of the water used in outdated agriculture is needed will let that filtered water to produce a more crop or yield in aquaponics than in soil. About 378 L of water added to an aquaponic system could be recirculated for numerous days, maybe 7 days or more, but that same water spread on a sundrenched thirsty soil garden would be gone by lunchtime, and the plants would be suffering.

5 Hydroponics and Fish [43, 46]

Aquaponics may be still largely unknown to the general popular, but it is a growing trend, which has the probable to feed the growing world population. This green technology couples aquaculture (production of fish) and horticulture (production of vegetables) while saving water, energy, and nutrients.

Aquaponics is the land-based assembly of fish in tanks shared through the recirculation of the water from the fish reservoirs through hydroponic systems to produce horticultural plants. The waste commodity from the fish is changed by a biofilter into soluble nutrients, which are absorbed by the plants and make it "clean" water to be resumed back to the fish. Thus, it yields a valuable fish protein with minimum pollution of freshwater reserves, at the same time generates horticultural (usually vegetable) plants.

The effects of deficit irrigation system on growth, yield, vitamin C content, and irrigation water use capability of hot pepper (*Capsicum annuum* cv. Battle) grown in soilless culture in China were studied by Ahmed et al. [47]. They grew hot pepper in a hydroponic system and subjected to four irrigation treatments: 100% of water-holding quantity (control) and 85, 70, and 55% of water-holding quantity, which were premeditated deficit irrigation treatments. Results manifested that deficit irrigation had an adverse effect on plant growth and yield. Rising irrigation deficiency exhibited a minimization in vegetative growth, fruit characters, and yield and a no powerful increase in irrigation water use capability and a corresponding minimization in the amount of irrigation water. Vitamin C content in fruit was powerfully decreased by deficit irrigation treatments at various ripening phases. The water-holding quantity of 100 and 85%, consideration, resulted in the maximum content of vitamin C obtained at the ripening phase. They carry out that "Battle" hot pepper is sensitive to deficit irrigation. Sourcing this cultivar with water at 85% of water-holding quantity could be a functional irrigation technique for the high value of vitamin C production as well as saving a large amount of water, which overcome the decrease in total fresh yield of fruit, especially in areas suffering from water deficit.

6 Irrigation Control in Hydroponics: Hydroponics Systems and Deficit Irrigation

Irrigation executive in hydroponics is more sensitive than in soil-based systems. In comparison to the soil, hydroponic systems typically have a much smaller roots zone size, higher water-holding quantity and are more applicable. Irrigation water has a lower moisture tension, higher hydraulic conductivity, and higher dissolved oxygen (DO) concentration in the nutrient solution. Typically, soil contains a multitude of materials that act as a buffer to chemical adjustments. In some hydroponics systems, the size of the roots zone is very small, e.g., 14 Lm^{-2} in rockwool in comparison with 500 Lm^{-2} in soil [48]. In the same way, the water and nutrient quantity are drastically diminished. Consequently, fertigation is the higher important executive component through which growers can control plant growth, yield, and quality in hydroponics [49]. Hydroponics commonly desire on-demand irrigation.

Plant water prerequisites can be determined either directly or indirectly. The higher common is an indirect way in which plant position is correlated with the climate (humidity, temperature, vapor pressure deficit or radiation). Direct measurements with sensors (e.g., a stem flow gauge) are now acceptable advanced and capable of fruitfully monitoring the plant. The irrigation system includes sharing constituents which cooperate with the substrate and climate.

6.1 *Irrigation in Hydroponics*

There are two points of view to irrigation in hydroponics that afford the support for making such systems work regularly. One of these sees the roots zone as a container that must be restocked each time a certain level of reduction has occurred. The other attitude is to treat the perfect system as a conduit of materials from a source to the roots surface of the plants. In improving irrigation, it is vital to understand both of these features of the system. It is commonly understood that the roots zone is a container or storage compartment where a number of important ingredients for plant growth and survival are stored.

Once any of the ingredients depleted, irrigation needs occur to resource this ingredient. It is also achievable (and, in fact, likely) that some appropriate elements may develop into excessive; then, irrigation is needed to flush out or dilute this element. In addition to its function as a container, the roots zone must also act as a conduit for these same materials. Nutrients that are near the roots surface applied to the plant, and the roots use active action to move materials into the plant. This depletes the nutrients in the actual surrounding of the roots. Thus, concentration grade is build up and drive nutrients to travel within the roots zone from sites of higher concentration to those of lower concentration. If this instruction does not work at a suitable rate, then the plant will be starved of appropriate nutrients or water, even though acceptable overall total amounts may be present in the container. This locality can be overcome by initiating irrigation to displace spent nutrient solutions from around the roots and replace it with fresh, excellently formulated nutrient solutions.

It should be noted that the plant discards water and ions selectively at a distinct standard, so it is achievable for more water to be used and ions left behind (or vice versa). This can cause salts to build up in the roots zone, and this buildup must be escaped as it is deleterious to plant growth. If low concentrations of ions are afforded, then it is achievable for the plants to run out of appropriate nutrients while there is still an acceptable source of water in the roots zone. Thus, irrigation systems in hydroponics achieve two assignments:

(1) They restore various elements that are in storage in the roots zone and
(2) They afford weight flow of such materials through the conduit.

Since weight flow is much quicker in moving materials than diffusion action, frequent irrigation can be used to move the needed elements to the roots surface. In hydroponic systems, this type of control is appropriate since all variables can be controlled; in soil-based systems, this is not achievable since the roots would develop into waterlogged and starved of oxygen.

In systems where the drainage water is recirculated, the container is effectively contributed to the tank where such water is collected. The large increase of yields in hydroponics over that in soil is due to several components, of which irrigation executive is of higher importance. Other components are the higher capability of water use, because water is applied directly to the roots, diminished evaporation, and prevention of water stress. Anyhow, hydroponic irrigation executive desires an accurate source and dynamic control because of the low water-holding quantity and the restricted possibility of nutrients due to the controlled size of substandard. In the present horticultural process, the hydraulic circumstances in the roots climate are such that plants are grown at very high-pressure heads. The reason for this is that plants are grown in low size substandard to decrease the cost and handling of growing media, so high irrigation frequencies are used.

Plant growth in hydroponics is related mainly to water, nutrient and oxygen source. Water and nutrient source can be regulated by the nutrient solutions together with the irrigation system and the irrigation frequency. In the same way, differences in O_2, CO_2, and ethylene in the roots zone have been shown to be significant by the growing medium and irrigation [50]. The roots zone in hydroponics is generated for plant growth, water, nutrient, and oxygen source only, but not for the abutment of the plants. The irrigation system should adjust the dose according to the prerequisite of the distinct hydroponic systems and substandard. Comparable high yields can be carried out as long as irrigation and fertilization are raised for the specific substrate. A number of authors describe a surplus of nutrient solutions of between 20 and 50% in the process in hydroponics. In an open irrigation system, each surplus leads to losses of water and nutrients. Closed irrigation systems weaken the surplus, but need expensive disinfection of the solution. Acceptable aeration of the roots zone is carried out. Roots need air, accordingly oxygen, for respiration.

The quantity of water needed is commonly dictated by the climatic conditions surrounding the aerial part of the plants and the amount of green leaves present in the plants. Under temperature, high humidity, and low light, the rate of water usage can be almost zero. Any production system must have the irrigation system range to accommodate the best rate of water usage; this occurs when the canopy is fully grown, the air is dry and hot (e.g., summer conditions), and there is appreciable air movement (as in a ventilated greenhouse). It is very important to estimate the higher rate of water use when an irrigation system is designed and installed since an unacceptable-range system will not be able to meet the needs of the plants and will result in production losses during the summer. Higher water use figures have been expressed for various plants.

The production of fertilizers is becoming advance expensive due to high prices of fossil fuels, and this may have long-term implications for nutrient use in agriculture in the future. Aquaponics uses waste commodity derived from animals and plants which are fed to the fish, and thus converted into valuable animal protein and fresh herbaceous plants. With the world's freshwater reserves controlled, aquaponics would appear to have appreciably probable for arid and similar climates. Disease and pest problems are minimized because of the development of an ecological balance. Production is commonly as good (or better) than with conventional hydroponic systems.

6.2 Water Conservation in Hydroponics

Essentially, hydroponic plants grow in water, with growing media material such as fibers from coconut shells or Styrofoam holding the plants in place. The roots float in water that affords all the nutrition the plants need. This water works double-times by both quenching the plants' thirst and feeding them. Cup to abutment plants as grow medium in higher hydroponic systems, water is recirculated. Excess water that is not absorbed by the plants is recollected. Every minute of water is reused over and over again, an unfeasibility in traditional, soil-based agriculture. Since it is recirculated and recycled, water is not ever discharged in aquaponics or hydroponics. In hydroponics, nitrate-rich water is presented in the hydroponically grown plants. These plants are cultivated in beds that sit on tubes occupied with water, and the water is improved by the nitrate harvested from the fish waste. The plants' bare roots droop through holes in the beds and hang in the nutrient-laden water. Water Conservation in Aquaponics Goldfish with mint aquaponics is a similar method which also raises fish while using far lesser water than used in hydroponics, where the continual reuse and recycling of water through naturally occurring biological action. Fish needs water restoration. The basic principle of aquaponics is to put waste to use and recycle it. The waste from fish produces natural bacteria that convert excess like ammonia into nitrate. This nitrate is absorbed by plants as a source of nitrogen. Nutrients are regularly added by fish waste, and water incomes to the plants. Fish in aquaponic tanks expels waste and releases ammonia into the water. Ammonia is deadly to fish in great concentrations, so it has to be detached from the fish tanks for fish to continue healthy. Ammonia-laden water is treated to harvest cooperative types of bacteria such as *Nitrobacter and Nitrosomonas*. *Nitrosomonas* transformed ammonia to nitrite, while *Nitrobacter* converts into nitrate. Both of these materials can be used as plant nutrient. The roots of the plants absorb nit standard, which acts as nutrient-rich plant food. These nit standard, which originate from fish manure, algae, and decomposing fish feed, would then transformed to toxic levels in the fish tanks and kill the fish. However, in its place, they serve as fertilizer for the plants. The plants' roots meaning as a biofilter—they band ammonia, nitrites, and phosphorus from the water. Then, that clean water is dispersed back into the fish tanks. Because fish waste is used as fertilizer, there's no essential for chemical fertilizers. The energy and money it would take to put those chemicals to work are secured. In fact, the only conventional agriculture method that is used to operate an aquaponics system is feeding the fish. Water loss happens only through evapotranspiration. The water that is transpired withdrawals through the leaves as water vapor. One purpose it helps is cooling the plant, and in that way, it can be crudely in comparison to how people sweat to keep cool. There is no way to eliminate evapotranspiration; it is a necessary function of living plants! You can, anyhow, keep it as efficient as biologically achievable by making sure that your temperature range is suitable for the plants you are growing. The exceptional case is any leakage. This might be a broken pipe or split tubing.

7 Growing Medicinal Plants in Hydroponic Systems

Medicinal plants are increasingly cultivated on a small scale in hydroponic systems. A lot of species are commonly grown in open field, which results in large year-to-year variability in both biomass production and content of active principles. Hydroponic technology may be applied to production high-standard plant material all year-round in consideration of the possibility to control growing conditions and to stimulate active metabolism by appropriate manipulation of nutrient nutrition.

The floating raft growing system for the greenhouse cultivation of Echinacea (*Echinacea angustifolia* DC) and basil (*Ocimum basilicum* L.) which are typically cultivated for their roots and leaves was studied [55]. They found that both species grew accelerated and healthy, and in two to four months, they accumulated large biomass with minimum contamination. Nevertheless, in Echinacea, the high biomass production was not correlated with great levels of cads, and the concentration of Echinacoside (the indicator compound used for value standardization) never extended the minimum standard (1% on a dry weight basis) for the industrial production of dry extract. In comparison, basil accumulated an acceptable content of rosmarinic acid. One additional advantage was the possibility to harvest also the roots system of basil, which contains higher levels of rosmarinic acid in comparison to the leaves.

The buildup of selected caffeic acid derivatives (CADS), inappropriate rosmarinic acid (RA), was explored in distinct tissues (leaves, roots, and plantlet shoots) plants grown either in vitro or hydroponic culture (floating system) under greenhouse conditions [51]. They used two cultivars with green leaves (Genovese and Superbo) and one with purple leaves (Dark Opal). They found that the content of rosmarinic acid RA ranged approximately from 4 to 63 mg/g DW, depending on the growing system. The total phenolic and RA levels were higher in sweet basil grown hydroponically than in soil-grown plants [52].

The hydroponic systems can enhance the yield and have shown a huge economic advantage [13]. This system is a rigorous way of controlling the temperature, humidity, light, irrigation, and fertilization of the plants that are growing in an inert substrate to attain superior growth allied with excellent plant quality and high bioactive substance. The risks of hydroponic cultivation of medicinal plant are low germination rate and ecological problems that can be overcome by controlling the excellent characters to obtain the best climate for germination. Having all aspects taken into consideration, hydroponic cultures can have relevant results like uniform yield, with a high percentage of bioactive materials and this kind of system can be the way to cultivate the medicinal plant) in economic purpose. Besides the economic and chemical advantages, the hydroponic system of cultivating the medicinal plant helps to protect spontaneous flora and the diversity of the species that can be found in the wilderness.

In order to guess the growth probable and volatile oil production of some medicinal plants, Azarmi et al. [11]. cultivated lemon verbena (*Lipia citriodora* Var *verbena*) and valerian (*Valeriana officinalis var. Common*) in various soilless and soil production systems that consisted of floating, aeroponic, growing media and soil in a research greenhouse [11]. They harvested plants after six months and measured the vegetative and roots characteristic and volatile oil content. They found that the concentration of volatile oil in both floating and aeroponic systems was higher compared to those of growing media and soil systems. In valerian, plants size of roots in the floating and aeroponic was powerfully higher and led to the increased total volatile oil content. No powerful difference in leaf area was found between floating and growing media systems. However, it was appreciably diminished in aeroponics and soil systems. Meanwhile, distinct systems had no powerful effect on the volatile oil concentration in lemon verbena. Anyhow, the total content of oil in the floating system was powerfully increased as the result of the higher fresh weight of the leaves. They recognized that both floating and growing media systems could be used for the production of both valerian and lemon verbena under the greenhouse. In general, distinct culture systems under greenhouse condition can be used for more efficient production of some medicinal and fragrant plants.

Hydroponic production of medicinal and fragrant plants is a new trend in the agricultural system, accordingly in organic and intensive agriculture. High productivity, cleaner, offseason, and cost-effective production, water use capability, acceptable aeration, abundant water possibility, drastic increase of photosynthetic probable and elimination of hard and labor consuming works such as cultivation, loosening, weeding, irrigation, have been defined as the main advantages of soilless production systems. All these advantages can be used to deviate from labor consuming, the expensive traditional culture of medicinal plants to industrial production in greenhouse hydroponic system [53].

The possibility and economic assistance of medicinal plants cultivation in the hydroponic system have been proved, and it has been shown that buildup of bioactive substance in soilless culture can also affect plant production in intensive plant production system [54]. Anyhow there is very little data applicable to the effect of distinct soilless culture methods on the growth probable and active constituents as volatile oil production in the medicinal plant.

7.1 Growing Medicinal Plants in Sand Culture, the Higher Commonly Used System, as a Hydroponic System

Medicinal plants cultivation is widespread, and the interest in this kind of cultures is growing at a quick pace. Sand culture as a hydroponic system getting very beloved because of the results it showed in recent process. The main plants cultivated in this kind are herbaceous plants, herbs or other plants used in the food industry. The hydroponic system can enhance the yield and have shown a huge economical advantage [13]. The risks of hydroponic cultivation of medicinal plant are ecological problems that can be overcome by controlling the excellent characters to obtain the best climate for germination and growth; another risk is that it may need specific nutrient solutions. Having all aspects taken in consideration, hydroponic cultures can have relevant results like uniform yield, with a high percentage of bioactive materials and this kind of system can be the way to cultivate the medicinal plant in economic purpose. Besides the economical and chemical advantages, the hydroponic systems of cultivating the medicinal plant help to protect spontaneous flora and the diversity of the species that can be found in the wilderness [55].

WHO estimated that 80% of the populations of developing countries rely on traditional medicines, highly plant drugs, for their elementary health care needs [56]. Likewise, recent pharmacopoeia still contains at least 25% drugs derivative from plants and many others which are synthetic equivalents built on model compounds isolated from plants. Demand for the medicinal plant is rising in both developing and advanced countries, and surprisingly, the bulk of the material traded is still from wild harvested sources on forest lands and only a very small number of species are cultivated. The expanding trade in the medicinal plant has severe implications on the survival of several plant species, with many under severe threat to develop into extinct.

8 Using Biostimulants for Production Medicinal Plant in Sand Culture

In recent years, new approaches have been proposed to enhance the sustainability of production systems for horticultural plants. A promising tool would be the use of materials and/or microorganisms defined as "biostimulants" that able to enhance plant quality characters, nutrient capability and abiotic stress tolerance. In this concern, plant biostimulants are diverse materials used to enhance plant growth [57–62].

Plant biostimulants formulations are, commonly, proprietary compositions based on complex organic materials, plant hormone-like compounds, amino acids, and humic acids [63]. The beneficial effect of natural biostimulants on specific aspects of plant growth, production and quality in distinct plants was studied [23, 64]. Specific plant biostimulants life such as increased roots and shoot growth, tolerance to abiotic stress, water uptake, minimization of transplant shock have also been expressed for *Coleus* and *Hedera* grew hydroponically [61, 62]. They used biostimulants extracted from willow trees.

9 Conclusion

Medicinal plants cultivated on a commercial scale to satisfy the large demand for natural remedies and to provide elemental natural materials for medicines, flavors, and cosmetics. Because of the rapid consumption of water resources and the increased stroke of climate change, ordinary irrigation practices can no longer be continued in many areas, especially in the Middle East. So, production of medicinal plants in hydroponics under greenhouse conditions advance faster. Some hydroponics systems matching other water control methods to further save water. Water management is not the only advantage of hydroponics, but it can also get rid of the use of pesticides, fertilizers, and herbicide, and gives more control to the farmers and they can plant their medicinal plants any time of the year. Under the system of deficit irrigation, as a potential technique for saving water, may advance water utilization and affect plant growth and may increase active constituents. In the situation of remodeling water productivity, there is an increasing interest of medicinal plants under the hydroponic systems with the application of deficit irrigation, whereby water supply is reduced diminished below maximum levels, and delicate stress is allowed with minimum effects on yield and sometimes increased the biosynthesis of secondary metabolites. Deficit irrigation can advance greater economic benefit than boost yields per unit of water for a given plant. Farmers are more willing to use water more efficiently, and more water-efficient cash-crop selection helps optimize returns.

References

1. Omar M, Moussa A (2016) Water management in Egypt for facing the future challenges. J Adv Res 7:403–412
2. Rorabaugh P (2017) Hydroponics page Introduction to Hydroponics and Controlled Environment Agriculture. Controlled Environment Agriculture Center, University of Arizona
3. FAO. Food and Agricultural Organization (2017). Statement on Water Scarcity in Agriculture. Meeting on 19–20 Apr 2017 at FAO headquarters in Rome, Italy
4. Raviv M, Lieth J (2008) Soilless culture: theory and practice. Working together to grow libraries in developing countries. Plant Science/Agriculture, ELSEVIER
5. Schroeder F, Lieth J (2002) Irrigation control in hydroponics. Chapter 11. In: Savvas D, Passam H (eds) Hydroponic production of vegetables and ornamentals. Embryo Publications, Athens Greece, pp 265–298
6. Lieth J, Oki I (2008) Irrigation in soilless production. Soilless culture, pp 117–156
7. Stemeroff J (2017) Irrigation management strategies for medical cannabis in controlled environments. M. Sc. Thesis, The University of Guelph, Canada
8. Wahed RA, Aslam Z, Moutonnet P, Kirda C, Tahir GR (1998) Scheduling for occasional omission of irrigation water for crop production in moisture deficit areas. Pakistan J Bio Sci 1:44–52
9. Falivene SG, Navarro JM, Connolly K (2015) Open hydroponics of citrus compared to conventional drip irrigation best practice: first three years of trialing and Australian experience. Acta Hortic 1065:1705–1712
10. Abdelghany A (2009) Study the performance of pulse drip irrigation in organic agriculture for potato crop in sandy soils. Ph.D. Thesis faculty of agriculture Cairo University Egypt

11. Azarmi F, Tabatabaie S, Nazemieh H, Dadpour M (2012) Greenhouse production of lemon verbena and valerian using different soilless and soil production systems. J Basic Appl Sci Res 2(8):8192–8195
12. Katsoulas N, Kittas C, Dimokas G, Lykas C (2006) Effect of irrigation frequency on rose flower production and quality. Biosyst Eng 93:237–244
13. Giurgiu RM, Morar GA, Dumitraş A, Boancă P, Duda BM, Moldovan C (2014) Study regarding the suitability of cultivating medicinal plants in hydroponic systems in controlled environment. Res J Agri Sci 46(2):84–92
14. Giurgiu RM, Morar G, Dumitraş A, Vlăsceanu G, Dune A, Schroeder F (2017) A study of the cultivation of medicinal plants in hydroponic and aeroponic technologies in a protected environment. Acta Hortic 1170:671–678
15. Chow Y, Lee LK, Zakaria NA Foo KY (2017) New emerging hydroponic system. In: International Malaysia-Indonesia-Thailand symposium on innovation and creativity (iMIT-SIC), vol 2, pp 1–4
16. Maucieri C, Nicoletto C, Junge R, Schmautz Z, Sambo P, Borin M (2018) Hydroponic systems and water management in aquaponics: a review. Italian J Agronomy 13:1012–1022
17. Waller P, Yitayew M (2016) Hydroponic irrigation systems. In: Irrigation and drainage engineering. Springer International Publishing Switzerland, pp 369–386
18. Khan F, Kurklu A, Ghafoor A, Ali Q, Umair M, Shahzaib M (2018) A review on hydroponic greenhouse cultivation for sustainable agriculture. Int J Agric Environ Food Sci 2(2):59–66
19. Valenzano V, Parente A, Serio F, Santamaria P (2008) Effect of growing system and cultivar on yield and water-use efficiency of greenhouse-grown tomato. J Hort Sci Biot 83(1):71–75
20. Al-Tawaha A, Al-Karaki G, Al-Tawaha A, Sirajuddin S, Makhadmeh I, Wahab P, Youssef R, Al Sultan W, Massadeh A (2018) Effect of water flow rate on quantity and quality of lettuce (*Lactuca sativa* L.) in nutrient film technique (NFT) under hydroponics conditions. Bulgarian J Agri Sci 24:793–800
21. Sardare MD, Shraddha VA (2013) A review on plant without soil-hydroponics. Int J Res Eng Technol 2(3):299–304
22. Fertinnowa (2017) Transfer of innovative techniques for sustainable water use in fertigated crops. Semi-closed hydroponic system. CORDI. EU research results
23. Anastasiou A, Ferentinos KP, Arvanitis KG, Sigrimis N (2005) DSS-Hortimed for on-line management of hydroponic systems. Acta Horti 691:267–274
24. Geerts S, Raes D (2009) Deficit irrigation as an on-farm strategy to maximize crop water productivity in dry areas. Agric Water Manage 96:1275–1284
25. Gruda N, Tanny J (2014) Protected Crops. In: Dixon G, Aldous DE (eds) Horticulture: plants for people and places, Volume 1. Springer, Dordrecht Heidelberg New York London, pp 327–406
26. Levidowa L, Zaccariab D, Maiac R, Vivasc E, Todorovicd M, Scardignoda A (2014) Improving water-efficient irrigation: Prospects and difficulties of innovative practices. Agric Water Manag 146:84–94
27. Lee J, Oh M (2017) Mild water deficit increases the contents of bioactive compounds in dropwort. Hort Environ Biot 58:458–466
28. Barzegar T, Lotfi H, Rabiei V, Ghahremani Z, Nikbakht J (2017) Effect of water-deficit stress on fruit yield, antioxidant activity, and some physiological traits of four Iranian melon genotypes. Iranian J Hort Sci (Special Issue):13–25
29. Soni P, Abdin MZ (2017) Water deficit-induced oxidative stress affects artemisinin content and expression of proline metabolic genes in Artemisia annua L. FEBS Open Bio 25;7(3):367–381
30. Khalid K (2006) Influence of water stress on growth, essential oil, and chemical composition of herbs (*Ocimum* sp.). Int Agrophysics 20:289–296
31. Moosavi SG, Seghatoleslami M, Fazeli M, Jouyban Z, Ansarinia E (2014) Effect of water deficit stress and nitrogen fertilizer on flower yield and yield components of marigold (*Calendula officinalis* L.). Int J Biosci 4:42–49
32. Hund A, Ruta N, Liedgens M (2009) Rooting depth and water use efficiency of tropical maize inbred lines, differing in drought tolerance. Plant Soil 318:311–325

33. Fischer R, Rees D, Sayre K, Lu Z-M, Condon A, Saavedra A (2009) Wheat yield progress associated with higher. Plant Soil 318:311–325
34. Tátrai Z, Sanoubar R, Pluhár Z, Mancarella S, Orsini F, Gianquinto G (2016) Morphological and physiological plant responses to drought stress in thymus citriodorus. Int J Agr 2016:1–8
35. Pagliarulo C, Hayden A (2002) Potential for greenhouse aeroponic cultivation of medicinal root crops. College of Agriculture and Life Sciences, The University of Arizona, The Controlled Environment Agricultural Center
36. Hayden A, Giacomelli G, Yokelson T, Hoffmann J (2004) Aeroponics: an alternative production system for high-value root crops. Acta Hort 629:207–213
37. Hayden A, Brigham L, Giacomelli G (2004) Aeroponic cultivation of ginger (*Zingiber officinale*) rhizomes. Acta Hort 629:397–402
38. Hayden A (2006) Aeroponic and hydroponic systems for medicinal herb, rhizome, and root crops. HortScience 41(3):536–538
39. Mairapetyan S, Alexanyan J, Tadevosyan A, Tovmasyan A, Stepanyan B, Galstyan H, Daryadar M (2018) The productivity of some valuable medicinal plants in conditions of water stream hydroponic. J Agr Sci Food Res 9:237–240
40. Mairapetyan S, Alexanyan J, Tovmasyan1 A, Daryadar M, Stepanian B, Mamikonyan V (2016) Productivity, biochemical indices and antioxidant activity of Peppermint (Mentha piperita L.) and basil (Ocimum basilicum L.) in condition of hydroponics. J Sci Technol Environ Inform 3:191–194
41. Daryadar M (2015) Water stream hydroponics as a new technology for soilless production of valuable essential oil and medicinal plant peppermint. Acad J Agri Res 3(10):259–263
42. Keat C, Kannan C (2015) Development of a cylindrical hydroponics system for vertical farming chow. J Agr Sci Tech B 5:93–100
43. Wilson G (2005) Greenhouse aquaponics proves superior to inorganic hydroponics. Aquaponic J. Issue #39 4th quarter
44. Woodruff J (2015) Aquaponic farming saves water, but can it feed the country? https://www.pbs.org/newshour/show/aquaponic-farming-saves-water-can-feed-country
45. Ray M (2017) Aquaponics: an interview with sweet water organics' world watch institute
46. Wilson AL (2004) Aquaponics research at RMIT University, Melbourne Australia. Aquaponic J. Issue #35 4th quarter
47. Ahmed A, Yu H, Yang X, Jiang W (2014) Deficit irrigation affects growth, yield, vitamin c content, and irrigation water use efficiency of hot pepper grown in soilless culture. Hort Sci 49(6):722–728
48. Strojny Z, Nelson PV, Willitz DH (1998) Pot soil air composition in conditions of high soil moisture and its influence on chrysanthemum growth. Sci Horti 73:125–136
49. Koriesh EM, Khalil AM, Abd El-Fattah YM, Attia K (2009) Application of one system of hydroponics in production of *Catharanthus roseus* L. G. Don. J. Agric Sci Mansoura Univ 34:6595–6615
50. Sonneveld C (1981) Items for application of macro-elements in soilless cultures. Acta Hort 126:187–195
51. Kiferle C, Lucchesini M, Mensuali-Sodi A, Maggini R, Raffaelli A, Pardossi A (2011) Rosmarinic acid content in basil plants grown in vitro and in hydroponics. Cent Eur J Biol 6:946–957
52. Sgherri C, Cecconami S, Pinzino C, Navari-Izzo F, Izzo R (2010) Levels of antioxidants and nutraceuticals in basil grown in hydroponics and soil. Food Chem 123:416–422
53. Hassanpouraghdam M, Tabatabaie S, Nazemiyeh H, Aflatuni A (2008) Essential oil composition of hydroponically grown *Chrysanthemum balsamita*. J Essent Oil-Bear Plants 11:649–654
54. Resh H (2012) Hydroponic food production: a definitive guidebook for the advanced home gardener and the commercial hydroponic grower, 7th edn. CRC, Inc., 560p
55. Maggini R, Kiferle C, Lucia G, Andrea R (2012) Growing medicinal plants in hydroponic culture

56. Beyene B, Deribe H (2016) Review on application and management of medicinal plants for the livelihood of the local community. J Resour Dev Manag 22:33–39
57. Brown P, Saa S (2015) Biostimulants in agriculture. Front Plant Sci 6:671
58. Bulgari R, Cocetta G, Trivellini A, Vernieri P, Ferrante A (2015) Biostimulants and crop responsesa review. Biol Agric Hortic 31:1–17
59. Calvo P, Nelson L, Kloepper JW (2014) Agricultural uses of plant biostimulants. Plant Soil 383:3–41
60. Colla G, Nardi S, Cardarelli M, Ertani A, Lucini L, Canaguier R, Rouphael Y (2015) Protein hydrolysates as biostimulants in horticulture. Sci Horti 196:28–38
61. Koriesh EM, Abd El-Fattah YM, Abo El-Soud IH, Khalil MF (2018) Effects of different nutrient solution formulations supplemented with willow bark or juvenile branches decoction on growth of *Coleus* plants. HortScience J Suez Canal Univ 7:11–19
62. Koriesh EM, Abo-El-Soud IH, Abd El-Fattah YM, Khalil M (2019) Comparison of nutrient solution formulations supplemented with willow extract on coleus (*Plectranthus scutellarioides,* (L.) r.br.) grown in sand culture. ii. active constituents (Under publication)
63. Povero G, Mejia JF, Tommaso D, Piaggesi A, Warrior P (2016) A systematic approach to discover and characterize natural plant biostimulants. Front Plant Sci 7:435
64. Saa S, Olivos-DelRio A, Castro S, Brown PH (2015) Foliar application of microbial and plant based biostimulants increases growth and potassium uptake in almond (*Prunus dulcis* [Mill] DA Webb). Front Plant Sci 6(87):1–10

Irrigation Management

Accurate Estimation of Crop Coefficients for Better Irrigation Water Management in Egypt

Samiha Ouda

1 Introduction

Egypt is characterized by having a semi-arid climate, where low or no rain events occur throughout crops' growing season. Thus, irrigation of the cultivated crops is needed for providing the best level of production. However, water is becoming a scarce natural resource and agriculture represents the major water consumption (about 85% of water resources is allocated to agriculture). Therefore, proper irrigation scheduling has to be employed by the farmers to attain water-saving measures. The misuse of water resources in Egypt is due to either low efficiency of irrigation or inadequate irrigation scheduling. Under these circumstances, loss of valuable water resources occurs, which result in higher production cost, in addition to negative environmental consequences.

Sustainability of water resources highly depends on matching water supply and demand, which also results in increasing crop water productivity per unit water. Moreover, knowledge of crop water requirements is crucial for water resources management and planning [1]. Crop water requirements consist of two components, namely evapotranspiration (ETo) and crop coefficients (Kc). ETo is the total amount lost from the field by both soil evaporation and plant transpiration [2]. Accurate estimation of ETo is an important factor to attain prop water management. Several equations were developed to calculate ETo, such as the temperature-based models ([3, 4] and [5]). Moreover, there are the radiation models, which are based on solar radiation ([6] and [7]). Furthermore, the combination models, which are based on the energy balance and mass, transfer principles and include the Penman [8], modified Penman [9], and FAO Penman-Monteith equation [10]. Earlier studies compared different ETo equations for their accuracy revealed that the Penman-Monteith equation

S. Ouda (✉)
Water Requirements and Field Irrigation Research Department; Soils, Water and Environment Research Institute, Agricultural Research Center, 9 El-Gamah Street, Giza 11211, Egypt
e-mail: samihaouda@yahoo.com

E.-S. E. Omran and A. M. Negm (eds.), *Technological and Modern Irrigation Environment in Egypt*, Springer Water,
https://doi.org/10.1007/978-3-030-30375-4_8

is the most accurate because of its detailed theoretical base and its accommodation of small time periods [11].

On the other hand, the Kc takes into account the relationship between atmosphere, crop physiology, and agricultural practices [10]. The concept of Kc was introduced by Jensen [12] and further developed by other researchers [10, 13]. Values of Kc for most agricultural crops increase from a minimal value at planting to a maximum Kc value near full canopy cover or pollination [10]. Kc is defined as the ratio between crop evapotranspiration (ETc) and ETo, from a well water (not limiting) reference surface [10]. Crop Kc plays an important role in the exact calculation of ETc and consequently water requirements. Thus, correct knowledge of ETc allows improving "water management by changing the volume and frequency of irrigation to meet the crop requirements and to adapt to soil characteristics" [1].

Furthermore, it was reported that the Kc is affected by all the factors that influence soil water status, for instance, the irrigation method and frequency [14], the weather factors, the soil characteristics, and the agronomic techniques that affect crop growth [15]. Consequently, the reported values of crop coefficients in the literature can vary significantly from the actual measured values in a location, if growing conditions differ from those where the said coefficients were experimentally obtained [16]. Ko et al. [17] observed that Kc values could be different from one region to the other. Thus, the available values of Kc published by Allen et al. [10] in the FAO paper 56 were done in experiments implemented in countries with different weather conditions, compared to the observed weather in Egypt. Furthermore, it is assumed that the different environmental conditions between regions allow variation in variety selection and crop developmental stage, which affect Kc [10]. Elevated air temperatures and water vapor pressure deficit over the growing seasons can cause temporal and transient leaf stomata closure [18], impeding plants to transpire at its full potential [17].

Using field experiments to develop values for Kc for the cultivated crops in Egypt is a time-consuming and highly expensive. Thus, using modeling to estimate Kc values can be with great value for water management in Egypt. A very famous irrigation scheduling model called Basic Irrigation Scheduling model (BISm) [19] can be used in this matter. The BISm model provided an easy method to determine Kc values for a large number of crops, as affected by the weather in a certain region, irrigation method, as well as planting and harvest dates.

Thus, the objective of this chapter was to calculate Kc values for 14 field crops, 7 fruit crops, and 13 vegetable crops. The calculation will be done for these crops grown under the weather conditions of the five agro-climatic zones in Egypt.

2 Agro-Climatic Zoning in Egypt

The current situation of water scarcity that faces Egypt creates challenges for agricultural scientists to manage water properly, taking into consideration water resources conservation. One of the suggestions to attain that is to divide Egypt into agro-climatic zones. A region can be divided into agro-climatic zones based on homogeneity in weather variables that have the greatest influence on crop growth and yield [20]. Thus, the agro-climatic zone is a "land unit in terms of major climate, superimposed on the length of the growing period, i.e., moisture availability period" [21]. As a result, crop's growth periods, crop's water requirements, and irrigation scheduling are dependent on weather conditions.

In Egypt, there were several attempts to develop agro-climatic zones. The earliest attempt was the division of Egypt into three main agro-climatic zones, i.e., Lower Egypt, Middle Egypt, and Upper Egypt. However, this classification was more administrative than climatic. Eid et al. [22] compared the annual ETo for each governorate. When the difference between ETo values of several governorates was less than 5%, they were grouped in one zone. Thus, they defined nine agro-zones in Egypt: (1) Coastal zone; (2) Central Delta; (3) East and West Delta; (4) Giza; (5) Minia; (6) Assuit and Sohag; (7) North Qena; (8) South Qena, and (9) Aswan. On the other hand, Medany [23] developed agro-ecological zones using regression equations to predict ETo for a certain zone using the average temperature and month number in the year. He distinguished six zones: (1) North Delta (Dakhlia, Gharbia, Damietta, and Kafr El-Sheikh); (2) West Delta (Alexandria and Beheira governorates); (3) Middle Delta (Ismailia, Kalubia, Minofia, Port-Said, and Sharkia governorates); (4) South Delta (Giza, Cairo, Beni Suef, and Fayoum governorates); (5) Middle Egypt (Sohag, Qena, Assuit, and Minia governorates); and (6) Upper Egypt region (Aswan governorate).

A more recent classification was published by Khalil et al. [24], where CROPWAT model [25] was used to calculate ETo using 10-year weather data (1997-2006) for 20 governorates in Egypt. Their results indicated that there were eight agro-climatic zones: (1) Alexandria and Damietta; (2) Dakahlia and North Sinai; (3) Kafr El-Sheikh and El-Gharbia; (4) Ismailia, El-Sharkia, and El-Monofia; (5) El-Kalubia, Beni Sweif, and El-Minia (6) Giza, Qena, Sohag, and El-Wadi El-Gedid; (7) El-Beheira and El-Fayoum; and (8) Assuit and Aswan. The problem with that classification is that it includes one governorate cultivated under rainfed (North Sinai) and another governorate cultivated using groundwater (El-Wadi El-Gedid), and the soil of these two governorates is sandy, whereas the soil of the other governorates is clay.

Furthermore, Noreldin et al. [26] used 30-year ETo data from 1985 to 2014 calculated by BISm model [19] to divide Egypt into seven agro-climatic zones in the Nile Delta and Valley governorates only. In that methodology, monthly means of weather data for 10 years were calculated for each governorate. Analysis of variance was used, and the means were separated and ranked using the least significant difference test ($LSD_{0.05}$). The results are distinguished seven agro-climatic zones (Table 1).

Lastly, Ouda, and Noreldin [27] used BISm model [19] to calculate ETo values for 20 years from 1995 to 2014 and 10 years from 2005 to 2014 and developed agro-climatic zones for Egypt in each time interval. Zoning using 10-year values of ETo resulted in five agro-climatic zones only and higher values of ETo in each zone, compared to 20-year and 30-year ETo values. Table 2 shows the five agro-climatic zones developed by Ouda and Noreldin [27].

Figure 1 illustrates the five agro-climatic zones developed by Ouda and Noreldin [27].

Table 1 Agro-climatic zones of Egypt as determined by statistical analysis

Zone number	Governorate	ETo (mm/day)
Zone 1	Alexandria	4.520
Zone 2	Damietta	4.687
	Kafr El-Sheik	4.677
	El-Dakahlia	4.700
Zone 3	El-Beheira	5.084
	El-Gharbia	5.063
Zone 4	El-Monofia	5.176
	El-Sharkia	5.246
	El-Kalubia	5.348
	Giza	5.410
	Fayoum	5.548
Zone 5	Beni Sweif	5.681
	El-Minia	5.740
	Assuit	5.810
	Sohag	5.881
Zone 6	Qena	6.002
Zone 7	Aswan	6.167
Mean		5.338
$LSD_{0.05}$		0.146

Source Noreldin et al. [26]

3 Calculation of ETo

The BISm model [19] was used to calculate the monthly values of ETo. The model calculates ETo using the Penman-Monteith equation as presented in the United Nations FAO Irrigation and Drainage Paper (FAO 56) by Allen et al. [10]. After calculating daily means per month, a cubic spline curve fitting subroutine is used to estimate daily ETo rates for the entire year.

The model contained a database for 64 crops (field crops, vegetables, and fruit trees), concerning planting and harvesting dates and morphological characteristics for the calculation of crop Kc values in each growth stage of each crop. Thus, to calculate the values of Kc for a certain crop, planting and harvest dates should be entered in the model to determine the Kc value as a percentage of the crop growing season. Snyder et al. [19] indicated that "the model assumed that as a crop canopy develops, the ratio of transpiration (T) to ET increases until most of the ET comes from T and evaporation (E) becomes a minor component. This occurs because the light interception by the foliage increases until the most light is intercepted before it reaches the soil. Thus, for field and row crops, crop coefficients generally increase until the canopy attain ground cover and reaches about 75%, and the light interception became near 80%. Regarding tree, the peak Kc is reached when the canopy has

Table 2 Agro-climatic zones of Egypt classification using 10-year time interval

Zone number	Governorate	ETo ($mm.day^{-1}$)
Zone 1	Alexandria	4.279
	Kafr El-Sheik	4.852
Zone 2	Damietta	5.123
	El-Dakahlia	5.344
	El-Beheira	5.192
	El-Gharbia	5.125
Zone 3	El-Monofia	5.800
	El-Sharkia	5.869
	El-Kalubia	5.964
	Giza	5.701
	Fayoum	5.587
Zone 4	Beni Sweif	6.139
	El-Minia	6.140
	Assuit	6.122
	Sohag	6.127
Zone 5	Qena	6.480
	Aswan	6.600
Average		5.673
Rang		2.321
$LSD_{0.05}$		0.217

Source Ouda and Noreldin [27]

reached about 63% ground cover". http://biomet.ucdavis.edu/irrigation_scheduling/bis/BIS.pdf.

The model used a two-stage method for estimating soil evaporation presented by Stroonsnjider [28] to estimate bare soil crop coefficients. Thus, the crop coefficient during initial growth (Kc_{ini}) is determined by the ETo rate and irrigation frequency using the bare soil evaporation model previously mentioned. The values for Kc_{mid} and Kc_{end} depend on the difference in (1) daily net radiation (Rn) and soil heat flux density (G); (2) crop morphology effects on turbulence; and (3) physiological differences between the crop and reference crop [19].

The model also calculates ETc and schedules irrigation by taking into consideration water depletion from the root zone. Therefore, it requires to input total water holding capacity and available water in the soil. The model then determines the time when irrigation needs to be applied and the required amounts [19].

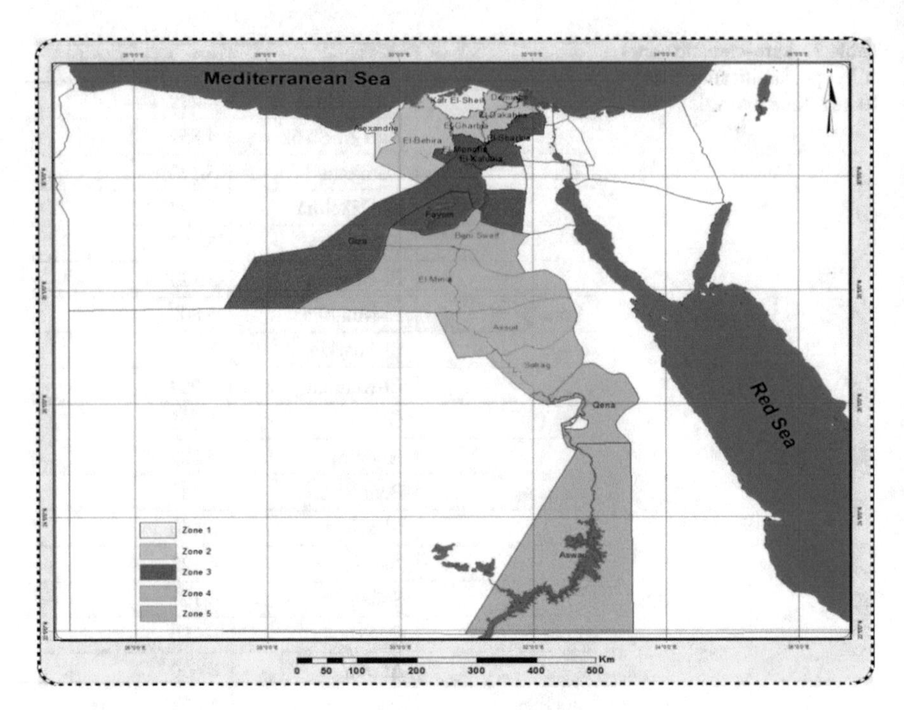

Fig. 1 Map of agro-climatic zones of Egypt using 10 years of ETo values. *Source* Ouda and Noreldin [27]

4 Crop Coefficients in the Agro-Climatic Zones of Egypt

4.1 Weather Descriptions of the Agro-Climatic Zones of Egypt

The weather elements in the five agro-climatic zones in 2016 are presented in Table 3. The results in that table indicated that there is an increasing trend in the value of solar radiation (SRAD) and maximum temperature (TMAX) from the first agro-climatic zone to the fifth agro-climatic zone. Morsy et al. [29] reported that the downward shortwave radiation decreases gradually from south to north according to the apparent position of the sun and reaches its maximum value in the summer season. Downward shortwave radiation has a strong gradient in winter, while it has a weak gradient in summer. They also added that the maximum temperature in Egypt increases gradually southward in all seasons following the apparent position of the sun. The minimum value of maximum temperature occurs in the winter, while it tends to reach its maximum value in the summer. One can notice that the gradient of maximum temperature intensifies during summer and spring seasons over the northern part of Egypt and declines in the winter season.

Table 3 Yearly averages of weather elements and ETo values in the five agro-climatic zones in 2016

Zone	SRAD (MJ/m^2/day	TMAX (°C)	TMIN (°C)	WS (m/s)	TDEW (°C)	ETo (mm/day)
Zone 1	19.3	26.4	18.4	2.7	14.0	5.1
Zone 2	19.4	27.7	18.2	2.6	13.4	5.4
Zone 3	20.4	30.3	15.3	2.2	9.4	6.1
Zone 4	22.1	28.6	14.1	2.1	4.8	6.9
Zone 5	22.5	33.1	16.8	2.0	2.9	7.8
Mean	20.8	29.2	16.6	2.3	8.9	6.3
Range	3.2	6.7	1.6	0.7	5.1	2.7

Furthermore, there is a decreasing trend in the value of the minimum temperature (TMIN), wind speed (WS), and dew point temperature (TDEW) from the first agro-climatic zone to the fifth agro-climatic zone. Morsy et al. [29] indicated that most of the middle region of Egypt have the lowest values of minimum temperature in all seasons, where the lowest value of minimum temperature occurs in the winter, while it tends to be at its maximum value in the summer. The gradient of minimum temperature intensifies during winter and autumn seasons over the northern part of Egypt and declines over the summer season. They also reported that wind speed increases in winter, compared to summer. The maximum core of wind speed is found over the southwest of Egypt, while the minimum core is found over Middle Egypt in all seasons. The strong gradient of wind over the northern part is according to Mediterranean depression in the winter season.

The net result of the interaction of these weather elements resulted in an increase in the value of ETo by 53% in the fifth agro-climatic zone, compared to the first agro-climatic zone.

The results in Table 3 are also shown that the range (the difference between highest and lowest values) was the highest for maximum temperature (6.7°C), followed by dew point temperature (5.1°C). The range between the lowest value of ETo in the first agro-climatic zone and the highest value in the fifth zone was 2.7 mm/day. It is expected that these high differences will have its implication on water consumptive use of the crops grown in these five zones.

4.2 Planting and Harvest Date for the Selected Crops

Recommended planting and harvest dates for the studied field crops are presented in Table 4. It worth mentioning that there is a suitable range, where a certain crop can be planted. However, for an easy calculation, a certain planting date was assumed to suitable for each crop.

Table 5 presents the recommended planting and harvest dates for fruit crops, as

Table 4 Planting and harvest dates for the selected field crops

Crop	Planting date	Harvest date	Season length (days)
Barley	15-Nov	1-Apr	138
Bean (faba)	25-Oct	25-Apr	152
Clover	15-Oct	1-Apr	169
Cotton	15-Mar	15-Aug	154
Flax	15-Nov	13-Apr	150
Lentil	25-Oct	25-Mar	152
Maize	15-Mar	1-Sep	110
Rice	15-May	16-Sep	125
Sorghum	15-May	1-Sep	110
Soybean	15-May	25-Aug	103
Sugar beet	15-Oct	12-Apr	180
Sugarcane	15-Feb	14-Feb	365
Sunflower	15-May	15-Aug	93
Wheat	15-Nov	18-Apr	155

Table 5 Planting and harvest dates for the selected fruits crops

Crop	Planting date	Harvest date	Season length (days)
Apple	15-Feb	14-Feb	365
Citrus	15-Feb	14-Feb	365
Date	15-Feb	14-Feb	365
Grape	15-Feb	14-Feb	365
Olive	15-Feb	14-Feb	365
Stone fruit	15-Feb	14-Feb	365

well as its season length.

Similarly, recommended planting and harvest dates for vegetable crops, as well as its season length are presented in Table 6.

4.3 *Kc for Crops Grown in the First Agro-Climatic Zones*

Table 7 shows the date of Kc growth stages and its values for the selected field crops in the first agro-climatic zone. It is also shown from the table that sugarcane is not cultivated in this zone. It is worth noting that Kc_{ini} starts from planting date and ends with the recorded date in the table.

Table 8 presents the dates of Kc growth stages and its value for the selected fruit crops in the first agro-climatic zone.

Table 6 Planting and harvest dates of selected vegetable crops

Crop	Planting date	Harvest date	Season length (days)
Cucumber	15-Mar	15-Jun	93
Eggplant	1-Apr	1-Aug	130
Onion	15-Nov	15-Apr	152
Peas	1-Sep	30-Nov	91
Pepper	1-Apr	8-Aug	130
Potato (winter)	1-Nov	1-Feb	93
Potato (summer)	1-Aug	28-Nov	120
Squash	15-Mar	15-Jun	93
Strawberries	1-Sep	15-May	257
Sweet potatoes	1-Oct	1-Mar	152
Tomato (winter)	1-Oct	1-Mar	152
Tomato (summer)	1-May	1-Sep	124
Watermelons	15-Feb	30-Jun	136

Table 7 Date of Kc growth stages and its values for the studied field crops in the first agro-climatic zone in 2016

Crop	Data of growth stages			Kc value		
	Kc_{ini}	Kc_{mid}	Kc_{end}	Kc_{ini}	Kc_{mid}	Kc_{end}
Barley	12-Dec	16-Jan	1-Apr	0.31	1.11	0.21
Bean (faba)	29-Nov	24-Dec	25-Apr	0.29	0.99	0.21
Clover	26-Oct	3-Dec	1-Apr	0.26	1.13	0.40
Cotton	7-Apr	23-Jul	15-Aug	0.30	0.93	0.46
Flax	10-Dec	21-Jan	13-Apr	0.31	1.10	0.25
Lentil	10-Nov	24-Dec	25-Apr	0.23	0.99	0.21
Maize	6-Jun	3-Jul	1-Sep	0.24	1.04	0.58
Rice	14-Jun	30-Jun	16-Sep	0.37	1.02	0.78
Sorghum	1-Jun	30-Jun	1-Sep	0.20	1.06	0.50
Soybean	4-Jun	30-Jun	25-Aug	0.24	1.11	0.39
Sugar beet	11-Nov	4-Jan	12-Apr	0.27	1.15	0.95
Sugarcane	–	–	–	–	–	–
Sunflower	2-Jun	25-Jun	15-Aug	0.24	1.09	0.37
Wheat	16-Dec	22-Jan	18-Apr	0.31	1.06	0.19

The date of Kc growth stages and its values for the selected field crops in the first agro-climatic zone is presented in Table 9.

Table 8 Date of Kc growth stages and its values for the studied fruit crops in the first agro-climatic zone in 2016

Crop	Data of growth stages			Kc value		
	Kc_{ini}	Kc_{mid}	Kc_{end}	Kc_{ini}	Kc_{mid}	Kc_{end}
Apple	15-Feb	16-Aug	14-Feb	0.55	1.05	0.80
Citrus	15-Feb	15-Jun	14-Feb	1.00	1.00	1.00
Date	15-Feb	15-Jun	14-Feb	0.95	0.95	0.95
Grape	15-Feb	28-Apr	14-Feb	0.45	0.80	0.35
Olive	15-Feb	15-Jun	14-Feb	0.80	0.80	0.80
Stone fruit	15-Feb	16-Aug	14-Feb	0.55	1.05	0.65

Table 9 Date of Kc growth stages and its values for the studied vegetable crops in the first agro-climatic zone in 2016

Crop	Data of growth stages			Kc value		
	Kc_{ini}	Kc_{mid}	Kc_{end}	Kc_{ini}	Kc_{mid}	Kc_{end}
Cucumber	1-Apr	27-Apr	15-Jun	0.26	0.85	0.85
Eggplant	30-Apr	10-Jun	1-Aug	0.23	0.91	0.86
Onion	30-Nov	24-Dec	15-Apr	0.30	1.19	0.54
Peas	19-Sep	13-Oct	30-Nov	0.22	0.98	0.98
Pepper	27-Apr	28-May	8-Aug	0.23	0.96	0.83
Potato (winter)	19-Nov	12-Dec	1-Feb	0.28	1.09	0.68
Potato (summer)	25-Aug	24-Sep	28-Nov	0.25	1.10	0.69
Squash	2-Apr	30-Apr	15-Jun	0.26	0.90	0.70
Strawberries	10-Oct	25-Dec	15-May	0.23	0.70	0.70
Sweet potatoes	31-Oct	8-Dec	1-Mar	0.26	1.10	0.69
Tomato (winter)	8-Nov	16-Dec	1-Mar	0.26	1.10	0.64
Tomato (summer)	1-Jun	2-Jul	1-Sep	0.25	1.10	0.65
Watermelon	14-Mar	24-Apr	30-Jun	0.29	1.00	0.75

4.4 Kc for Crops Grown in the Second Agro-Climatic Zones

Table 10 shows the date of Kc growth stages and its values for the selected field crops in the second agro-climatic zone. The table showed that the date of every Kc growth stage was similar in the first and second agro-climatic zones. However, the values of Kc were different. It can be noticed from the table that sugarcane is not cultivated in this zone.

Table 11 indicates the date of Kc growth stages and its values for the selected fruit crops in the second agro-climatic zone. The table showed that the date and the value of every Kc growth stage were similar in the first and second agro-climatic zones.

The results in Table 12 are indicated the Kc values for the studied vegetable crops

Table 10 Date of Kc growth stages and its values for the studied field crops in the second agro-climatic zone in 2016

Crop	Data of growth stages			Kc value		
	Kc_{ini}	Kc_{mid}	Kc_{end}	Kc_{ini}	Kc_{mid}	Kc_{end}
Barley	12-Dec	16-Jan	1-Apr	0.30	1.11	0.20
Bean (faba)	29-Nov	24-Dec	25-Apr	0.28	0.99	0.20
Clover	26-Oct	3-Dec	1-Apr	0.26	1.15	0.40
Cotton	7-Apr	23-Jul	15-Aug	0.28	0.93	0.46
Flax	10-Dec	21-Jan	13-Apr	0.30	1.10	0.25
Lentil	10-Nov	24-Dec	25-Apr	0.23	0.99	0.20
Maize	6-Jun	3-Jul	1-Sep	0.23	1.04	0.58
Rice	14-Jun	30-Jun	16-Sep	0.35	1.02	0.78
Sorghum	1-Jun	30-Jun	1-Sep	0.20	1.06	0.50
Soybean	4-Jun	30-Jun	25-Aug	0.23	1.11	0.39
Sugar beet	11-Nov	4-Jan	12-Apr	0.26	1.15	0.95
Sugarcane	–	–	–	–	–	–
Sunflower	2-Jun	25-Jun	15-Aug	0.23	1.09	0.37
Wheat	16-Dec	22-Jan	18-Apr	0.31	1.08	0.19

Table 11 Date of Kc growth stages and its values for the studied fruit crops in the second agro-climatic zone in 2016

Crop	Data of growth stages			Kc value		
	Kc_{ini}	Kc_{mid}	Kc_{end}	Kc_{ini}	Kc_{mid}	Kc_{end}
Apple	15-Feb	16-Aug	14-Feb	0.55	1.05	0.80
Citrus	15-Feb	15-Jun	14-Feb	1.00	1.00	1.00
Date	15-Feb	15-Jun	14-Feb	0.95	0.95	0.95
Grape	15-Feb	28-Apr	14-Feb	0.45	0.80	0.35
Olive	15-Feb	15-Jun	14-Feb	0.80	0.80	0.80
Stone fruit	15-Feb	16-Aug	14-Feb	0.55	1.05	0.65

in the second agro-climatic zone, where the date of every growth stage was similar to its counterpart in the first agro-climatic zone.

4.5 *Kc for Crops Grown in the Third Agro-Climatic Zones*

Table 13 shows the date of Kc growth stages and its values for the selected field crops in the third agro-climatic zone. The results in the table showed that similar dates for Kc growth stages in the first and second agro-climatic zones were found,

Table 12 Date of Kc growth stages and its values for the studied vegetable crops in the second agro-climatic zone in 2016

Crop	Data of growth stages			Kc value		
	Kc_{ini}	Kc_{mid}	Kc_{end}	Kc_{ini}	Kc_{mid}	Kc_{end}
Cucumber	1-Apr	27-Apr	15-Jun	0.25	0.85	0.85
Eggplant	30-Apr	10-Jun	1-Aug	0.22	0.91	0.86
Onion	30-Nov	24-Dec	15-Apr	0.29	1.19	0.54
Peas	19-Sep	13-Oct	30-Nov	0.21	0.98	0.98
Pepper	27-Apr	28-May	8-Aug	0.22	0.99	0.83
Potato (winter)	19-Nov	12-Dec	1-Feb	0.27	1.09	0.68
Potato (summer)	25-Aug	24-Sep	28-Nov	0.24	1.10	0.69
Squash	2-Apr	30-Apr	15-Jun	0.25	0.90	0.70
Strawberries	10-Oct	25-Dec	15-May	0.22	0.70	0.70
Sweet potatoes	31-Oct	8-Dec	1-Mar	0.25	1.10	0.69
Tomato (winter)	8-Nov	16-Dec	1-Mar	0.25	1.10	0.64
Tomato (summer)	1-Jun	2-Jul	1-Sep	0.24	1.10	0.65
Watermelon	14-Mar	24-Apr	30-Jun	0.28	1.00	0.75

Table 13 Date of Kc growth stages and its values for the studied field crops in the third agro-climatic zone in 2016

Crop	Data of growth stages			Kc value		
	Kc_{ini}	Kc_{mid}	Kc_{end}	Kc_{ini}	Kc_{mid}	Kc_{end}
Barley	12-Dec	16-Jan	1-Apr	0.30	1.11	0.18
Bean (faba)	29-Nov	24-Dec	25-Apr	0.27	0.99	0.19
Clover	26-Oct	3-Dec	1-Apr	0.24	1.15	0.40
Cotton	7-Apr	23-Jul	15-Aug	0.28	0.95	0.50
Flax	10-Dec	21-Jan	13-Apr	0.30	1.10	0.25
Lentil	10-Nov	24-Dec	25-Apr	0.22	0.99	0.19
Maize	6-Jun	3-Jul	1-Sep	0.21	1.03	0.58
Rice	14-Jun	30-Jun	16-Sep	0.31	1.02	0.78
Sorghum	1-Jun	30-Jun	1-Sep	0.18	1.06	0.50
Soybean	4-Jun	30-Jun	25-Aug	0.21	1.11	0.39
Sugar beet	11-Nov	4-Jan	12-Apr	0.25	1.15	0.95
Sugarcane	–	–	–	–	–	–
Sunflower	2-Jun	25-Jun	15-Aug	0.21	1.09	0.37
Wheat	16-Dec	22-Jan	18-Apr	0.31	1.08	0.17

with different values of Kc. Table 13 also indicates that sugarcane is not cultivated in this zone.

Concerning the studied fruit crops, the date and value of each Kc growth stage were similar to what was obtained in the first and second agro-climatic zones (Table 14).

A similar trend was found for vegetable crops, and it is presented in Table 15, where the date of each Kc growth stage was similar in the first, second, and third agro-climatic zones. Also, the values of Kc for the selected crops were different.

Table 14 Date of Kc growth stages and its values for the studied fruit crops in the third agro-climatic zone in 2016

Crop	Data of growth stages			Kc value		
	Kc_{ini}	Kc_{mid}	Kc_{end}	Kc_{ini}	Kc_{mid}	Kc_{end}
Apple	15-Feb	16-Aug	14-Feb	0.55	1.05	0.80
Citrus	15-Feb	15-Jun	14-Feb	1.00	1.00	1.00
Date	15-Feb	15-Jun	14-Feb	0.95	0.95	0.95
Grape	15-Feb	28-Apr	14-Feb	0.45	0.80	0.35
Olive	15-Feb	15-Jun	14-Feb	0.80	0.80	0.80
Stone fruit	15-Feb	16-Aug	14-Feb	0.55	1.05	0.65

Table 15 Date of Kc growth stages and its values for the studied vegetable crops in the third agro-climatic zone in 2016

Crop	Data of growth stages			Kc value		
	Kc_{ini}	Kc_{mid}	Kc_{end}	Kc_{ini}	Kc_{mid}	Kc_{end}
Cucumber	1-Apr	27-Apr	15-Jun	0.23	0.85	0.85
Eggplant	30-Apr	10-Jun	1-Aug	0.20	0.91	0.86
Onion	30-Nov	24-Dec	15-Apr	0.29	1.19	0.54
Peas	19-Sep	13-Oct	30-Nov	0.20	0.98	0.98
Pepper	27-Apr	28-May	8-Aug	0.20	0.99	0.83
Potato (winter)	19-Nov	12-Dec	1-Feb	0.27	1.09	0.68
Potato (summer)	25-Aug	24-Sep	28-Nov	0.22	1.10	0.69
Squash	2-Apr	30-Apr	15-Jun	0.23	0.90	0.70
Strawberries	10-Oct	25-Dec	15-May	0.21	0.70	0.70
Sweet potatoes	31-Oct	8-Dec	1-Mar	0.24	1.10	0.69
Tomato (winter)	8-Nov	16-Dec	1-Mar	0.24	1.10	0.64
Tomato (summer)	1-Jun	2-Jul	1-Sep	0.21	1.10	0.65
Watermelon	14-Mar	24-Apr	30-Jun	0.26	1.00	0.75

4.6 *Kc for Crops Grown in the Fourth Agro-Climatic Zones*

Table 16 shows the date of Kc growth stages and its values for the selected field crops in the fourth agro-climatic zone. The table also showed that both the date of Kc stages and its value were different in the fourth agro-climatic zone, compared to the third agro-climatic zone. Furthermore, rice is not cultivated in this zone, as it is prohibited by the law and sugarcane is suitable to be cultivated in this zone.

Table 17 indicates that both the date of Kc stages and its value were similar in the fourth agro-climatic zone to its counterpart in the third agro-climatic zone.

Table 16 Date of Kc growth stages and its values for the studied field crops in the fourth agro-climatic zone in 2016

Crop	Data of growth stages			Kc value		
	Kc_{ini}	Kc_{mid}	Kc_{end}	Kc_{ini}	Kc_{mid}	Kc_{end}
Barley	13-Dec	17-Jan	1-Apr	0.29	1.11	0.18
Bean (faba)	30-Nov	25-Dec	25-Apr	0.26	0.99	0.18
Clover	27-Oct	4-Dec	1-Apr	0.26	1.15	0.40
Cotton	8-Apr	24-Jul	15-Aug	0.24	0.93	0.45
Flax	11-Dec	22-Jan	13-Apr	0.29	1.10	0.25
Lentil	11-Nov	25-Dec	25-Apr	0.21	0.99	0.18
Maize	7-Jun	4-Jul	1-Sep	0.19	1.03	0.58
Rice	–	–	–	–	–	–
Sorghum	2-Jun	1-Jul	1-Sep	0.16	1.06	0.50
Soybean	5-Jun	1 Jul	25-Aug	0.19	1.11	0.39
Sugar beet	12-Nov	5-Jan	12-Apr	0.23	1.15	0.95
Sugarcane	18-Apr	21-Oct	14-Feb	0.4	1.25	0.75
Sunflower	3-Jun	26-Jun	15-Aug	0.20	1.09	0.37
Wheat	17-Dec	23-Jan	18-Apr	0.29	1.08	0.17

Table 17 Date of Kc growth stages and its values for the studied fruit crops in the fourth agro-climatic zone in 2016

Crop	Data of growth stages			Kc value		
	Kc_{ini}	Kc_{mid}	Kc_{end}	Kc_{ini}	Kc_{mid}	Kc_{end}
Apple	15-Feb	16-Aug	14-Feb	0.55	1.05	0.80
Citrus	15-Feb	15-Jun	14-Feb	1.00	1.00	1.00
Date	15-Feb	15-Jun	14-Feb	0.95	0.95	0.95
Grape	15-Feb	28-Apr	14-Feb	0.45	0.80	0.35
Olive	15-Feb	15-Jun	14-Feb	0.80	0.80	0.80
Stone fruit	15-Feb	16-Aug	14-Feb	0.55	1.05	0.65

Table 18 Date of Kc growth stages and its values for the studied vegetable crops in the fourth agro-climatic zone in 2016

Crop	Data of growth stages			Kc value		
	Kc_{ini}	Kc_{mid}	Kc_{end}	Kc_{ini}	Kc_{mid}	Kc_{end}
Cucumber	2-Apr	28-Apr	15-Jun	0.21	0.85	0.85
Eggplant	1-May	11-Jun	1-Aug	0.19	0.91	0.86
Onion	30-Nov	25-Dec	15-Apr	0.28	1.19	0.54
Peas	20-Sep	14-Oct	30-Nov	0.18	0.98	0.98
Pepper	28-Apr	29-May	8-Aug	0.19	0.99	0.83
Potato (winter)	20-Nov	13-Dec	1-Feb	0.25	1.09	0.68
Potato (summer)	26-Aug	25-Sep	28-Nov	0.20	1.10	0.69
Squash	3-Apr	30-Apr	15-Jun	0.21	0.90	0.70
Strawberries	11-Oct	26-Dec	15-May	0.19	0.70	0.70
Sweet potatoes	31-Oct	9-Dec	1-Mar	0.22	1.10	0.69
Tomato (winter)	9-Nov	17-Dec	1-Mar	0.22	1.10	0.64
Tomato (summer)	1-Jun	3-Jul	1-Sep	0.20	1.10	0.65
Watermelon	15-Mar	25-Apr	30-Jun	0.24	1.00	0.75

Concerning vegetable crops, both the date of Kc stages and its value were different in the fourth agro-climatic zone, compared to the third agro-climatic zone (Table 18).

4.7 *Kc for Crops Grown in the Fifth Agro-Climatic Zones*

Table 19 shows the date of Kc growth stages and its values for the selected field crops in the fifth agro-climatic zone. Both the date of Kc stages and its value were different in the fifth agro-climatic zone, compared to the third agro-climatic zone. Three crops are not cultivated in the fifth agro-climatic zone, namely cotton, rice, and sugar beet. Furthermore, the climate of this zone is suitable to grow sugarcane.

It is clear from Table 20 that both the date of Kc stages and its value were similar in the fifth agro-climatic zone and different from its counterpart in the third agro-climatic zone.

Regarding vegetable crops, the results in Table 21 are indicated that both the date of Kc stages and its value were similar in the fifth agro-climatic zone, compared to the fourth agro-climatic zone.

Table 19 Date of Kc growth stages and its values for the studied field crops in the fifth agro-climatic zone in 2016

Crop	Data of growth stages			Kc value		
	Kc_{ini}	Kc_{mid}	Kc_{end}	Kc_{ini}	Kc_{mid}	Kc_{end}
Barley	13-Dec	17-Jan	1-Apr	0.26	1.11	0.17
Bean (faba)	30-Nov	25-Dec	25-Apr	0.23	0.99	0.17
Clover	27-Oct	4-Dec	1-Apr	0.21	1.15	0.40
Cotton	–	–	–	–	–	–
Flax	11-Dec	22-Jan	13-Apr	0.26	1.10	0.25
Lentil	11-Nov	25-Dec	25-Apr	0.19	0.99	0.17
Maize	7-Jun	4-Jul	1-Sep	0.18	1.03	0.58
Rice	–	–	–	–	–	–
Sorghum	2-Jun	1-Jul	1-Sep	0.15	1.06	0.50
Soybean	5-Jun	1-Jul	25-Aug	0.18	1.11	0.39
Sugar beet	–	–	–	–	–	–
Sugarcane	18-Apr	21-Oct	14-Feb	1.25	1.25	0.75
Sunflower	3-Jun	26-Jun	15-Aug	0.18	1.09	0.37
Wheat	17-Dec	23-Jan	18-Apr	0.26	1.08	0.16

Table 20 Date of Kc growth stages and its values for the studied fruit crops in the fifth agro-climatic zone in 2016

Crop	Data of growth stages			Kc value		
	Kc_{ini}	Kc_{mid}	Kc_{end}	Kc_{ini}	Kc_{mid}	Kc_{end}
Apple	15-Feb	16-Aug	14-Feb	0.55	1.05	0.80
Citrus	15-Feb	15-Jun	14-Feb	1.00	1.00	1.00
Date	15-Feb	15-Jun	14-Feb	0.95	0.95	0.95
Grape	15-Feb	28-Apr	14-Feb	0.45	0.80	0.35
Olive	15-Feb	15-Jun	14-Feb	0.80	0.80	0.80
Stone fruit	15-Feb	16-Aug	14-Feb	0.55	1.05	0.65

5 Water Consumptive Use for Crops in the Agro-Climatic Zones

Table 22 shows the water consumptive use for the selected field crops in the five agro-climatic zones of Egypt. The results in this table indicated that the lowest value of water consumptive use could be found in the first agro-climatic zone and the highest value can be found in the fifth agro-climatic zone. This result can be attributed to the higher value of ETo in the fifth zone compared to the first agro-climatic zone.

Table 23 indicates that the percentage of increase in the water consumptive use was higher in the second agro-climatic zone with low percentage, compared to the first

Table 21 Date of Kc growth stages and its values for the studied vegetable crops in the fifth agro-climatic zone in 2016

Crop	Data of growth stages			Kc value		
	Kc_{ini}	Kc_{mid}	Kc_{end}	Kc_{ini}	Kc_{mid}	Kc_{end}
Cucumber	1-Apr	27-Apr	15-Jun	0.20	0.85	0.85
Eggplant	30-Apr	10-Jun	1-Aug	0.18	0.91	0.86
Onion	30-Nov	24-Dec	15-Apr	0.25	1.19	0.54
Peas	19-Sep	13-Oct	30-Nov	0.16	0.98	0.98
Pepper	27-Apr	28-May	8-Aug	0.18	0.99	0.83
Potato (winter)	19-Nov	12-Dec	1-Feb	0.23	1.09	1.09
Potato (summer)	25-Aug	24-Sep	28-Nov	0.18	1.10	0.69
Squash	2-Apr	30-Apr	15-Jun	0.20	0.90	0.70
Strawberries	10-Oct	25-Dec	15-May	0.17	0.70	0.70
Sweet potatoes	31-Oct	8-Dec	1-Mar	0.20	1.10	0.69
Tomato (winter)	8-Nov	16-Dec	1-Mar	0.20	1.10	0.64
Tomato (summer)	1-Jun	2-Jul	1-Sep	0.19	1.10	0.65
Watermelon	14-Mar	24-Apr	30-Jun	0.22	1.00	0.75

Table 22 Water consumptive use (mm) for the selected field crops in the five agro-climatic zones of Egypt in 2016

Crop	Zone 1	Zone 2	Zone 3	Zone 4	Zone 5
Barley	327	337	359	403	475
Bean (faba)	356	367	387	433	512
Clover	566	591	629	700	825
Cotton	798	854	995	1092	–
Flax	399	415	452	505	585
Lentil	348	358	378	424	503
Maize	561	598	679	765	846
Rice	696	739	835	–	–
Sorghum	581	620	709	790	884
Soybean	557	595	678	752	839
Sugar beet	546	569	617	683	–
Sugarcane	–	–	–	2594	2970
Sunflower	502	536	614	676	750
Wheat	385	401	437	487	563

Table 23 Percentage of increasing the value of the water consumptive use of field crops compared to the value of the first agro-climatic zone in 2016

Crop	Zone 1	Zone 2	Zone 3	Zone 4	Zone 5
Barley	–	3	10	23	45
Bean	–	3	9	22	44
Clover	–	4	11	24	46
Cotton	–	7	25	37	0
Flax	–	4	13	27	47
Lentil	–	3	9	22	45
Maize	–	7	21	36	51
Rice	–	6	20	–	–
Sorghum	–	7	22	36	52
Soybean	–	7	22	35	51
Sugar beet	–	4	13	25	
Sugarcane	–	–	–	–	14
Sunflower	–	7	22	35	49
Wheat	–	4	14	26	46

zone with a low percentage. The value of the percentage of the difference between the water consumptive uses in the first agro-climate became higher when the comparison was made with the other zones. This result can be attributed to higher values of weather elements and ETo in the second, third, fourth, and fifth agro-climatic zones as compared to the first agro-climatic zone.

A similar trend was observed for fruit crops, where there was a gradual increase in the value of water consumptive use (Table 24) and the percentage of increase in water consumptive use (Table 25) from the first to the fifth agro-climatic zone.

Concerning vegetable crops, its water consumptive use (Table 26) and its percentage of increase in water consumptive use took an increasing trend, where the highest value was in the fifth agro-climatic zone (Table 27).

Table 24 Water consumptive use (mm) for the selected fruit crops in the five agro-climatic zones of Egypt in 2016

Crop	Zone 1	Zone 2	Zone 3	Zone 4	Zone 5
Apple	1676	1776	2001	2247	2561
Citrus	1865	1978	2236	2505	2849
Date	1772	1879	2125	2379	2706
Grape	1198	1278	1470	1647	1851
Olive	1492	1582	1789	2004	2279
Stone fruit	1686	1787	2011	2258	2575

Table 25 Percentage of increase in the value of water consumptive use of fruit crops compared to the value in the first agro-climatic zone in 2016

Crop	Zone 1	Zone 2	Zone 3	Zone 4	Zone 5
Apple	–	4	14	26	46
Citrus	–	6	19	34	53
Date	–	6	20	34	53
Grape	–	6	20	34	53
Olive	–	7	23	37	55
Stone fruit	–	6	20	34	53

Table 26 Water consumptive use (mm) for the selected vegetable crops in the five agro-climatic zones of Egypt in 2016

	Zone 1	Zone 2	Zone 3	Zone 4	Zone 5
Cucumber	386	414	487	530	582
Eggplant	591	631	728	802	886
Onion	492	512	556	619	725
Peas	291	305	324	366	414
Pepper	667	713	825	909	1006
Potato (winter)	211	215	223	233	289
Potato (summer)	460	480	521	595	674
Squash	386	414	487	529	581
Strawberries	587	617	680	743	845
Sweet potatoes	375	385	394	436	525
Tomato (summer)	642	684	779	866	967
Tomato (winter)	351	360	368	408	490
Watermelons	604	647	763	830	909

6 Comparison Between Measured and Estimated Values of Water Consumptive Use

Published papers concerning measured water consumptive use for crops grown in Egypt were collected for some of the studied crops, and it is presented in Table 28. These papers contained experiments conducted between 1990 and 2014. Therefore, different values of water consumptive use existed in the same agro-climatic zone, but different experiments. To ease the presentation, these values were summarized in the lower and upper limit. Furthermore, Ouda and Zohry [30] estimated water consumptive use for some of these crops in 2015, and the values are presented in Table 28. The estimated values in 2016 are also presented in Table 28.

The results in the table showed that although the measured values were in earlier years than the estimated value, the estimated values lay in a reasonable range,

Table 27 Percentage of increase in the value of water consumptive use of vegetable crops compared to the value in the first agro-climatic zone in 2016

	Zone 1	Zone 2	Zone 3	Zone 4	Zone 5
Cucumber	–	7	26	37	51
Eggplant	–	7	23	36	50
Onion	–	4	13	26	47
Peas	–	5	11	26	42
Pepper	–	7	24	36	51
Potato (winter)	–	2	6	10	37
Potato (summer)	–	4	13	29	47
Squash	–	7	26	37	51
Strawberries	–	5	16	27	44
Sweet potatoes	–	3	5	16	40
Tomato (winter)	–	7	21	35	51
Tomato (summer)	–	3	5	16	40
Watermelons	–	7	26	37	50

compared to the calculated values. The results also indicated that the value of water consumptive use was higher in 2016, compared to 2015. It can be attributed to the observed increasing trends in weather elements in the past few years.

7 Conclusions

Proper irrigation water management under the prevailing conditions of water scarcity in Egypt required the knowledge of exact values of water consumptive use of the cultivated crops. This can be attained by the accurate estimation of the values of ETo and crop Kc. Although the calculation of ETo can be easily implemented, the calculation of Kc required more efforts to be calculated. This chapter provided the date and the values of the Kc for 33 crops (field crops, fruit trees, and vegetable crops) to contribute to irrigation water management in Egypt.

8 Recommendations

It is recommended to establish field experiments to measure Kc values for the studied crops to verifying the estimated values by the BISm model.

Table 28 Comparison between measured and estimated values of water consumptive use (mm) in the agro-climatic zones of Egypt

		Water consumptive use (mm)			
Crop	Zone	Measured	Estimated (2015)[a]	Estimated (2016)	Reference
Barley	3	306–330	346	359	Abd El-Rahman et al. [31]
Bean (faba)	3	355–366	375	387	Ouda et al. [32a]
Cotton	2	609–720	792	854	Abu–Grab et al. [33]
	3	876–899	830	995	Ouda et al.[34b]
	4	1005–1056	905	1092	El-Sayed [35]
Maize	1	548–555	535	561	Abd El-Lattif et al. [36]
	2	421–534	579	598	Abu-Garb et al. [33]
					Taha 2012 [37]
					El-Sharkawy et al. [38]
	3	541–621	637	679	Khalil and Mohamed [39]
					Abd El-Maksoud et al. [40]
					Farrag et al. [41]
	4	523–625	643	705	Badawi et al. [42]
Sugar beet	3	534–565	577	617	Personal commutations
Sunflower	1	447–462	474	502	Darwesh et al. [43]
	3	527–536	530	614	Abdou et al. [44]
					Khalil [45]
Wheat	1	419–428	363	385	Kassem [46]
	2	334–384	385	401	Moussa et al. [47]
					Taha [37]
	3	381–464	409	437	Khalil et al. [48]
					Ouda et al. [49]
					Abdou et al. [44]
	4	479–494	431	487	Noreldin and Mahmoud [50]

[a]*Source* Ouda and Zohry [30]

References

1. Katerji N, Rana G (2008) Crop evapotranspiration measurement and estimation in the Mediterranean region. ISBN 978 8 89015 2412. INRA-CRA, Bari
2. Gardner FP, Pearce RB, Mitchell RL (1985) Physiology of crop plants. Iowa State University Press, Ames, USA
3. Thornthwaite CW (1948) An approach toward a rational classification of climate. Geograph Rev 38:55–94
4. Blaney HF, Criddle WD (1962) Determining consumptive use and irrigation water requirements. In: USDA Technical Bulletin 1275, US Department of Agriculture, Beltsvill
5. Hargreaves GH, Samani ZA (1985) Reference crop evapotranspiration from temperature. Appl Engine Agric 1(2):96–99
6. Priestley CHB, Taylor RJ (1972) On the assessment of the surface heat flux and evaporation using large-scale parameters. Mon Weather Rev 100:81–92
7. Makkink GF (1957) Testing the Penman formula by means of lysimeters. J Inst Water Eng 11:277–288
8. Penman HL (1948) Natural evaporation from open water, bare soil, and grass. Proc Roy Soc London A193:120–146
9. Doorenbos J, Pruit WO (1977) Guidelines for predicting crop water requirements. FAO irrigation and drainage paper, 24
10. Allen RG, Pereira LS, Raes D, Smith M (1998) Crop evapotranspiration-guidelines for computing crop water requirements-FAO irrigation and drainage paper 56; FAO: Rome, Italy, p D05109
11. Valipour M (2014) Analysis of potential evapotranspiration using limited weather data. Appl Water Sci. https://doi.org/10.1007/s13201-014-0234-2
12. Jensen ME (1968) Water consumption by agricultural plants: In: Plant water consumption and response. Water deficits and plant growth, vol II. Academic Press: New York, NY, USA, pp. 1–22
13. Reddy KC, Arunajyothy S, Mallikarjuna P (2015) Crop coefficients of some selected crops of Andhra Pradesh. J. Inst. Eng. (India): Ser. A, 96, 123–130
14. Wright JL (1982) New evapotranspiration crop coefficients. J Irrig Drain Div ASCE 108:57–74
15. Annandale JG, Stockle CO (1994) Fluctuation of crop evapotranspiration coefficients with weather: a sensitivity analysis. Irrig Sci 15:1–7
16. Tarantino E, Onofrii M (1991) Determinazione dei coefficienti colturali mediante lisimetri. Bonifica 7:119–136 (in Italian)
17. Ko J, Piccinni G, Marek T, Howell T (2009) Determination of growth-stage-specific crop coefficients (Kc) of cotton and wheat. Agric Water Manag 96:1691–1697
18. Baker JT, Gitz DC, Payton P, Wanjura DF, Upchurch DR (2007) Using leaf gas exchange to quantify drought in cotton irrigated based on canopy temperature measurements. Agron J 99:637–644
19. Snyder RL, Orang M, Bali K, Eching S (2004) Basic irrigation scheduling (BISm). http://www.waterplan.water.ca.gov/landwateruse/wateruse/Ag/CUP/Californi/Climate_Data_010804.xls
20. FAO (1979) Yield response to water. In: Doorenbos J Kassam A (eds) FAO irrigation and drainage Paper No. 33. Rome
21. FAO (1983) Land evaluation of rain-fed agriculture. Soil Bull. 52, FAO, Rome, 237p
22. Eid HM, El-Marsafawy SM, Ouda SA (2006) Assessing the impact of climate on crop water needs in Egypt: the CROPWAT analysis of three districts in Egypt. CEEPA Discussion Papers No. 29
23. Medany M (2007) Water requirement for crops in Egypt. Central Laboratory of Agricultural Climate. Agricultural Research Center, Egypt
24. Khalil F, Ouda SA, Osman N, Khamis E. (2011) Determination of agro-climatic zones in Egypt using a robust statistical procedure. In; Proceeding of the 15th international conference on water technology, Alexandria, Egypt. 30 May–2 June

25. FAO (1992) CROPWAT, a computer program for irrigation planning and management. FAO Irrigation and Drainage Paper No. 26. Rome
26. Noreldin T, Ouda S, Amer A (2016) Agro-climatic zoning in the Nile Delta and Valley to improve water management. J Water Land Dev Water Land Dev 31(X-XII):113–117
27. Ouda S, Noreldin T (2017) Evapotranspiration data to determine agro-climatic zones in Egypt. J Water Land Dev 32(I-III):79–86
28. Stroosnijder L (1987) Soil evaporation: test of a practical approach under semi-arid conditions. Neth J Agric Sci 35:417–426
29. Morsy M, Sayad T, Ouda S (2017) Present and future water requirements for crops. In: Future of food gaps in Egypt: obstacles and opportunities. Springer Publishing House. ISBN 978-3-319-46942-3
30. Ouda S, Zohry A (2017) Water Requirements for prevailing cropping pattern: In: Cropping pattern to overcome abiotic stresses: water, salinity, and climate. Springer Pulishing house
31. Abd El-Rahman MFS, Khalil FA, Anton NA (2012) Effect of irrigation scheduling and nitrogen fertilization on barley yield and water use efficiency. J Soil and Agric Eng 7(3):234–239
32. Ouda SA, AbouElenin R, Shreif MA (2010a) Increasing water productivity of faba bean grown under deficit irrigation at middle Egypt. In: Proceeding of the 14th international conference on water technology, Egypt
33. Abu-Grab OS, Kaoud EE, Abdel-Maksoud HH (1994) Effect of some chemicals as transpiration—suppressants on yield, water consumptive use, chlorophyll, and minerals contents for some field crops. J Agric Sci 19(3):875–889
34. Ouda SA, Abd El-Baky H, AbouElenin R, Shreif MA (2010b) Simulation of the effect of irrigation water management on cotton yield at two locations in Egypt. In: Proceeding of the 14th international conference on water technology, Egypt
35. El-Sayed AEM (2016) Response of Egyptian cotton to alternative systems of irrigation and different rates of splitting of NPK under two planting dates. Ph.D. Thesis. Assuit University
36. Abd El-Latif KM, Osman EAM, Eid TAA (2012) Available soil moisture depletion levels and some foliar sprayed organic acids effects on yield and some crop water relations of drip-irrigated maize. J Soil Agric Res 8(11):267–273
37. Taha A (2012) Effect of climate change on wheat and maize crops grown in new reclaimed land under fertigation practice. Ph.D. Thesis. College of Agriculture, Tanta University, Egypt
38. El-Sharkawy AF, Bondok MY, Abdel-Maksoud HH (2008) Irrigating maize crop via some different ETo estimating formulae and consequent influence on crop water use and water use efficiency. Minufiya J Agric Res 33(4):941–954
39. Khalil FAF and Mohamed SGA (2006) Studies on the inter-relationship among irrigation and maize varieties on yield and water relations using some statistical procedures. Anal Agric Sci 44(1):393–406
40. Abdel-Maksoud HH, Ashry MRK, Youssef KMR (2008) Maize yield and water relations under different irrigation and plant density treatments. J Agric Sci 33(5):3881–3893
41. Farrag FRM, El-Akram MFI, Abdou SMM, El-Masry AA (2011) Water management of maize crop under liquid ammonia gas fertilization. Minufiya J Agric Res 36(4):1133–1149
42. Badawi MI, Mokhtar NY, Noreldin T (2017) Evaluation of induced microclimate modification via changing planting dates, and/or irrigation methods on maize water productivity in two locations in Egypt. J Soil Sci Agric Eng 8(4):163–169
43. Darwesh RKh, El-Mansoury MAM, El-Shamy MA (2016) Effect of irrigation scheduling on sunflower/forage cowpea intercropping pattern, growth, yield and its components. J Soil Sci Agric Eng 7(2):135–146
44. Abdou SMM, Farrag RMF, El-Akram MFI, Ashry MRK (2011) Water relations and yield of wheat under different N-fertilizer forms and scheduling irrigation. J Soil Sci Agric Eng 2(1):25–41
45. Khalil FAF (2007) Effect of some agricultural practices on productivity and water use efficiency for sunflower. Minufiya J Agric Res 32(1):283–296
46. Kassam A (2016) Impact of climate change on wheat under water stress in North Nile Delta. Ph.D. Thesis. Mansoura University

47. Moussa AM, Abdel- Maksoud HH (2004) Effect of soil moisture regime on yield and its components and water use efficiency for some wheat cultivars. Annals Agric Sci 49(2):515–530
48. Khalil FA, El-Shaarawy GA, Hassan YM (2007) Irrigation scheduling for some wheat cultivars through pan evaporation norms and its effect on growth, yield and water use efficiency. Fayoum J Agric Res Dev 21(1):222–233
49. Ouda SA, Khalil FA, AbouElenin R, Shreif MA, Benli B, Qadir M (2008) Using Yield-Stress model in irrigation management for wheat grown in Egypt. J Appl Biol Sci 2(1):57–65
50. Noreldin T, Mahmoud MSM (2017) Evaluation of some wheat genotypes under water stress conditions in upper Egypt. J Soil Sci Agric Eng 8(4):218–224

Vermicomposting Impacts on Agriculture in Egypt

Archana Singh, El-Sayed E. Omran and G. S. Singh

1 Introduction

Municipal solid waste (MSW) generation in Egypt has augmented drastically around >36% since year 2000, that touch 20.5 million tons per year [1, 2] (Table 1). The MSW generation is 35 million tons per year and around 60% of the total MSW is organic fraction [3]. The MSW generation predicted about 300 g–800 g/day/capita, with a growth 3.4% annually. Additionally, around 6.2 million tons per year industrial waste generated with 0.2 million tons of hazardous waste and 23 million tons per year agricultural waste produced [4]. In Egypt, post-harvest waste of six largest crops produces more than 10.5×10^6 tons of agricultural solid waste annually [5]. Merely, 40% of this waste presently used and 60% of it discarded [4]. Around, 18.7 million animals in Egypt produced 14.7 million tons waste per year [5]. Approximately, 60% of the entire municipal solid waste generated through domestic activities and the remaining 40% municipal solid waste generated through commercial purposes, streets and gardens, service institutions and other sectors. In cities, villages and towns, waste generation rates per individual differ from lower than 0.3 kg in low socio-economic groups and rural areas, and around more than 1 kg in upper class living in urban centres. In a country, on an average around 50–60% food waste, 10–20% paper waste and 1–7% metals, cloth, glass and plastic wastes and at last the remaining waste is basically inorganic matter and others. Municipal solid waste finally included the waste in which about 8% composted waste, 2% recycled, 2% landfilled and 88% is dumped in open dumps. After the swine flu, government decided to recycle the organic waste using earthworms to get rid of swine flu caused by waste.

A. Singh · G. S. Singh (✉)
Institute of Environment and Sustainable Development, BHU, Varanasi 221005, India
e-mail: gopalsingh.bhu@gmail.com

E.-S. E. Omran
Soil and Water Department, Faculty of Agriculture, Suez Canal University, 41522 Ismailia, Egypt
e-mail: ee.omran@gmail.com

E.-S. E. Omran and A. M. Negm (eds.), *Technological and Modern Irrigation Environment in Egypt*, Springer Water,
https://doi.org/10.1007/978-3-030-30375-4_9

Table 1 Municipal solid waste generation in Egypt [2]

Region	Governorate	MSW (ton/day)	MSW (million ton/year)
Greater Cairo	Cairo, Giza, Helwan, Qalyubia	25,000	9.51
Alexandria and Matruh	Alexandria, Marsa Matruh	6300	1.45
Canal, Sinai and Red Sea	Ismailia, North Sinai, Port Said, Red Sea, South Sinai, Suez	2650	0.933
Delta	Beheira, Dakahlia, Damietta, Gharbia, Kafr El-Sheikh, Monufia and Sharqia	17,700	6.43
Upper Egypt	Al-Minya, Aswan, Asyut, Beni Suef, Fayoum, Luxor, New Valley, Qena and Sohag	5950	2.13

Earthworms might handle the waste with more profits and clean the environment. There is a nice prospect for vermicomposting adaptation by municipal waste systems in country operation [6].

Some of the waste could be used in the generation of organic fertilizers, animal fodder, production of food, energy production and other resolutions. Vermiculture is also an important process to transform organic waste into vermicompost. By following appropriate training and cognizance, vermiculture could be opted in all villages. This operation can be adopted by rich as well as rural individuals to produce enriched and good quality vermicompost that will enrich the soil, like sandy and reclaimed soil, and produce organic matter and biofertilizers in the form of proteins, enzymes, hormones, humus matter, vitamins, sugars and synergistic components that make it useful as soil. Earthworms have been originated 20 million years ago on earth. Their role is only to turn over the soil; the organic matter is recycled through dead bodies of organisms. In the ancient culture of Greece and Egypt, an earthworm's role in soil was valued. The earthworm's importance was firstly recognized by Egyptians. In 69-30 BC, Cleopatra gave the statement about the earthworms "Earthworms are sacred". And, she knew the importance of worms to fertilize the Nile valley crops fields after yearly floods. Even there was a rule of death if someone removed earthworms from Egypt. In fact, farmers were not permitted to touch the earthworm as they are believed as the god of fertility. And ancient Greeks thought earthworm's importance in soil fertility. Aristotle, a Greek philosopher, stated earthworm as "the intestines of the earth". In addition, historical information given by Jerry Minnich that the last Ice Age around 10,000 years ago, populations of earthworm have been destroyed in many glaciers and other extreme climatic conditions [7]. Many existing species were neither productive and nor fertile. In suitable environments, such as Nile valley, earthworms played a significant role in sustainable agriculture. However, the Nile's fertility is well recognized and also having the enriched alluvial deposits transported by yearly floods, these deposited materials were mingled and stabilized by valley-dwelling earthworms. USDA accessed that earthworms constitute around 120 tonnes of their castings in one acre annually in the Nile food plain in 1949 [8].

In Egypt, vermiculture is recognized from Cleopatra period. The dependence on Green Revolution along with the huge scale machinery and damming of the Nile River eradicated the ecosystem in which the worms (mainly *Eisenia fetida*) flourished. In Egypt, some of the most fertile lands were found. The Nile River provides the sediments to settle in nearby areas to make the soil nutrient-rich and also provided the water critical for agricultural purposes. In current years, Aswan High Dam has almost eradicated the yearly flood which leads to loss of enriched deposits resulting in a requirement of organic sources for agricultural land for an increase in productivity in Egypt. Charles Darwin (1809–1882) observed the earthworms for >40 years and published a book on earthworm "The Formation of Vegetable Mould through the Action of Worms". He stated that in books there is doubt on other animals which play important role as earthworm in world history.

For 3000 years, the Egypt civilization was flourished because of two reasons: (i) Nile river, which carried a copious amount of water to the drylands of that region; (ii) earthworm population in billions transformed the annual deposited organic matter brought through annual floods into nutrient-rich soil for higher food production in any region. These earthworms were thought to be night crawlers that would travel and spread into whole Europe and finally came in Western Hemisphere with before settlers [9].

From another prospect, waste handling properly, like organic sources in megacities such as Cairo will decrease the environmental effect on public as well as on government. Any sort of effort to clean the streets is commendable, and organic compost availability from various sources will show great result in agriculture in Egypt. As most of the soils due to modern agricultural practices leads to poor soil quality having poor organic material, the profits of waste conversion into compost improved soil quality as in Middle East and North Africa countries.

2 Vermicomposting

It is an eco-biotechnological process in which "vermi" word has been taken from Latin which means worm and thus the vermicomposting means composting process using earthworms to enhance the waste conversion process and production of a better product. It is a mesophilic process using the micro-organisms and earthworms which are effectively work on 10–32 °C. The vermicomposting process comparatively transforms the waste faster than in composting, because the organic matter passes through earthworm gut, a bioreactor that forms casting as a result. These castings are rich in microbial actions and plant growth regulators and stimulated pest repellence [10].

Earthworms are capable of converting garbage into worthy product through its biological actions [11]. The final goal is to achieve vermicompost in short period, and it is achieved through higher earthworm density throughout the method. And, if the aim is to produce high earthworm biomass, then less number of earthworms are needed to maintain the reproductive rates. It is identified that maximum of the extracellular enzymes can be attached to humic matter which is formed during the composting or vermicomposting method, irrespective of organic substrate used, but the information about these enzymes are very less [12].

2.1 Vermicomposting Process: Requirements

The first requirement is the earthworm application composting process,

(i) One thousand of adult earthworms weighing one kilogram.
(ii) One kilogram of earthworms used to convert 5 kg of organic wastes in a day. Around, 10 kg of adult earthworms can have converted one-ton waste in a month.
(iii) Approximately, 2000 earthworms can be applied in 1 m^2.
(iv) In ideal situation, 1000 earthworms and their progenies can convert one ton of organic waste into more biofertilizer yield in a year. This potential method is used to reduce the GHG emissions.

2.2 Vermicomposting Species

Some of the features should be believed to be used in the vermicomposting process. Those features are an ability of organic waste colonization, organic matter degradation, consumption, digestion and assimilation; have an ability to bear environmental factors in a wide range, higher reproduction rates, high productivity of cocoons that do not have long hatching period; and their time interval of *maturation* after hatching into an adult should be faster. It should be strong, tolerant and survive handling. Some of the species of earthworm have these characteristics and they are litter species which include *Eisenia fetida, E. andrei, Perionyx excavatus, Eudrilus eugeniae, Amynthas cortices and A. gracilis* and Pheretimas known as *P. hawayana, Eisenea hortensis and E. veneta*; *Lampito mauritii* "Mauritius Worm".

2.3 *Vermicomposting: Important Factors*

Basically, earthworms need five principle components: suitable environment called substrate bedding, a food source, appropriate moisture content (>50% water content by weight), proper aeration and temperature maintenance [10].

2.3.1 Bedding

It is any material or substrate which provides a stable habitat. And this system has following features:

(i) **Absorbency more**: moist system is needed for earthworms because of their skins for survival. Hence, the bedding material has to be highly absorbing material having high water retention to earthworms flourishment.
(ii) **Good bulking potential**: Earthworms need O_2 to survive, and hence, the bedding must be having good porosity and material should not be too dense or tightly packed and then porosity is reduced. Porosity depends on various factors like particle shape and size, texture and rigidity as well as strength of the structure.
(iii) **Low protein and/or nitrogen content** (high carbon: nitrogen ratio)

Earthworms depend on bedding material for their food by decomposition. High nitrogen levels lead to faster decomposition and its linked heating, often fatal to micro-organisms involved in the process [13].

2.3.2 Moisture

The material should provide appropriate moisture for earthworm survival. They breathe through skin pores which must be moistened to enable respiration. Earthworms can also live in submerged water for months as aquatic ancestors and they will die if they dry out [14]. The moisture content should be maintained at 45–60% range in a conventional composting system. On the other hand, vermicomposting requires 70–90% moisture content range.

In this range, various scientists observed somewhat dissimilar ideals: there is a direct relationship in the moisture content and earthworms' rate of growth [15]. For example, *E. andrei* grown in pig manure matured in 65–90% moisture content, and optimum is 85% and above moisture content growth of earthworm increased and resulted in increased biomass. Higher moisture content range like 90% speeds up the sexual maturity, and in lower 65–75% range, clitellum is developed after 44 days not in all worms. In Canada, scientists observed that 75–80% moisture content range is optimum for best growth and reproductive growth results in different substrate feedstocks [13]. With *E. fetida,* the moisture content ranged from 50 to 80% for adults and juvenile grows in lower range 65–70%. And clitellum development occurred at

60–70% range while clitellum occurrence takes place later in 55–60% moisture content range. *E. fetida* are more tolerant species to moisture content they grown on 50% moisture content up to one month [16]. On the other hand, earthworms are grown in different cattle manure at faster rates having 90% moisture content with maximum biomass and maximum mean weight of six hundred earthworms after twelve weeks [17].

2.3.3 Aeration

Earthworm is an aerobic animal and requires oxygen for its survival. If greasier or high moisture content with less aeration in substrate causes lack of oxygen and the system become anaerobic, then it leads to death of worms rapidly. They can also be killed by noxious compounds (e.g. ammonia) which can form by microbes in these conditions (anaerobic).

2.3.4 Temperature

It is an important factor for vermicomposting process. Various ranges and their results on earthworms are as follows:

(i) **Low temperatures**

It is commonly known that temperature for earthworm survival is >10 °C and >20 °C for an efficient vermicomposting process. *E. fetida* can survive at lowest temperature 0 °C, but they cannot reproduce and take much food at this temperature.

(ii) **Effects of freezing**

Eisenia species can also survive at frozen bedding but they will die due to non-consumption of food. However, studies at the Nova Scotia Agricultural College proved that cocoons of the *Eisenia* remained viable in deep freezing state [13].

(iii) **High temperatures**

Earthworms have lived at 30 °C range in the mid 30s, and above 35 °C, worms will leave their habitats, if they will not leave that place they will die quickly. At warmer temperatures (20 °C), they prefer reproduction. Environmental factors also have impact on earthworm survival and growth in municipal solid waste [18]. Earthworms achieved maximum growth rate at 19.7 °C which is 0.0459 g/g day, at 20, 15 and 25 °C the fastest growth rate achieved at 34, 53 and 27th day which was 0.0123, 0.0068 and 0.0138 g/g day.

Soil organism activities follow a specific seasonal fluctuation. Optimal temperature and moisture the cycle is related to a peak activity happens in the spring in which the temperature and moisture become appropriate after cold winter temperatures. At the lower temperatures, fungal activity may continue at high levels in whole winter in litter. Decomposition in winter undergoes at highest rates under snow in the litter. In summer, due to moisture limitation, activities of earthworms lower down as compared in the winters. In the rainy season, like in Mediterranean climates, a peak two of activity has been seen in the fall. If substrate lacks organic matter, then these peaks have not been seen. The *E. fetida* have been observed by different scientists at different moisture content and temperatures in the laboratory. These species have maximum weight in horse manure and activated sludge with 70% and 85% moisture content having temperature between 20 and 29 °C [19]. They have shown best growth in diverse animal and vegetable wastes at 25–30 °C with 75–90% moisture content, and these conditions may vary according to the substrate [20].

2.4 Earthworm Categories

Earthworm is a polyphagous annelid which plays an important role in soil system. While all species of earthworm decomposes plant-derived organic material, they varied in the way of degradation of residues. Earthworms can be categorized into three types according to their habitats and ecological functions;

(1) Anecic means “out of earth” in Greek which is defined as the species those burrow the soil in deep mineral layers of the soil and carry food from the surfaces. These species have important role in pedogenesis, e.g. nightcrawler [10].
(2) Endogeic means “within the earth” in Greek which is defined as burrowing the soil but not in deep only in shallow burrows. These species are limited to plant litter layer on soil surface, comprised of organic material or wood, and rarely penetrate soil in deep. Main role of these species is shredding of organic material into fine granules to facilitate enhanced microbial action.
(3) Epigeic means upon the earth in Greek which is defined as species living on organic substrate not in the soil, usually used in vermiculture or vermicomposting. They play an important role in the promotion of microbial actions in organic matter decomposition which finally involves humification into complex and stable vague colloids having phenolic compounds. e.g. *Eisenia fetida* (tiger worm).

2.5 Egypt: Native Earthworms

The earthworm diversity is affected by soil properties, climate and availability of organic sources in locality such as land use history. The earthworm communities found lower in biomass with lower pH, less fertility, less fertile litter or highly degraded soil. The most important factors of soils influencing the earthworm biomass are pH, C/N ratio, moisture content and Al, Ca, Mg, organic matter, silt and coarse sand [21]. Europe is an original habitat for maximum of the collective species of earthworm like, red worm or red wiggler (*Lumbricus rubellus*): tiger worm or manure worm known as *Eisenia fetida*: and field worm known as *Allolobophora calignosa.*

The major species of Egypt are nightcrawler and field worm whose living habitats are manure or compost heaps, both prefer woodland and grassland margins. The main types of species are *Alma nilotico and A. stuhlmannt.*

The Nile basin is sectioned into three Obligataete subregions: lower Nile, belongs to delta to Kartoum having *Alma nilotico* and *A. stuhlmannt*, upper Nile belongs from Kartoum to central and east Africa with *A. emini*, and Ethiopian sub-region having *Eudrilus eugeniae.* And, it involves *Allolboplora (Aporrectodea) caliginosa*, linked to the *aquatic Eiseniella tetraedra* in spring nearby the St. Catherine monastery in south Sinai, *Allolboplora (Aporrectodea) rosea (Eisenia rosea)* on the slopes of the Mountain of Moses, and near Monastery. *Allolobophoru jassyensis* is located in the Delta and *Eiseniella tetraedra* in Sinai [22]. The earthworm shortage in biomass results in arid soils of the Egypt and most of the cultivated land is arid. In Egypt, in arid condition, almost rainless earthworms are not able to move in favourable place to survive in these stress conditions. Because of poor soil conditions, earthworms are less in density (e.g. orchard and forest). Moreover, in arable lands, favourable conditions are transient. These favourable conditions are:

1. An uninterrupted soil.
2. A steady and satisfactory water supply.
3. A well soil quality (to increase the obtainability of water).
4. A steady and passable supply of organic matter.

Various species can grow in sand dunes soils in Egypt like *Aporrectodea caliginosa* that can live in sand dunes soils but their biomass reduced with augmented fractions of sand and coarser gravels. In 1965 through sampling five species found from 14 localities in Beheira governorate and nearby areas by El-Duweini and Ghabbour. Those five species are: *Gordiodrilus* sp., *Pheretima califonica*; *Pheretima Elongate*; *Allolbophora caliginoosa, f. trapezoids* and *Eisenia rosea f. Biomastoides.*

3 Compost

It is a microbial degradation of organic matter refuse in controlled environment. The final product is used as soil amendment called compost. It stimulates microbial activities in soils required for plants growth, disease resistance, water holding capacity and filtration and also prevents erosion of soil. It can be used as amendment to enhance physicochemical and biological properties of soil. But, generally, the macro-nutrient value of compost is not high as other fertilizers. It enriches soil by increasing organic matter and also increases water retention ability. By using compost nutrients cannot leach easily and released to plants at slower rate to increase crop production. It can reduce fungal diseases in soil which is important for golf and nursery industries.

3.1 Compost Versus Vermicompost

Composting generally clearly explained as aerobic transformation process in which an organic substrate can be transformed in different organic product which can be used as soil amendment without having negative impacts on plants growth, which is the adequate method for pre-treatment and management of organic wastes. The most traditional method of composting comprised of bio-oxidation of organic material as it circulated by thermophilic stage (45–65 °C) in which earthworm species releases heat, CO_2 and water, whereas vermicompost is an aerobic process having nutrients in available forms for plants, viz. nitrates, exchangeable P, K, Ca and Mg. It has greater potential as media for plants growth in horticultural and agricultural industries. It can be used as soil additives or as horticultural media which enhances seed germination and improves seedling growth rates and development.

However, composting and vermicomposting are having differences particularly temperature ranges, microbial groups' presence which involved in the process such as thermophilic bacteria in composting and mesophilic bacteria and fungi in vermicomposting. The two processes have different paths of waste conversion. Vermicompost is finer granules than the compost. The plants will uptake finer nutrients more readily than coarser granules. Vermicompost also has plant growth regulators [23].

4 Gap Analysis of Current Scenario of on-Farm Along with Organic Waste Management Practices in Egypt

Here, are two chief bases of organic matter in Egypt for composting farm waste and urban wastes. To attain these materials, one should know the waste management practices of this area.

4.1 On-Farm Organic Waste

Agricultural wastes are derived from agriculture which are discarded or intended to be discarded, and a discarded biomass creates a problem for society and industries. It has been predicted that off farm disposed of plants and animal wastes are about 27 and 12 million tonnes in a year, respectively. Crop residues burning causes problem in a country like rice straw wastes. In Egypt, around 360 ha rice is cultivated in 2008, with 6 million tons of straw production. It depends on farmer how they will dispose the waste. The most common method of disposition is dumping the waste at municipal waste sites, dumping the waste in a desert or by burning it. The grower thinks that waste is free and difficult to handle but after that farmer realized the value of waste after collection and arrangement for transport. They usually burn the biomass for their agricultural benefits from ash as minerals or get rid of insects or diseases by burning the surface of the soil. This practice is well known and they did not realize the demerits of this practice. Farmers are not easily convinced for organic recycling because of various constraints like the absence of appropriate technology, long duration, more labour, more space need, costly and investment requirements. In rural areas, various constraints are there in solid waste management. Those constraints are related to the financial matter, institutional and environmental conditions, planning and legal limitations and deficiency of technology. Two methods recommended for rice straw treatment to collect the dispersed waste for compost method usage and other methods are to use the rice straws as virgin material in fibre processing for crop exportation.

4.1.1 Weak Points in Rice Straw System in Egypt

There is an economic problem which causes extreme scarcity of raking, combining and baling machines and fewer trucks for transportation. Additionally, the management problems make the transportation more difficult between farms and markets. In contrast, agricultural cooperation must provide the GIS facility to know road status that will be used in the transportation.

4.2 Urban Wastes

The governmental system involving Cairo and Giza having purgative and prettification authorities with four main systems work in solid waste management earlier privatization trend. These agencies involved in municipal solid waste management regulating the private service delivery. Instead, the formation of such entities was not valuable and faced problems. The other system is informal waste collectors known as Zabbaleen system which gives door to door service in return for fees in a month. There is private sector at third group which has been introduced in big cities and local towns. The third operator has a licence or local municipality that gives a chance to operate in an assigned area. In addition, the last organiser is NGO which balance solid waste.

4.3 Municipal Solid Waste

The solid waste concentration of collected solid waste (i.e. waste not collected and dumped in disposal sites but rather dumped on roads and empty lands) was calculated about 9.7 million tonnes for the year 2000, with a total volume of 36,098,936 m^3 [6]. This solid waste can be divided into municipal waste, industrial waste, agriculture waste, waste from cleaning waterways and healthcare waste. Domestic waste contributed 60% of the total municipal waste quantity, with leftover around 40% being produced by commercial firms, service institutions, streets, hotels, gardens and other showbiz sector. In Egypt, 16 landfills exist; 7 in the Greater Cairo Region, 5 in the Delta governorates and 4 in Upper Egypt. Their capacities range between 0.5 and 12 Mt in a day. They are usually operated by private entities. New landfills currently, 53 sites have been found and 56 composting plants constructed in the country is under progress.

4.3.1 Overview of Solid Waste Management Problem in Egypt

In Egypt, the solid waste management problem is increasing at an alarming rate. And, its negative consequences on environment, health of individuals and on Egypt, mainly manpower productivity and tourism are becoming quite clear and severe. In the cities, such as Cairo and Alexandria, the problem touched such fractions that they involved considerable government and the actions of prudent in all terms (short, long and medium).

In spirit, the problem is explained in the National Waste Management Strategy 2000 as follows:

"The present systems could not satisfy the served community needs with its various strata for a reasonably accepted cleansing level, as well as in reducing the negative health and environmental impacts, or in improving the aesthetic appearance".

The apparent symptoms for this difficulty are:

(i) Waste accumulation at different locations will cause various activities such as rodents and insects danger, environmental pollution, rotten and bad smell, frequent open burning which will have adverse impact on human health and on environment.
(ii) Non-effective treatment, handling and recycling cause health risks.
(iii) Predominant open dumping of solid waste and unselective dumping result in various environmental and health risks.

4.3.2 Main Factors Contributing to Soil Waste Management Problem

The key factors involved in solid waste generation in Egypt can be concise as follows:

(i) Sustainable actions were not taken in past, and the issues were not proper.
(ii) Actions taken in the past were not always sustainable, and the issues were not dealt with an inclusive and combined way.
(iii) Reliability and accuracy of quantities of solid wastes, production rates, information about composition were not there. Various efforts have been made to measure the problem, but these efforts were hard and inclusive to achieve.
(iv) With weak procedures of implementation, laws cannot be followed.
(v) The SWM participation in Egypt was minimum in previous decades till the private sectors were more engaged.

(vi) Non-effective recycling methods, for example, the mixing of all types of wastes altogether without any planning to support sorting at source. However, non-toxic and toxic wastes were mixed altogether.
(vii) Lack of public awareness and improper actions regarding the handling and disposal of solid waste.

5 Fertilizer Status in Egypt

Fertilizers application for crops cultivation is a routine in modern agriculture practices and main need of this era is high yield and production in intensive agro-ecosystem. Fertilizers are needed in priority in modern practices but they are main polluters of air, soil and water and also greenhouse gases GHG emissions with changing climate.

In the ancient period, fertilizers have been used. In contrast, soluble salts in huge amounts in the soil cause delayed seed germination, reduce plant growth and also reduce groundwater and soil use. The environment is adverse through the application process of fertilizers. Presently, Egypt is at ranked country for fertilizers consumption in agro-ecosystem. The total amount of fertilizers production is about 2 million of which 32% is exported. Such chemical fertilizers cause harmful effects on human health and environment of Egypt, which need to be resolved through some alternative methods such as organic agriculture which is one of the strongest and appropriate for land reclamation of Egypt. However, farm waste recycling and composting is an alternative method for restoring soil fertility which has low organic content. Fruits or grains harvest which is small part of plants and return back the remaining residues of plant after composting to the soil leads to add minerals. Hence, displacement of chemical fertilizers will clean the environment and also produce less GHG emissions, and therefore, the farmers can displace through the clean development mechanism (CDM) of Kyoto protocol.

The food demand and other services of agriculture are rising because of population escalation and development of living standards in Egypt. Continuous efforts increase crop production and quality. Fertilization in a proper way is one of the most important agricultural practices which results in improved agriculture [24].

The fertilizers production rate increased from the last epochs. The production of fertilizers is about 2 million tonnes, and from this total amount, thirty-two per cent is exported. The rest amount of fertilizers after the production is about 43%. Hence, Egypt imported less amount and 43% of total consumption is compensated. Fertilizers export and consumption increased. This increment is because of the aggravation for intensification of agriculture in horizontal and vertical expansion. Through the statistics of FAO, it can be estimated that most type of fertilizer used are phosphorus and nitrogen in accordance with Egypt. FAO has shown that in 2010 there is increase in nitrogen fertilizers from 2008 (1,721,105 tonnes Nitrogen) and phosphorus (229,911 tonnes). This has touched 60 and 61%, respectively, compared to

2002 year for nitrogen and phosphorus. Hence, the fertilizer demand is probably increase in next future years.

5.1 Vermicompost as Fertilizers in Egypt

The good quality of vermicompost generation is important for farmers to cultivate the more valuable crops. The production factors effects like characteristic variability of organic substrate, vermicomposting duration and different parameters treated as maturity indicators, for quality assessment of vermicompost important aspects in developing guidelines to be considered. The marketing of vermicompost increases by consideration of vermicompost indicators as utilization, generation and commercialization. The feedstock substrate was examined to check the compatibility with various earthworm species like paper mill industries sludge, water hyacinth, paper waste, cattle manure, crop waste, sewage sludge and so forth.

Vermicomposting concept is not new in which earthworms are used to stabilize the organic waste in both developed and developing countries. The investment in establishment of vermicomposting system has to be proven a hurdle for the large-scale projects because of expensive earthworms usage. Three aspects give the economic sustainability to the system. First is a provision of waste stabilization sustainably which can produce income at low cost. In second aspect, the formation of soil conditioner which can be sold as vermicast. The third and last aspect is to produce protein in the form of protein meal, valuable source of vitamins, amino acids, minerals and long-chain fatty acids for chicken and fish. Agricultural waste recycling and composting is a substitute method for mineral fertilizers use. The compost use enhanced in traditional agriculture will reduce the use of fertilizers and leads to good quality generation with less polluted effects. An organic agriculture is an option with a significant prospect for land reclamation in Egypt at broad level. More production of organic fertilizers will induce the chances of farmers to be connected with organic farming. The chapter concentrates on needs of fertilizers in Egypt and the potential of vermicompost as fertilizer for organic farming in Egypt.

5.1.1 Potentiality of Vermicompost as a Source of Fertilizer in Egypt

Urban waste for the year 2005 has range of 15–16 million tonnes, biodegradable matter in the wastes was 50–60% and regular collection ability was 70%. In Egypt, it has been predicted that urban wastes contributed about 1.99 million tonnes compost generated per year having around 21,000 tonnes N, 5000 tonnes P and 10,640 tonnes K [6]. Solid waste management inaptly and poor-quality compost generation are main restraint in manipulating as huge quantity of plant nutrients for crop production augmentation. In contrast, in Egypt, agricultural waste could generate about four times compost material in comparison to urban wastes, suppose that hundred per cent of it is organic matter and all of it is reachable to the grower. There are different merits of this waste, which are space availability and direct association to the field. This reduces the collection requirement and transportation need. The N, P and K concentration which can be generated from agricultural wastes are almost four times of urban wastes. The whole composted material is almost 10 million tonnes with both organic matter, involving around 10,000 tonnes of N, 20,000 tonnes of P and 41,000 tonnes of K. N fertilizer attained from organic wastes could conserve up to 5.9% of that consumed in 2008; while >10% of P fertilizers used in 2000 should be conserved.

5.2 Vermicomposting of Agricultural Wastes

Crop residues and cattle shed wastes used in vermicomposting not only produce fertilizer but also act as the best culture medium for growth of large-scale earthworm production. The composting capability and growth rate of *E eugenaie* were assessed by applying various mixtures of crop residues and cattle dung, under laboratory conditions. In terms of nutrient increment have been seen in the end product as best results in vermicompost bed in comparison to composting without earthworms. However, vermicompost application has shown higher concentration of total N, available P, exchangeable K and Ca content. And the end product has revealed the relatively lower C: N ratio and stabilised product comparatively. A noticeable quantity of biomass of earthworm and cocoons were generated in various treatments. Moreover, feedstock quality, applied in that study was having importance to determine the growth parameters of earthworm, e.g. individual biomass, cocoon numbers and growth rate. Various results have shown that crop residues can be used as an efficient culture media for *E. eugenaie* production at large-scale level for land restoration practices sustainably at low inputs [25].

5.3 Urban Waste Vermicomposting

Many countries have tradition of home composting and is suggested as one of the best option for waste management by the European Union Policy. Merits of this option are waste should not be transported and home gardens will benefit with nutrients and humus. Additionally, it also has prospect of education for enhancing environment awareness. Adoption for this method has limitations of space to conduct this process, knowledge deficiency regarding correct process of composting. This involves substrate assortment which are efficient for home composting and efficient process situation. There is a probability of vermicompost production from each home in the city as Cairo. For the process, the earthworms in an appropriate amount kept in a double basket system with one perforation inside, organic resources should be used without any kind of odours or smell. However, the system is not widely used, but due to proper awareness and public support should be applied. This can initiate the monetary value for the poor families and generate appropriate amount of vermicompost which can be used for agricultural activities directly. Furthermore, it has following benefits:

(i) Money saving and environment protection;
(ii) Reduction of household garbage disposal expenses;
(iii) It is odourless and tempts some pests when keep food wastes into waste container;
(iv) It increases water and electricity savings that is consumed by disposal units of kitchen sink;
(v) It generates a high-quality soil amendment called vermicompost;
(vi) It needs less space, labour and maintenance;
(vii) It reproduces free earthworms for fishing as fish bait.

Various options for Zabbaleen combination into the international companies interviews of staff members with CID members were done to explore, increases the local-world conflict issue and the possible involvement of public-private partnership. The Zabbaleen execute the segregation system hence it is known as sub-contractors, which separates organic waste from non-organic waste. They are involved in household waste collection, whereas medical and industrial waste and landfill management could be targeted by multinational companies. Stations were made for transportation where a main fraction of non-organic waste has been found and given to existing businessman. Organic waste can be received by Zabbaleen from other companies as input for their recycling businesses and small community-based composting facilities are initiated. In this way, the conventional informal Zabbaleen system has been changed into new privatized large-scale waste collection system for mutual benefits to both sides. In spite of these recommendations, current improvements have explained the implausibility of local world partnerships. Despite international companies favours the Zabbaleen system training as waged employees, they allow them to find landfill sites for organic waste for pig-raising actions [26].

5.4 *Vermicomposts Effect on Plant Growth*

Earthworms have an important role on chemical, physical and biological properties of soils it is a well-known fact and also enhances plants growth and crop productivity in natural as well as in managed ecosystems. These helpful effects have contributed to soil structure and properties of soil and made mineral nutrients avail to plants and biologically active secondary metabolites act as plants growth regulators.

Compost formed by *Eisenia fetida* through strongly have impact on soil fertility through enhancing nutrients availability, improves structure of soil and water retention. Earthworms have highly recommended for organic matter decomposition and generate various bioactive humic substances. Those substances are gifted with hormone activity which enhances growth and plant nutrition. Humic acids (HAs) embrace with main proportion of humic components.

In a study, it was precisely mentioned that vermicompost biologically influences the growth of plants certainly and generates gradual increase in overall plant productivity, which is not dependent on uptake of nutrients. Blending of elevated amount of vermicompost-derived humic acids with container media to increase growth of plant, and higher amounts generally declined growth, in a same manner of plants responses to growth detected by applying vermicompost into container media with all the required mineralized nutrition. Plant treatments having 50–500 mg/kg humic acids enhanced growth of plants, but declined gradually if the amount of humic acids increased 500–1000 mg/kg concentration. However, growth of plants has been seen partially increased because of elevated rates of uptake of nitrogen through plants, in most of such experiments. Moreover, this does not include the other mechanisms contribution by which humic acids may impact on plants growth. There is other clarification for the hormone action of humic acids in the experiments. Other scientists have extracted plant growth regulators like indole acetic acid, gibberellins and cytokinins from vermicompost in the form of aqueous solution and explained that this solution has a gradual effect on plants growth. Those components might be relatively momentary in the soils. Moreover, there is high probability appears like plants growth regulators which are relatively momentary, were adsorbed onto the humates and deed in aggregation with them to impel plants growth [27]. In the horticulture industry, vermicompost has been endorsed as a possible alternative culture media component. Vermicompost addition blend in media of 10 and 20% volume has progressive impacts on plans growth. The utmost growth increase on seedling has been seen when the step of plug of the bedding of the plant crop cycle. Growth surges up to forty per cent which has seen in dry shoots tissue and leaf area of tomato, marigold and green pepper plants. The augmented vigour unveiled was also sustained when the seedlings plugs were transferred into containers with commercial standard potting feed stocks without vermicompost. In addition, there were merits seemingly resulting from nutritional constituent of vermicompost. All of the lumps were generated without the additional fertilization input. During the plugs generation stage vermicompost mediated commercial potting media has a potent role and used without additional fertilizer applied by famers [28].

6 Environmental Impacts of Current on-Farm and Urban Organic Waste Management Practices

This work benefits the agriculture producers usually and for organic farming producers precisely. In organic farmer interest, the affirmative effects on environment have been seen by applying as such process of vermicompost generation, it is essential to understand how vermicompost help to reduce the greenhouse gases generation, and as a result supports to mitigate the global warming.

6.1 Emissions from Vermicompost

It has been determined that composting is an important origin of CH_4 and N_2O. With increased deviation of decomposable waste from landfill into the composting sector, it is essential to measure methane and nitrogen dioxide emissions released through composting and in all stages. The investigation concentrated on the end step of a two-step composting method and evaluated the production and release of methane and nitrogen dioxide linked with two different composting methods: automatically turned windrow and vermicomposting. The automatic turned windrow system was described by methane release with lesser amount of nitrogen oxide.

Moreover, the vermicomposting method releases gradual amount of nitrogen dioxide and methane only in traces amount. Huge nitrogen oxide rates through vermicomposting explained to strongly nitrifying circumstances in the beds in use in combination with the occurrence of de-nitrifying bacteria within the gut of earthworm [29].

Various other reports from different countries said that any possible release of GHG emission through earthworms from soils or through vermicomposting method is very small in comparison of the GHG emissions (N_2O and CO_2 and CH_4) as documented because of chemical fertilizers manufacture, landfills, manure heaps, crop residues in soil, lagoons and pigs manure and cattle manure. Although nitrogen dioxide is generated from these sources, there is no explanation for recommending environment-friendly and energy efficient processes for vermicompost generation and compost should be limited due to their importance to generate greenhouse gases. The GHG (nitrogenous) gases generated at world level in agriculture should be evaluated release from all sources before vermicomposting publicly fated in a sensational way [30].

Current study has found that vermicomposting types certainly can produce substantial concentration of nitrogen dioxide. These preliminary findings have revealed that there is still need of more research in this direction to be conducted to comment on vermicomposting and to give any statement. Subsequently, the emissions concentration through composting be contingent on the particular composting process used and how properly the method is accomplished, it is not possible to comment about the quantity of emissions by vermicomposting contributed in climate change. Maximum investigations on emission through composting have been done in developing countries in which the situations varied from the countries of present study targeted. Nonetheless, various environmental agencies have summarized that when composting process finished properly, it produced less concentration of GHG emissions [31]. GHG emissions through three different domestic waste treatment process in Brisbane, Australia, have been investigated Chan et al. [32], in which gas samples were received monthly from the backyard composting bins which were 34 in numbers in year 2009 from January to April. Be around in this investigation duration the composting (aerobic) bins emits less concentration of methane which was 2.2 mg $m^{-2}h^{-1}$ in comparison to anaerobic composting bins 9.5 mg $m^{-2}h^{-1}$ and the vermicomposting bins have 4.8 mg $m^{-2}h^{-1}$. The vermicomposting bins have emitted lower nitrogen dioxide concentration that was 1.2 mg $m^{-2}h^{-1}$ compared to other gases 1.5–1.6 mg $m^{-2}h^{-1}$. GHG emissions in total includes both nitrogen oxide and methane were 463, 504 mg with 694 mg CO2 e $m^{-2}h^{-1}$ through vermicomposting, aerobic composting and anaerobic composting respectively, with the nitrogen dioxide contributed about more than 80% in the total budget. The GHG emissions varied gradually with duration and were controlled by moisture content, temperature and composition of waste, indicated towards the capacity to mitigate GHG emission by proper management of the composting systems. The result recommends that domestic composting offers an effective and viable supplementary waste management process particularly centralized facility for cities with lower population density as Australian cities. GHG emissions in the course of maturation process, the windrow method of composting, were categorized by methane emission. GHG emission by vermicomposting emits nitrogen oxide predominantly compared to methane, explained that aerobic condition was efficiently maintained in the vermicomposting beds to constrain methane production. The global warming potential (GWP) during the maturation process of vermicomposting was predicted about 30-folds more than that for windrow composting system. The GHG emissions by these types of composting need further evaluation. The decomposition fraction reduces as the anaerobic to aerobic in vermicomposting through worms leads to gradual reduction in methane and volatile sulphur components which were swiftly release through the traditional (microbial) composting method. Hence, vermicomposting method has merit over the traditional composting process using organic waste as this (vermicomposting) process not permitted the GHG (CH_4) formation. Methane is 20–25 folds stronger than carbon dioxide at molecules level. Earthworm has a lead role in the reduction strategy of greenhouse gas and mitigation during the organic waste disposal at world level and landfills emit CH_4 formed by slow anaerobic degradation of organic waste in many years. Moreover, study in Germany has observed that earthworms generate a one-third part of nitrous oxide gas in vermicomposting process. Molecule by molecule N:O ratio is 296-folds stronger GHG than CO_2. There is further need to explore in this field [33].

6.2 Total Emissions from Waste Sector in Egypt

In 2000, total generation of GHG concentration was about 193 megatons of CO_2 equal to 1*. Total emissions in 1990 were around 117 megaton of carbon dioxide equivalent 1. The mean of GHG emissions is around 5% in a year. In this aspect, the total GHG generation was predicted around 288 megaton of CO_2 equivalent. In Egypt, particular GHG gases involved in 2000 have concentration around 2.99 megaton of CO_2 per individual.

In 1990, the total GHG emission of CO_2, N_2O, haloflourocarbons, perflourocarbons, methane and sulphur hexafluoride (except land use change emissions) in the world was calculated around 29,910 megaton of carbon dioxide. In Egypt, in 1990 the total emissions measured were about 117 megatons of CO_2 and rest were methane and nitrogen oxide [34]. In 1990, Egypt contributed about 0.4% of GHG emissions of the total GHG emissions in world having emissions of N_2O, CO_2 and methane through manure, agricultural activities, and burning of field biomass. In addition, the sub-categories of the sources of GHG emissions were methane production from aerobic wastewater treatment plants, NO_2 emissions through domestic wastewater and emissions from incineration of wastes, these all were not included in the 1990 figures. However, the updated figures for these actions were utilized in solid waste production and waste water production in the 2000. Based on these figures, the total world emissions were discussed in detail for 2000 year with concentration 33,017 megaton of CO_2 equivalent, and Egypt shared about 0.58% of total emission in world in year 2000 [2].

6.3 *Emissions from Agricultural Wastes*

Crop residues incorporated around 0.4 Tg nitrogen in a year globally and have role in N_2O emissions, using IPCC default released factor of 1.25% of applied residue N emitted as N_2O. Moreover, the default emission factor is relied relatively on less no. of observation. Current researches have shown that emission factor for crop residues can fluctuate noticeably with residue quality, specifically the C: N ratio and the concentration of mineralized N. Usually, emissions in higher amount keep on involving of residue with lessen C/N ratios. It can be summarized that earthworm burrowing activities increased the N_2O generation through crop residues by 18 times; this earthworm impact is mainly independent of bulk density; but earthworm species has particular effects on N_2O emissions and stabilized residues in soil organic matter. Moreover, earthworm-mediated N_2O emissions mostly occurred due to residue incorporation into the soil and vanished when ploughing of residue was simulate in the soil. Irrespective of earthworm actions, farmers might reduce direct N_2O emissions from crop residues with a relatively low C:N ratio by kept it on above for few weeks before ploughing it in the soil [6]. Moreover, studies at ground have affirmed this impact, and possible trade-offs to other (indirect) emissions of N_2O should be noticed before this can be suggested [35].

In the last three years, an inclusive research programme has been developed in Ohio State University on vermicomposting. This has involved studies which evaluated the vermicompost impacts on seed germination, growth, fruiting and flowering of vegetable plants. viz. bell peppers and tomatoes, and also on broad range it included petunias, marigolds, bachelor's button, chrysanthemums, impatiens, sunflowers, and poinsettias. A continuous vogue in all.

The trial experiments that have been observed are the best plant growth responses, with all required provided nutrients, formed when vermicompost created a small fraction (10–20%) of the total volume of the container medium mixture, with higher fraction of vermicompost in the plant growth medium not always increased plants growth. Some of the responses of plant growth in horticultural medium, replaced with a sort of dilutions of vermicompost, were same to those which were as in composts were utilized [23].

6.4 Analysis of the Egyptian Context and Applicability of Vermiculture as a Means of Greenhouse Gas Emission Reduction

Egypt-relevant ministries in collaboration with related governorates in the waste sector have made many plans and programmes in the last ten years to enhance the method of collection, reuse and waste recycling, hitherto there are various obstacles in the achievement of these goals. These aims involved financial restraints for GHG mitigation by waste sector; the gradual reliance on external financial support, as grants and loans at concession, and this slows down the implementation and complicates the process plan; lacking of public awareness of the economic profits by reuse and waste recycling leads to the funding institutions doubt for consideration of waste management activity as a feasible option, the requirement of technology transfer and high investments for some waste treatment options, like anaerobic digestion; weak enforcement of existing laws ;and violation in regulation of waste handling.

7 Mitigating Greenhouse Gas from the Solid Wastes

In a non-Annex I country, Egypt is not needed to fulfil any particular emission decrease or restraint in terms of commitments beneath UNFCCC, or the Kyoto protocol. Moreover, extenuation events are already underway. Egypt is totally conscious about the GHG generation reduction, mainly through major generators, is the only calculate that could confirm the global warming vindication of global warming and climate change. Six foremost criteria have been elected for vindication measures prioritization in the waste sector in accordance to Second National Communication of Egypt. These includes investment costs; payback periods; greenhouse gases emission reductions potentials; duration of implementation; priority in national strategies/programmes; and role in sustainable development. Vindication options summarized from a multi-criteria analysis were merged for each sub-sector in order to produce a number of situations for solid waste and wastewater. The lowermost GHG emission scenario was chosen for implementation in the period of 2009–2025 [6]. Various obstacles related to achieve the targets of these programmes are:

- Though the financial support for vindication of greenhouse gas emissions from the Egypt waste sector has increased gradually in the past years, it also shows a clear restraint in the application of the intended programmes.
- There is dependency on external financial support, as grants and concessionary loans, complicated the process plan and decrease implementation rate.

- The lack of public awareness about the financial benefits of vindication options in the waste field results in unwillingness of funding institutions to study waste management activity as an economically feasible option.
- Transfer of technology shows another obstacle chiefly in anaerobic digestion technologies as it requires high capital investment and skills to function appropriately. Few technologies are designed on site-specific bases, which are not ideal for other areas. Further, highly intensive studies and local skilled experts are required for showing the aptness and applicability of the technology in accordance to different local conditions in Egypt.
- All agencies of the waste sector are moderate to limited environmental management knowledge and the coordination mechanism with EEAA is not properly founded. Also, waste sector privatization lacks clear vision for partnership, specifically in regard with public–private partnership.
- Weak implementation of present laws and regulations to achieve the planned programmes.

8 Benefits of Vermicompost to Save Water

To the many benefits of composting, add another for water saving and water conservation. Using vermicompost can save water where soil moisture is retained, thus limiting the amount of irrigation required and minimizing run-off. Vermicompost is suitable for reducing water shortage's harmful effects (25). The vermicompost application significantly risen the microspores number and reduced the macropores number ($P \leq 0.05$) and also resulted to the improved in the water holding capacity (26). Compost, rice straw and sawdust to clay loam-textured soil augmented the hydraulic conductivity, pore size, and water holding capacity of the soil (27). The quantity of water retained by soils for plant use is considered as water holding capacity (26). Vermicompost can also lighten heavy clay soils and enhance the ability of sandy soils to hold water, aiding you to manage your water resources further. When vermicompost is applied as a thin layer to bare soils, it is an effective barrier to soil moisture evaporation, a practice known as top or side dressing. Vermicompost also reduces plants' needs for water by increasing how much water can be held by the soil (soil's water holding capacity). Improving the soil's water holding capacity by putting vermicompost enabled all crops with reduced periods of water stress during summer droughts. So, save money on water by improving water penetration and storage and reducing run-off and evaporation.

9 Conclusions

Egypt is a country in which the solid waste is upsurging in an abrupt manner. The agricultural and municipal solid wastes are generated through urban and rural areas, respectively. To combat the problem of waste generation and GHG emissions through waste burning vermicomposting is one of the best methods. The organic waste conversion into value-added products plays a significant role in balancing the circular economy by making money from organic waste. Vermicomposting covers many aspects like solid waste reduction, as biofertilizers have positive effects on plants growth. It also reduces GHG emissions, and so from every point of view, vermicomposting is a better option to manage Egypt waste generation. And, with these above points, vermicomposting promotions needed further. And, more aspects need to be considered for vermicomposting adoption, like substrate composition, earthworm species and pre-composting.

References

1. SWEEPNET (2010) Country report on the solid waste management in Egypt. The regional solid waste exchange of information and expertise network in Mashreq and Maghreb countries. http://www.sweep-net.org/ckfinder/userfiles/files/countryprofiles/rapport-Egypte-en.pdf
2. EEAA (2010) Solid waste management in Egypt, Egyptian Environmental Affairs Agency. http://www.eeaa.gov.eg
3. Tawfik A, El-Qelish M (2012) Continuous hydrogen production from co-digestion of municipal food waste and kitchen wastewater in mesophilic anaerobic baffled reactor. Bioresour Technol 114:270–274
4. EEAA (2011) Egyptian Environmental Affairs Agency (EEAA), Annual report. http://www.eeaa.gov.eg
5. Abou El-Azayem MGM, Abd El-Ghani SS (2010) Economic return of recycling the agricultural wastes in Egypt and Spain. J Am Sci 6:960–970
6. Mahmoud M, Yalhia E (2011) Vermiculture in Egypt: current development and future potential. Ph.D. Agro Industry and Infrastructure Officer Food and Agriculture Organization (FAO/UN), Cairo, Egypt
7. Rodale JI (1977) Organic gardening and farming. Rodale Books, Emmaus, Pa., © 1959. Rodale Books, Emmaus, Pa., [1977, © 1959]
8. Tilth (1982) Earthworms—surprising partners in the creation of fertile soils. Tilth producers quarterly. J Org Sustain Agric 8(1 & 2) (Soil supplement)
9. Burton M, Burton R (2002) International wildlife encyclopedia, vol. 6. Marshall Cavendish, New York, p 734
10. Munroe G (2007) Manual of on-farm vermicomposting and vermiculture. Organic Agriculture Centre of Canada. http://www.allthingsorganic.com/How_To/01.asp
11. Nagavallemma KP, Wani SP, Stephane L, Padmaja VV, Vineela C, Babu RM, Sahrawat KL (2004) Vermicomposting: recycling wastes into valuable organic fertilizer. Global theme on agroecosystems report no. 8. International Crops Research Institute for the Semi-Arid Tropics. P. Ecoscience Research Foundation, Andhra Pradesh, India, Sahrawat 502324. http://www.erfindia.org
12. Benítez E, Nogales R, Masciandaro G, Ceccanti B (2000) Isolation by isoelectric focusing of humic-urease complexes from earthworm (Eisenia fetida)-processed sewage sludges. Biol Fertil Soils 31:489–493

13. GEORG (2004) Feasibility of developing the organic and transitional farm market for processing municipal and farm organic wastes using large-scale vermicomposting. The good earth organic resources group limited Sackville, Nova Scotia
14. Sherman R (2003) Raising earthworms successfully. North Carolina Cooperative
15. Dominguez J, Edwards CA (1997) Effects of socking rate and moisture content on the growth and maturation of Eisenia Andrei (Oliogochaeta) in pig manure. Soil Biol Biochem 29(3–4):743–746
16. Reinecke AJ, Venter JM (1987) Moisture preferences, growth and reproduction of the compost worm *Eisenia fetida* (oligochaeta). Biol Fertil Soils 135–141
17. Gunadi B, Edwards, CA, Blount IVC (2003) The influence of different moisture levels on the growth, fecundity and survival of Eisenia fetida (Savigny) in cattle and pig manure solids. Eur J Soil Bio 39:19–24
18. Hou J, Qian Y, Liu G, Dong R (2005) The influence of temperature, pH and C/N ratio on the growth and survival of earthworms in municipal solid waste. Agricultural Engineering International: CIGR Ejournal VII. Manuscript FP 04 014
19. Kaplan DL, Hartenstein R, Neuhauser EF, Malechi MR (1980) Physicochemical requirements in the environment of the earthworm *Eisenia fetida*. Soil Biol Biochem 12:347–352
20. Edwards CA (1988) Breakdown of animal, vegetable and industrial organic wastes by earthworms. In: Edwards CA, Neuhauser EF (eds) Earthworms in waste and environmental management. SPB Academic Publishing, The Hague, pp 21–31
21. Ghafoor A, Hassan M, Alvi ZH (2008) Biodiversity of earthworm species from various habitats of district Narowal, Pakistan. Int J Agri Biol 10:681–684
22. Ghabbour S (2009) The Oligochaeta of the Nile basin revisited. In: The Nile origin, environments, limnology and human use. Monographiae biologicae, vol. 89
23. Atiyeh RM, Subler S, Edwards CA, Bachman G, Metzger JD, Shuster W (2000) Effects of vermicomposts and composts on plant growth in horticultural container media and soil Pedobiologia 44(5):579–590
24. FAO (2005) Fertilizer use by crop in Egypt. Land and Water Development Division and Plant Nutrition Management Service, Food and Agriculture Organization of the United Nations
25. Suthar S (2008) Bioconversion of post-harvest crop residues and cattle shed manure into value-added products using earthworm *Eudrilus eugenaie* (Kinberg). Ecol Eng 32:206–214
26. Fahmi WS (2005) The impact of privatization of solid waste management on the Zabaleen garbage collectors of Cairo. Environ Urbanization 17(2):77
27. Atiyeh RM, Lee S, Edwards CA, Arancon NQ, Metzger JD (2002) The influence of humic acids derived from earthworm-processed organic wastes on plant growth. Biores Technol 84:7–14
28. Bachman GR, Metzger JD (2008) Growth of bedding plants in commercial potting substrate amended with vermicompost. Biores Technol 99(8):3155–3161
29. Hobson AM, Frederickson J, Dise NB (2005) CH_4 and N_2O from mechanically turned windrow and vermicomposting systems following in-vessel pre-treatment. Waste Manag 25(4):345–352
30. Edwards CA (2008) Can earthworms harm the planet? Bio Cycle 49(12):53
31. IGES (2008) Climate change policies in the Asia-Pacific re-uniting climate change and sustainable development. Institute for Global Environmental Strategies
32. Chan YC, Sinha RK, Wang W (2010) Emission of greenhouse gases from home aerobic composting, anaerobic digestion and vermicomposting of household wastes in Brisbane (Australia). Waste Manage Res. (http://wmr.sagepub.com)
33. Daven JI, Klein RN (2008) Progress in waste management research. Nova Publishers, p 392
34. EEAA (1999) First national communications of Egypt. Egyptian Environmental Affairs Agency, IPCC, UNFCCC
35. Rizhiya E, Bertora C, Vliet PC, Kuikman PJ, Faber JH, Groenigen JWV (2007) Earthworm activity as a determinant for N_2O emission from crop residue, soil biology and biochemistry 39(8):2058–2069

Irrigation Water Use Efficiency and Economic Water Productivity of Different Plants Under Egyptian Conditions

Nesreen H. Abou-Baker

1 Introduction

Egypt is confronting increasing water demand and will be suffering from water shortage till the year 2030, caused by an agricultural approach which concentrates on increasing farming production together with climatic change and scarcity in both of rainfall and other water resources [1, 2]. Surface water resources and most groundwater reservoirs are completely exploited. Ideal water administration is a fundamental prerequisite for satisfying future needs in Egypt depends on improved use and proficient utilization of rainfall and present water resources [2]. Supplemental irrigation (SI) system could be a chance to improve crop production and provision of population food needs [3].

Progressive drought is one of the common limiting factors lead to increasing the relative electrolyte leakage, proline, and malondialdehyde and caused a huge reduction in leaf relative water content, chlorophyll parameters, photosynthesis as well as crop production, inflicting nutritional and countries economics insecurity [4, 5].

Functions of total applied water with crop production like IWUE and EWP were developed and used in evaluation and selection of the suitable crops to cultivation, as well as estimation of irrigation water rates for maximizing production without reducing profit under restricted water resources [6, 7].

N. H. Abou-Baker (✉)
Agricultural and Biological Research Division, Soils and Water Use Department, National Research Centre El Behoos St., Dokki, Giza 12622, Egypt
e-mail: nh.abou-baker@nrc.sci.eg

E.-S. E. Omran and A. M. Negm (eds.), *Technological and Modern Irrigation Environment in Egypt*, Springer Water,
https://doi.org/10.1007/978-3-030-30375-4_10

2 IWUE and EWP Definition

Optimizing water is the main challenge for increasing crop productivity and maximizing water use efficiency. Irrigation water use efficiency (kg m^{-3}) is considered one of the factors used to estimate the performance of agricultural production systems and defined as a weight of yield (kg) for each unit of supplied irrigation water (m^3) or biomass accumulation over water consumed. It could be defined the IWUE as the quantity of yield produced by one cubic meter of water supplied and was calculated using the following formula: IWUE = grain, seeds, ears, pods, oil or biological yield (kg $fed.^{-1}$)/total water applied (m^3 $fed.^{-1}$) [8, 9]. This formula is also called irrigation water productivity (IWP) as reported by El-Nady and Hadad [10]. It is viewed as one of the parameters that used to assess the performance of cultivation systems. The same crop may record under the same condition and may give both of high and low value of IWUE given the part of yield (grain, seeds, pods, ears, or biological yield) that entered into the equation. One meter cubic of water can produce from 0.82 to 1.66 kg grains, 1.037 to 2.037 kg ears, and 1.773 to 3.443 kg biological yield, respectively [11].

At the photosynthetic scale in plant leaves, the instantaneous water use efficiency is defined as the ratio of net carbon dioxide assimilation rate to transpiration or stomatal conductance [4, 12, 13].

Many investigators calculate the water use efficiency WUE (or crop water Productivity as it called in some research papers) by divided the yield over crop evapotranspiration or water consumptive use [6, 14–17] or the effective actual evapotranspiration ET_{rEF} [18], irrespective of application uniformity Eu and/or leaching requirement Lr. This could raise the net value of WUE and it did not communicate with total water amount that was added actually.

In other studies, water consumptive use (WCU) was calculated as fallow: WCU = $\sum_{i=1}^{I=3} D_i \times D_{bi} \times \frac{\theta_2 - \theta_1}{100}$ Hansen et al. [19], then crop water use efficiency (CWUE) as it called by El-Nady and Hadad [10] and El-Nady and Abdallh [20] and was calculated by divided the yield on water consumptive use. Whereas, WCU = Water consumptive use in the effective root zone (0.60 m). D_i = Soil layer depth (m), D_{bi} = Soil bulk density (kg m^{-3}) for the 0.60 m soil depth, θ_1 = Soil moisture percentage before irrigation (%), θ_2 = Soil moisture percentage (%) at 48 h after irrigation, and I = Number of soil layers.

Beshara et al. [15] reported that water use efficiency values of wheat grains were 1.89, 1.85, and 1.83 kg m^{-3} by using water consumptive use (m^3 $fed.^{-1}$) and 1.38, 1.29, and 1.25 kg m^{-3} by using total applied irrigation water (m^3 $fed.^{-1}$) as a dominator of the equation. These results are in close agreement with those obtained by El-Nady and Abdallh [20] on maize and Abdel-Aziz [16] on cucumber and pepper yields. They calculated irrigation water use efficiency IWUE by using total irrigation water applied that produced low values compared with water use efficiency WUE that is calculated by using water consumptive use or actual evapotranspiration, while both of them take the same trend. The maximum values of WUE and IWUE were 22.2 and 19.9 kg m^{-3} for cucumber and 6.2 and 5.1 kg/m^3 for pepper yield, respectively.

Economic water productivity (EWP) or the gross economic return of each water cubic meter was expressed in gross income per gross water supplied in m^3 computed from the estimated irrigated area, obtainable yield and from the seasonal price of the main product and byproduct. The next equation shows the calculation method: EWP = Gl/GIWR [21] where Gl is gross income from the sale of grains (Egyptian pound, USD, Euro or the suitable currency); GIWR is gross irrigation water requirement (m^3) [22]. The mean EWP for forage cactus clones that grown in Brazil was 18 R$ mm^{-1} of the effective actual evapotranspiration [18].

Increasing the crop production per unit of water added did not communicate with increasing the farmer's profit; this may be due to the nonlinearity of production with the crop yield price [7, 21]. Consequently, biomass production outcome for each unit of applied irrigation water has been used extensively as a measure of EWP. There is need to be assessed on the EWP parameters based on the anticipated price before planting directly because seasonal price fluctuations will give illusion data consequently wrong decision particularly in countries which suffering from fluctuation in market prices [7]. It should be considered when comparing EWP values to decide that the used price in the calculation if it producer or customer price. There are some factors like net present value, benefit-to-cost ratio, and internal rate of return used in the economic evaluation of irrigation water management procedures [23]. Especially, in some plant kinds that can reserve water like cactus species, the water retained by the plant must be considered [18].

3 Effect of Experimental Location on IWUE

It is essential to enhance location-specific agronomic management to maximize IWUE values as possible. In the same location (North China Plain), 28 growing seasons of wheat were established under water shortage condition, and IWUE was raised from past to present. This may be referred to the significant increase in soil organic matter, N contents, and reducing soil evaporation over time consequently, increasing wheat yield and harvest index [24].

In Egypt, IWUE values ranged from 0.67 to 1.66 kg m^{-3} and from 0.60 to 2.2 kg m^{-3} for wheat grown in El-Arish region and Alexandria Governorate, respectively [8, 22].

Data presented in Table 1 showed the mean values of EWP (USD/kg) for wheat grown in Egypt at various locations and different harvest years (USD/kg) using producer prices (USD) that reported by FAO [25]. The differences between EWP values could be attributed to the variation in location, harvest year, and increasing grain prices over time.

Table 1 Mean values of wheat producer prices which grown in Egypt at various harvest years and calculated EWP (USD/kg)

Harvest year	Producer prices of wheat in the harvest year (USD/kg)[a]	EWP	References of yield and irrigation rates data which used in calculation
2009–2010	0.31	0.37	Abd-Eladl et al. [8]
2010–2011	0.36	0.51	Abd-Eladl [22]
2012–2013	0.40	0.63	El-Nady and Borham [26]

[a]Producer Prices Annual of most crops from 1991 to 2015 is recorded in FAO [25]

4 Effect of Some Management Procedures on IWUE

This part of the chapter reviews selected research from Egypt or around the world and summarizes some treatments that affect IWUE values.

4.1 Irrigation Rate and Frequency

Rainfall in Egypt ranged between 0 and 200 mm year^{-1} in the desert and the north coastal region, respectively, and the harvested rain was used mainly for irrigation [1]. A significant decrease in IWUE values of wheat grains (1.97, 1.61 and 1.19 kg m^{-3}) as soil moisture level decreased (80, 65 and 50%) of field capacity [27]. Hassanli et al. [28] showed that IWUE could be enhanced by improving irrigation technology, irrigation scheduling, and agronomic practices which led to yield increase. The value of EWP was higher when irrigated with rain than rain + SI treatment [22].

The IWUE increased gradually in the order 100% > 130% > 70% of wheat water requirement, this may be due to the enhancement in IWUE did not due to high amount of water, and low yields caused by water stress did not concomitant to low IWUE values. Mathematically, increasing irrigation applied leads to increase in the denominator of the calculation equation [yield (kg fed.$^{-1}$)/total water applied (m^3 fed.$^{-1}$)] and then decreases the result. As for the viewpoint of plant nutrition, plants can optimize their use of irrigation water, preserve the dry matter accumulation, and maximize their survival chance under stress conditions [8, 18].

Although the values of IWUE were greater at rainfed and decreased by the addition of SI to wheat [22] and forage cactus [18], the severe drought reduced IWUE of maize grown in North China Plain [24]. Drought stress led to increase in the relative electrolyte leakage, proline, abscisic acid, and malondialdehyde and caused a huge reduction in stomatal conductance, leaf relative water content, chlorophyll parameters, photosynthesis, transpiration inflicting crop production, nutritional and countries economics insecurity [4, 5, 29, 30].

By sacristy of water additions, deficit irrigation technique aims to augment water productivity for various crops without causing extreme yield reductions; however, it should be combined between field experiments of various plant genotypes at several growth stages and modeling of crop water productivity [31].

IWUE values increased gradually in the order 402 mm > 456 mm > 358 mm, the minority of IWUE values under 358 mm treatment may be referred to low yield produced by high water stress, but their majority under 402 mm treatment compared to 456 mm means better water usage and low water losses which concomitant to high yield [9]. In this concern, Abou Hadid [32] reported that 1 m^3 of irrigation water produces 5.65 kg potato.

The higher irrigation water use efficiency values were produced at the lowest rate and decreased by increasing irrigation amount and/or increasing interval between irrigations. Application of 3202, 2722, 2241 (m^3 $fed.^{-1}$) to maize resulted in 0.94, 1.14, and 1.5 IWUE values, respectively [11].

Linear regression with strong relationship ($R2 = 0.90$–0.94) relating IWUE of tomato plants with irrigation rates supplied was detected. The IWUE values increased by 30.3% under application of 40% ETc are compared with 100% ETc [33]. The highest irrigation rate (1263 m^3 $fed.^{-1}$) has the lowest IWUE and EWP values. Application of 758 and 1010 m^3 $fed.^{-1}$ increased IWUE by 14.1 and 19.2% and raised EWP by 15.6 and 20.3%, respectively, compared with the highest water application rate [7].

Sunflower plants that irrigated at 25% of available soil moisture depletion recorded higher values of IWUE than 50%, because the first rate increased water availability to plant, while under the second rate, the plants are suffering from water stress and required more energy to absorb water that influenced negatively on the yield. Irrigation of sunflower by 6742, 5858, 5549, 5518, and 5498 m^3 ha^{-1} produced IWUE values 0.47, 0.46, 0.53, 0.62, and 0.56 kg seed m^{-3} and 0.19, 0.20, 0.22, 0.25, and 0.21 kg oil m^{-3}, while EWP values calculated based on the price of seed yield were 0.63, 0.73, 0.66 \$ m^{-3}, respectively. The highest value of net income and net returns were 9268.7 and 5128.7 LE (1\$ = 5.5 LE in 2008–2009) produced by escaping irrigation at the age of 60 days after sowing of sunflower plants (2299 m^3 $fed.^{-1}$) [34].

Irrigation frequency is one of the essential factors in irrigation management, particularly under stress condition, whereas it affects IWUE and crop yield. The values of IWUE are different under the short distance between irrigations, medium, and long period between irrigations. The lowest irrigation frequency caused water stress, in contrast to high frequency. This effect is based mainly on soil properties especially the soil texture [11, 35].

4.2 *Irrigation System*

Water harvesting should be revised to provide plant production. It can reserve over half of the lost water and improve IWUE by concentrating rainwater to the planted crops by using cisterns as it is widely spread in North Egypt [36, 37].

Irrigation of maize plant with a subsurface drip irrigation system (0.2 m depth) gave higher IWUE value than surface one [35], while values of IWUE of the grape plant that grown under drip irrigation system and low head bubbler one exceeded that under gated pipe system by 129.7 and 96.9%, respectively [38]. Yield, yield attributes, IWUE, and leaf area index in the relay intercropping treatments were affected by limited single drip irrigation of wheat and maize because this practice saved water used for irrigation [39].

Values of IWUE of sunflower that irrigated by sprinkler system ranged from 0.52 to 0.82 kg seed m^{-3} and from 0.21 to 0.29 kg oil m^{-3} [40].

Figure 1 illustrates the interaction between three different irrigation methods (alternate furrow (AFI), fixed furrow (FFI), and conventional furrow (CFI) irrigation systems) and four various fertilizers combinations [NPK (F1), NPK + Zn (F2), NPK + Zn + B (F3), and NPK + Zn + B + Fe (F4)]. The obtained results indicated that FFI and AFI are ways to save water, which save about 24 and 13% of irrigation water, respectively. The interaction AFI × F4 was the best in maize yield production, but it gave low IWUE value (1.64 kg m^{-3}) compared with FFI × F1 (1.81 kg m^{-3}) [20].

Although highest soybean seed yield was obtained by traditional furrow irrigation method followed by the surge and alternate irrigation, the highest IWUE values were obtained by alternate irrigation followed by the surge and the next was the furrow method under two different tillage methods (conventional or no-tillage) as illustrated in Fig. 2. These may be attributed to the highest amount of irrigation water that was applied with traditional furrow irrigation method, while the alternate and surge systems saved about 29 and 15% of water applied compared with furrow system [10].

Figure 3 illustrates the effect of intelligent irrigation technique (IIT) (Hunter Pro-C (H)) and irrigation control technique (ICT) under both of surface (SDI) and subsurface drip irrigation (SSDI) systems on yield and IWUE values of cucumber and pepper. The values of IWUE were ranged between 11.6 and 19.9 kg m^{-3} for

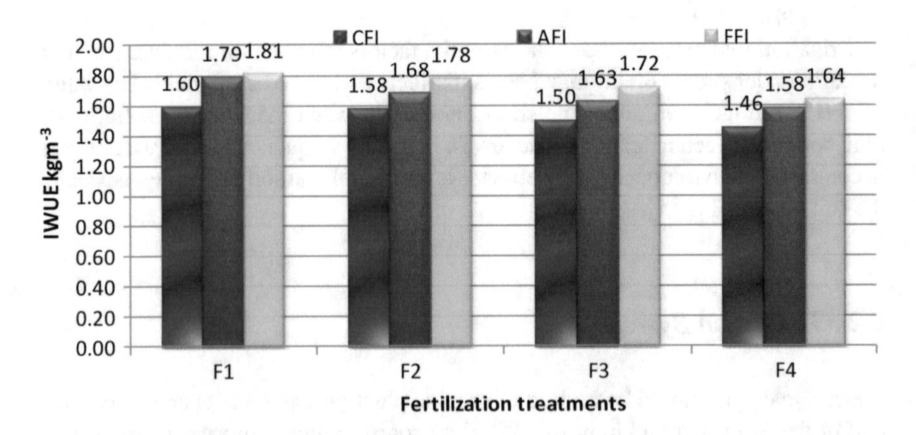

Fig. 1 Irrigation water use efficiency as affected by the interaction between irrigation systems and fertilizers combinations

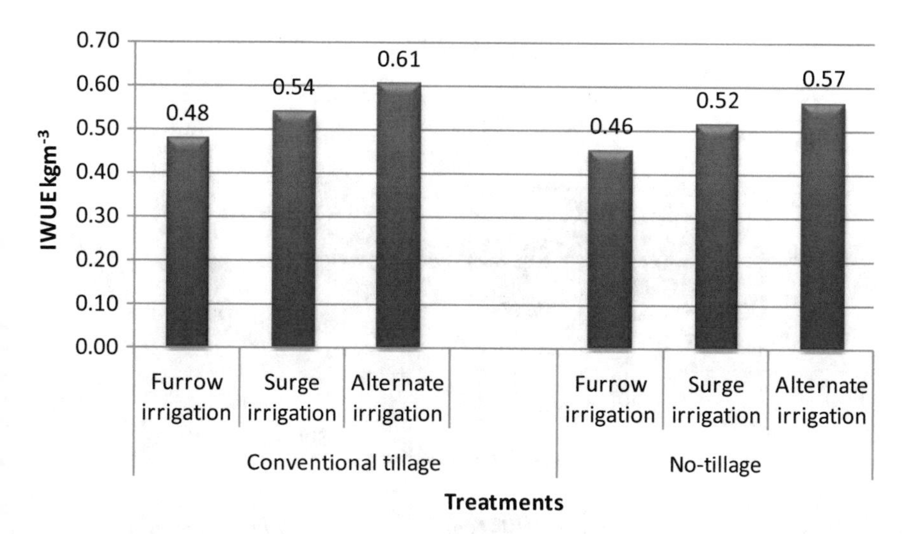

Fig. 2 Effect of the furrow, surge, and alternate irrigation systems under both of conventional tillage and no-tillage

cucumber, 3.1 and 5.1 kg m^{-3} for pepper under ICT × SDI and IIT × SSDI interactions. The integration between the IIT system and SSDI introduced smart control in adding the water requirement based upon plant status consequently increasing yield and decreasing water applied at the same time [16].

In contrast, Cai et al. [41] reported that the irrigation system (mobile rain shelter) may not be appropriate because the mild water stress did not significantly reduce maize yield compared with no water stress or severe water stress.

4.3 Soil Treatments

Concerning the effect of fertilization treatments irrespective of irrigation levels on IWUE, organic matter (OM) promotes the hydraulic conductivity, porosity, and soil moisture constants and plays a major role in enhancing IWUE, while bulk density takes an opposite trend [8, 26, 42]. Water use efficiency values were influenced by fertilization treatments (biofertilization, mineral, and combination between mineral and organic), and the highest value was produced by biofertilization treatment compared with mineral and mineral-organic combination applied [26]. The combination of NP and OM resulted in the most efficient use of irrigation water, whereas OM + NP treatment produced higher IWUE than those for sole application of N, NP, and OM [43]. Values of IWUE calculated by grain yield of wheat were 1.2, 1.6, 2.2, 0.6, 0.7, and 0.7 for rain treatment, rain + NPK, rain + OM, SI, SI + NPK, and SI + OM, respectively [22]. Splitting the recommended nitrogen rate of wheat—in urea fertilizer form—to 4, 3, and 2 equal doses resulted in IWUE values 1.42, 1.29,

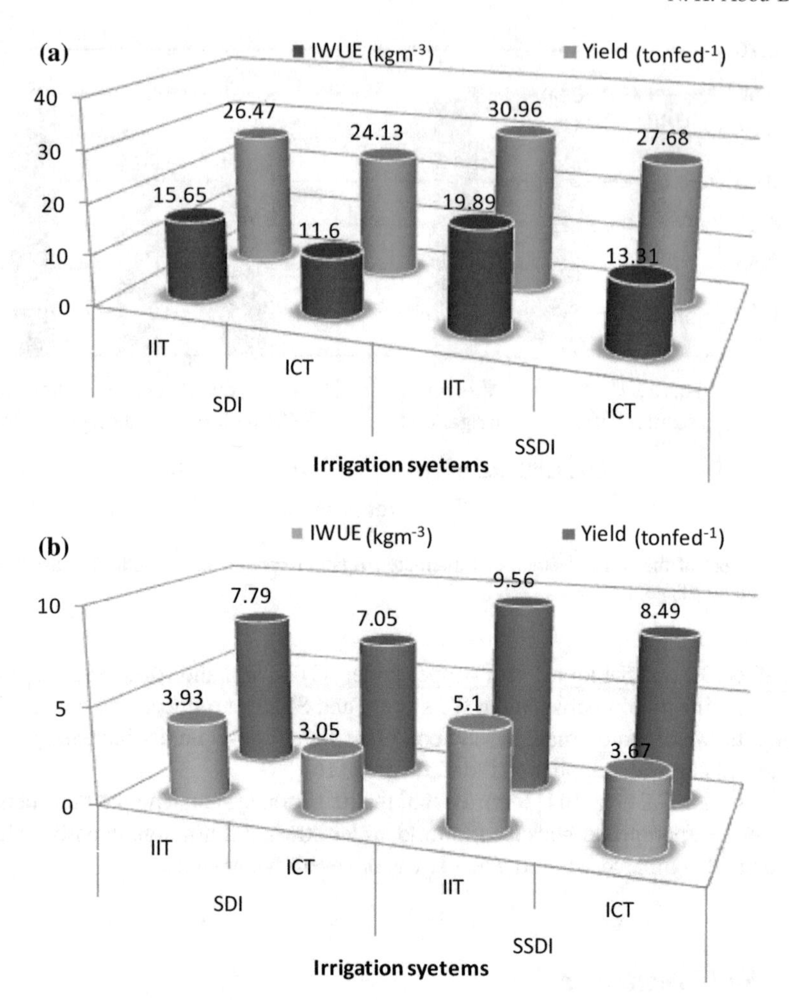

Fig. 3 Yield and IWUE values of cucumber (**a**) and pepper (**b**) as affected by intelligent irrigation technique (IIT) [Hunter Pro-C (H)] and irrigation control technique (ICT) under both of surface (SDI) and subsurface drip irrigation systems (SSDI)

and 1.20 kg grain m^{-3}, respectively [15]. Treating sandy soils with compost led to an increase in IWUE of maize plants as follows and revealed that IWUE of wheat decreased gradually in the rank (50% compost + 50% NPK) > (100% compost) > (full recommended NPK) > control as illustrated in Fig. 4 [8].

Irrespective of irrigation practice (rainfed or SI), the addition of OM increased EWP values of the maize plant, followed by mineral fertilization and the next is the control (Table 2). Compost application increased IWUE for wheat and sequenced maize at all applied irrigation treatments. Concerning the effect of fertilization treatments irrespective of irrigation levels, data showed that IWUE of wheat declined in

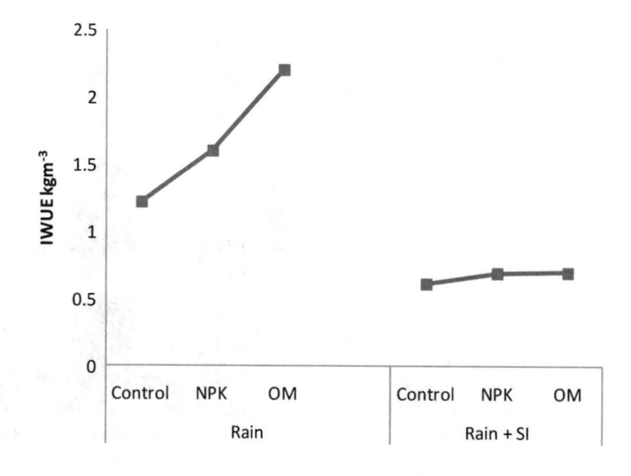

Fig. 4 Effect of mineral and organic fertilization under rain and supplemental irrigation (SI) on IWUE values of maize yield

Table 2 Average values of maize grain yield and EWP as affected by mineral and organic fertilization under rain and supplemental irrigation (SI)

Irrigation treatments	Fertilization treatments	Av. grain yield kg fed.$^{-1}$	Av. EWP Eur m^{-3}
Rain	Control	552	0.34
	NPK	724	0.45
	OM	998	0.62
Rain + SI	Control	992	0.17
	NPK	1118	0.19
	OM	1130	0.20

the order (7.5 tons fed.$^{-1}$ compost + 50% NPK of recommended doses) > (15 tons compost tons fed.$^{-1}$) > (full recommended NPK) > control [8].

The IWUE values increased gradually in the rank: (full recommended NPK) > (50%compost + 50%NPK) > (75%compost + 25%NPK) > (50%compost + 50%NPK), but with adding 1980 m^3 fed.$^{-1}$ only IWUE values arranged in the order (full recommended NPK) > (75%compost + 25%NPK) > (50%compost + 50%NPK) > (50%compost + 50%NPK). This means that under low amount of water, application of (75%compost + 25%NPK) superior than applied (50%compost + 50%NPK) as reported by Abou-Baker [44]. These ratios were calculated based on nitrogen recommendation of Egyptian ministry of agriculture for maize plants.

Increasing application rate of nitrogen and potassium humate (15 kg N + 2 kg K-humate fed.$^{-1}$, 30 kg N + 4 kg K-humate and 45 kg N + 6 kg K-humate fed.$^{-1}$) produced an increase in IWUE of sunflower crops 0.61, 0.68, and 0.81 kg seed m^{-3} or 0.23, 0.25, and 0.28 kg oil m^{-3}, compared with unamended plots (0.46 kg seed m^{-3} and 0.18 kg oil m^{-3}). Improving water holding capacity of the sandy soil and the depression in actual evapotranspiration with humate application led to enhancing both of the seed and oil yields consequently increasing IWUF values [40].

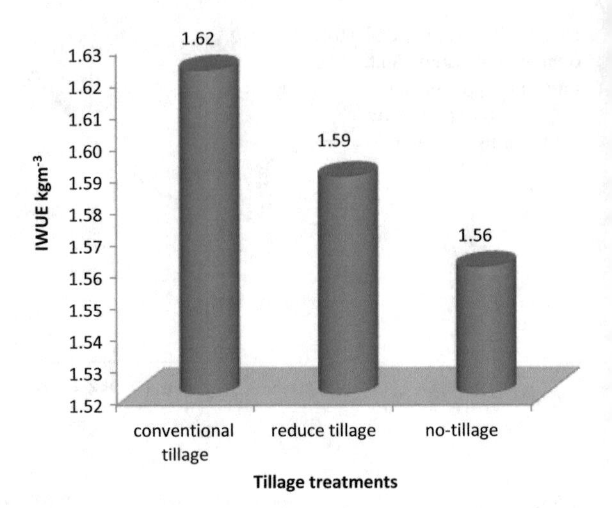

Fig. 5 Effect of conventional, reduce, and no-tillage on IWUE of wheat grains

Statistically, IWUE was affected significantly by the application of different hydrogel sources (acrylamide-hydrogel HA and straw based-hydrogel HS) with control (H0) which produced IWUE values; 1.29, 1.24, and 1.05 kg m^{-3} [11].

Water use efficiency values of wheat plants (calculated by water consumptive use) were influenced by tillage practices (conventional tillage, reduce tillage, and no-tillage), and the highest value was produced by conventional tillage, while the lowest one recorded under the no-tillage system (Fig. 5). This may be due to increasing wheat grain yield and decreasing the water consumptive use as a result of enhancing soil physical properties and the good distribution of roots under conventional tillage [26]. In another study, there is no significant difference between conventional tillage and no-tillage tillage systems on reducing applied water and increasing water productivity [10].

Water use efficiency values of wheat plants are enhanced with raising biofertilizer application rates (250, 500, and 750 gm $fed.^{-1}$), in concomitant with increasing yield with the higher rate of biofertilizer [26].

4.4 Plant Treatments

Crop selection, grafting technique, and a foliar of some intelligent compounds are the most important items in the plant section of this chapter because their influence on IWUE and EWP values is direct. In various recent studies, grafted tomato onto other interspecific tomato hybrid resulted in higher yield and WUE (12.1%) compared with ungrafted plants [33, 45, 46]. Also, Luffa rootstock produced higher leaf area, dry weight, and instantaneous WUE, when grafted with cucumber or with its scion [47]. Regardless of the irrigation regime, mini-watermelon and sweet pepper yield and their WUE values were higher in grafting treatment than in non-grafted plants

[48, 49]. These findings may be attributed to (1) vigorous rhizosphere of rootstocks compared to self-rooted plants, (2) enhancing plants nutrient status, (3) reducing transpiration rate, and (4) increasing the net carbon dioxide assimilation rate or due to all mentioned reasons [45, 48, 50]. The decline of water use efficiency by the plant was caused by a fast increase in both the transpiration process and stomatal conductance [13].

Cultivar selection and intercropping maize with peanut that known as less water requirement crop are the most important practices affecting plant yield and IWUE value [51, 52]. Average crop water productivity (16.6 kg mm^{-1}) and profitability of the continuous corn rotation were also higher than in the other crop rotations: corn wheat, corn wheat sorghum, and corn wheat sorghum soybean [17].

Plants can improve its adaptive mechanisms to regulate the developmental processes, alleviate stress impacts, and survive under drought conditions such as increasing proline, auxin, and cytokinin excretions [53]. The abscisic acid regulates crop vegetative and contributes to water stress adaptation of crops [30]. Spraying bean plants with 300 and 600 ppm Si three times within a growth period (30, 60, and 80 days after sowing bean) resulted in higher IWUE compared with those splashed with distilled water [7]. So, it can be recommended that spraying plants that were grown under abiotic stress with abscisic acid, proline, silicate, salicylic, auxin, and cytokinin may lead to alleviating stress and increasing growth, thus increasing IWUE and EWP under stress conditions.

5 Comparison Between Some of IWUE and EWP Values Under Both Egyptian and Worldwide Conditions

As for wheat plants, in Tunisia, the best water consumption of wheat was produced by adding three irrigations over rainfed, whereas it reached 384.6 mm with productivity 348.7 kg/dunum of grains and IWUE 1.53 kg m^{-3} [54]. In India, Singh et al. [55] reported that the greatest IWUE was recorded with pre-sowing irrigation up to field capacity and ranged from 1.13 to 1.28 kg m^{-3}. In China, Zhang et al. [56] reported that IWUE for grains of winter wheat was increased from 0.97 to 1.1 kg m^{-3} by supplemental irrigation. In Egypt, the IWUE values of wheat ranged from 0.61 to 2.26 kg m^{-3} [22] and from 0.67 to 1.66 kg m^{-3} [8], which is higher than that recorded by Zwart and Bastiaanssen [57] who reported that globally measured average IWUE value per unit water depletion is 1.09 kg m^{-3} and the range is very wide (0.6–1.7 kg m^{-3}) for wheat grains. This may be due to the progress in wheat cultivars, fertilization, and soil water and plant management for a long while or due to using compost [22].

As for maize plants, in Iran, adding 75% of maize water demands raised the soil wetting front, net present value, benefit-to-cost ratio, and internal rate of return by 110, 37.7, 6.14, and 192% compared with application 100% and 55% treatments, respectively. Thus, 25% of water quantity was saved by using this economic water-saving technique [23]. In Egypt, the values of IWUE were greater (1.19 kg maize ears m^{-3}) at lower water rates (1980 m^3 $fed.^{-1}$) [44]. The globally measured range is narrow (1.1–2.7 kg m^{-3}) for maize grains [57].

Concerning bean seeds, IWUE values of bean plant grown in Ismailia governorate (Egypt) ranged from 0.69 to 1.07 kg m^{-3} (based on seed yield), and from 1.29 to 2.06 (based on stover yield), their EWP values ranged between 0.58 and 0.88 Eur/m^3 [7]. In another study, been plants that grown in El-Nubaria produced the greatest IWUE value (9.75 kg bean pods m^{-3}) at lowest water rate (574 m^3 $fed.^{-1}$) [44].

Values of EWP (that calculated by using producer prices in Egypt at harvest year (USD/kg) as reported by FAO [25] were ranged from 0.19 to 0.43 USD m^{-3} for grape plant [38], from 0.6 to 1.7 USD m^{-3} for bean [7, 44], from 0.20 to 0.32 USD m^{-3} for sunflower [34, 40], from 0.12 to 0.6 USD m^{-3} for maize [8, 11, 20, 35], from 0.29 to 0.36 USD m^{-3} for soybean [10], from 2.40 to 2.93 USD m^{-3} for cucumber [16], and from 0.77 to 1.34 USD m^{-3} for potato [9].

6 Conclusion

Egypt characterizes by plenty of space cultivable soil, but water scarcity is a limiting factor in reclamation processes. Consequently, improving IWUE and EWP is the most important than increasing yield for each land unit. The equations that calculate IWUE and EWP did not estimate the retained in the plant. In the future, more efficient irrigation management must be adopted. It should increase farmers' and decision makers' understanding of the scientific communities' results and apply it, especially in concerning of agronomic management effects on IWUE and EWP across various, plant types, genotypes, soil properties, climate conditions, irrigation management (systems, rates, frequencies and timing), soil amendments, plant practices, etc.

Irrigation management (supplemental irrigation, irrigation systems, rates and frequencies), tillage methods, organic amendments, hydrogel application, grafting technique, cultivar selection, crop rotation, as well as spraying of abscisic acid, proline, silicate, salicylic, auxin, and cytokinin are the fundamental practices have a high potential for maximizing the benefits of water unit.

7 Recommendations

All efforts should come together to improve the agricultural management that reflects on the increase in both of IWUE and EWP especially in arid and semi-arid regions than decreasing the demand for water in the agricultural sector.

References

1. Abdel-Shafy HI, El-Saharty AA, Regelsberger M, Platzer C (2010) Rainwater in Egypt: quantity, distribution and harvesting. Mediterr Mar Sci 11(2):245–258
2. Abdin AE, Afify A, Adel A (2011) Comparative analysis of Egyptian water policy and water framework directive. In: Junier S, El Moujabber M, Trisorio-Liuzzi G, Tigrek S, Serneguet M, Choukr-Allah R, Shatanawi M, Rodríguez R (eds) Dialogues on Mediterranean water challenges: rational water use, water price versus value and lessons learned from the European water framework directive. CIHEAM, Bari, pp 169–179
3. Abderrazzak B, Daoui K, Kajji A, Dahan R, Ibriz M (2013) Effects of supplemental irrigation and nitrogen applied on yield and yield components of bread wheat at the Saïs region of Morocco. Am J Exper Agric 3(4):904–913
4. Kumar P, Rouphael Y, Cardarelli M, Colla G (2017) Vegetable grafting as a tool to improve drought resistance and water use efficiency. Front Plant Sci 8:1130. https://doi.org/10.3389/fpls.2017.01130
5. Liao T, Wang Y, Xu CP, Li Y, Kang XY (2017) Adaptive photosynthetic and physiological responses to drought and rewatering in triploid *Populus* populations. Photosynthetica. https://doi.org/10.1007/s11099-017-0704-5
6. Zhang H, Oweis T (1999) Water–yield relations and optimal irrigation scheduling of wheat in the Mediterranean region. Agric Water Manag 38(3):195–211
7. Abou-Baker NH, Abd-Eladl M, Eid TA (2012) Silicon and water regime responses in bean production under soil saline condition. J Appl Sci Res 8(12):5698–5707
8. Abd-Eladl M, Abou-Baker NH, El-Ashry S (2010) Impact of compost and mineral fertilization irrigation regime on wheat and sequenced maize plants. Minufiya J Agric Res 35(6):2245–2262
9. Eid TA, Ali SMM, Abou-Baker NH (2013) Influence of soil moisture depletion on water requirements, yield and mineral contents of potato. J Appl Sci Res 9(3):1457–1466
10. El-Nady MA, Hadad WM (2016) Water use efficiency of soybean under different tillage and irrigation methods. Egypt J Soil Sci 56(2):295–312
11. Ibrahim MM, Abd-Eladl M, Abou-Baker NH (2015) Lignocellulosic biomass for the preparation of cellulose-based hydrogel and its use for optimizing water resources in agriculture. J Appl Polym Sci 132(42):42652 (1–12)
12. Bacon MA (2009) Water use efficiency in plant biology. Blackwell Publishing Ltd., Oxford, p 344
13. Podlaski S, Pietkiewicz S, Chołuj D, Horaczek T, Wiśniewski G, Gozdowski D, Kalaji HM (2017) The relationship between the soil water storage and water-use efficiency of seven energy crops. Photosynthetica 55(2):210–218
14. Howell TA (2001) Enhancing water use efficiency in irrigated agriculture. Agron J Abst 93:281–289
15. Beshara AT, Borham TI, Saied MM, El-Hassanin AS (2013) Impact of soil moisture depletion and splitting the recommended nitrogen fertilizer rate on water requirements and water use efficiencies of wheat crop in North Delta. Egypt J Soil Sci 53(2):249–266
16. Abdel-Aziz AA (2016) Effect of intelligent irrigation technique on water use efficiency for cucumber and pepper crops in New Salhia Area. Egypt J Soil Sci 56(4):761–773
17. Schlegel AJ, Assefa Y, Dumler TJ, Haag LA, Stone LR, Halvorson AD, Thompson CR (2016) Limited irrigation of corn-based no-till crop rotations in west central Great Plains. Agron J 108(3):1132–1141
18. Morais JE, Silva TG, Queiroz MGD, Araujo GG, Moura MS, Araújo Júnior GDN (2017) Hydrodynamic changes of the soil-cactus interface, effective actual evapotranspiration and its water efficiency under irrigation. Revista Brasileira de Engenharia Agrícola e Ambiental 21(4):273–278
19. Hansen VW, Israelsen DW, Stringharm QE (1979) Irrigation principle and practices, 4th edn. Wiley, New York
20. El-Nady MA, Abdallh AM (2013) Corn yield response to some irrigation methods and fertilization with macro and micronutrients. Egypt J Soil Sci 53(3):347–360

21. Araya A, Habtu S, Haile M, Sisay F, Dejene T (2011) Determination of local barley (*Hordeum vulgare* L.) crop coefficient and comparative assessment of water productivity for crops grown under the present pond water in Tigray, Northern Ethiopia. Momona Ethiop J Sci 3(1):65–79
22. Abd-Eladl M (2014) Evaluation of supplemental irrigation and fertilization as integrated management of wheat under rainfed conditions in Egypt. J Genet Environ Resour Conserv 2(1):54–64
23. Karandish F (2016) Improved soil–plant water dynamics and economic water use efficiency in a maize field under locally water stress. Arch Agron Soil Sci 62:1311–1323
24. Zhang X, Qin W, Chen S, Shao L, Sun H (2017) Responses of yield and WUE of winter wheat to water stress during the past three decades—a case study in the North China Plain. Agric Water Manag 179:47–54
25. FAO (2017) FAOSTAT—producer prices, annual. http://www.fao.org/faostat/ar/#data/PP/visualize
26. El-Nady MA, Borham TI (2013) Influence of different tillage systems and fertilization regimes on wheat yield, wheat components and water use efficiency. Egypt J Soil Sci 53(4):579–593
27. El-Kommos F, Dahroug AA, Nour El-Din Y (1989) Evapotranspiration, water use efficiency and dry matter of wheat as affected by some soil conditioners at different soil moisture levels. Egypt J Soil Sci 29(4):387–400
28. Hassanli AM, Ahmadirad S, Beecham S (2010) Evaluation of the influence of irrigation methods and water quality on sugar beet yield and water use efficiency. Agric Water Manag 97(2):357–362
29. Beis A, Patakas A (2015) Differential physiological and biochemical responses to drought in grapevines subjected to partial root drying and deficit irrigation. Eur J Agron 62:90–97
30. Killi D, Bussotti F, Raschi A, Haworth M (2017) Adaptation to high temperature mitigates the impact of water deficit during combined heat and drought stress in C3 sunflower and C4 maize varieties with contrasting drought tolerance. Physiol Plant 159:130–147
31. Geerts S, Raes D (2009) Deficit irrigation as an on-farm strategy to maximize crop water productivity in dry areas. Agric Water Manag 96:1275–1284
32. Abou Hadid AF (2006) Water use efficiency in Egypt. In: Hamdan I, Oweis T, Hamdallah G (eds) Water use efficiency network. Proceedings of the expert consultation meeting, 26–27 Nov, ICARDA, Aleppo, Syria, pp 39–57
33. Ibrahim A, Wahb-Allah M, Abdel-Razzak H, Alsadon A (2014) Growth, yield, quality and water use efficiency of grafted tomato plants grown in greenhouse under different irrigation levels. Life Sci J 11(2):118–126
34. El-Henawy AS, Soltan EM (2013) Irrigation water management for sunflower production at North Nile Delta Soils. Egypt J Soil Sci 53(1):1–8
35. Mehanna HM, Hussein MM, Abou-Baker NH (2013) The relationship between water regimes and maize productivity under drip irrigation system: a statistical model. J Appl Sci Res 9(6):3735–3741
36. Oweis T, Prinz D, Hachum A (2001) Water harvesting: indigenous knowledge for the future of the drier environments. ICARDA, Aleppo
37. Oweis T, Hachum A (2006) Water harvesting and supplemental irrigation for improved water productivity of dry farming systems in West Asia and North Africa. Agric Water Manag 80(1):57–73
38. Tayel MY, El-Gindy AA, El-Hady MA, Ghany HA (2007) Effect of irrigation systems on: II-yield, water and fertilizer use efficiency of grape. J Appl Sci Res 3(5):367–372
39. Zhang BC, Huang GB, Feng-Min LI (2007) Effect of limited single irrigation on yield of winter wheat and spring maize relay intercropping. Pedosphere 17(4):529–537. Project supported by the National Key Basic Research Special Foundation (NKBRSF) of China (No. G2000018603) and the National High Technology Research and Development Program (863 Program) of China (No. 2002AA2Z4191)
40. El-Kotb HMA, Borham TI (2013) Combination impacts of irrigation regimes and n&k-humate rates on sunflower productivity and water use efficiency in sandy soil. Egypt J Soil Sci 53(1):119–133

41. Cai Q, Zhang Y, Sun Z, Zheng J, Bai W, Zhang Y, Liu Y, Feng L, Feng C, Zhang Z, Yang N, Evers JB, Zhang L (2017) Morphological plasticity of root growth under mild water stress increases water use efficiency without reducing yield in maize. Biogeosci Discuss 1–24. https://doi.org/10.5194/bg-2017-103
42. Fusheng L, Jiangmin Y, Mengling N, Shaozhong K, Jianhua Z (2010) Partial root-zone irrigation enhanced soil enzyme activities and water use of maize under different ratios of inorganic to organic nitrogen fertilizers. Agric Water Manag 97(2):231–239
43. Fan T, Stewart BA, Payne WA, Yong W, Luo J, Gao Y (2005) Long-term fertilizer and water availability effects on cereal yield and soil chemical properties in Northwest China. Soil Sci Soc Am J 69:842–855
44. Abou-Baker NH (2015) Organic fertilization, pollution and irrigation. LAP LAMPERT Academic Publishing. ISBN: 978-3-659-37159-2
45. Schwarz D, Rouphael Y, Colla G, Venema JH (2010) Grafting as a tool to improve tolerance of vegetables to abiotic stresses: thermal stress, water stress and organic pollutants. Sci Hortic 127:162–171
46. Al-Harbi A, Hejazi A, Al-Omran A (2016) Responses of grafted tomato (*Solanum lycopersicon* L.) to abiotic stresses in Saudi Arabia. Saudi J Biol Sci 24(6):1274–1280. https://doi.org/10.1016/j.sjbs.2016.01.005
47. Liu S, Li H, Lv X, Ahammed GL, Xia X, Zhou J et al (2016) Grafting cucumber onto luffa improves drought tolerance by increasing ABA biosynthesis and sensitivity. Sci Rep 6:20212. https://doi.org/10.1038/srep20212
48. Rouphael Y, Cardarelli M, Colla G, Rea E (2008) Yield, mineral composition, water relations, and water use efficiency of grafted mini watermelon plants under deficit irrigation. HortScience 43:730–736
49. López-Marín J, Gálvez A, del Amor FM, Albacete A, Fernandez JA, Egea-Gilabert C et al (2017) Selecting vegetative/generative/dwarfing rootstocks for improving fruit yield and quality in water stressed sweet peppers. Sci Hortic 214:9–17. https://doi.org/10.1016/j.scienta.2016.11.012
50. Khah EM, Katsoulas N, Tchamitchian M, Kittas C (2011) Effect of grafting on eggplant leaf gas exchanges under Mediterranean greenhouse conditions. Int J Plant Prod 5:121–134
51. Liu ZJ, Hubbard KG, Lin XM, Yang XG (2013) Negative effects of climate warming on maize yield are reversed by the changing of sowing date and cultivar selection in Northeast China. Glob Change Biol 19:3481–3492
52. Lu H, Xue J, Guo D (2017) Efficacy of planting date adjustment as a cultivation strategy to cope with drought stress and increase rainfed maize yield and water-use efficiency. Agric Water Manag 179:227–235
53. Bielach A, Hrtyan M, Tognetti VB (2017) Plants under stress: involvement of auxin and cytokinin. Int J Mol Sci 18(7):1427
54. Shawa F, Shaabouni Z, Baazez A (1993) Supplemental irrigation and its effect on wheat crop productivity in the semi-arid region of Tunisia. Damascus (Syria) Ministry Agriculture and Agrarian Reform, 41 p
55. Singh PN, Singh G, Dey P (1996) Effect of supplemental irrigation and nitrogen application to wheat on water use, yield and uptake of major nutrients. J Ind Soc Soil Sci 44(2):198–201
56. Zhang Jianhua, Sui Xiangzhen, Li Bin, Baolin Su, Jianmin Li, Dianxi Zhou (1998) An improved water-use efficiency for winter wheat grown under reduced irrigation. Field Crops Res 59:91–98
57. Zwart SJ, Bastiaanssen WGM (2004) Review of measured crop water productivity values for irrigated wheat, rice, cotton and maize. Agric Water Manag 69(2):115–133

Irrigation System Design

Improving Performance of Surface Irrigation System by Designing Pipes for Water Conveyance and On-Farm Distribution

Hossam Al-Din M. Hiekal

1 Introduction

Egypt is an arid country with a high rate of population growth and escalating living standards. The natural and geographical conditions of Egypt are not auspicious in terms of freshwater resources availability [1]. Egypt faces significant challenges due to its limited water resources by enforcement policies to improve the performance of existing irrigation systems and its development. However, agriculture is the major consumer of water. Following are the measures applied to agriculture among a complete package of water saving techniques, and one of these techniques is the use of modern irrigation systems in newly reclaimed land [2]. The improvement of irrigation systems is one of the most essential attempts in Egypt to implement more efficient irrigation technologies. This chapter presents an overview of the hydraulics of surface irrigation system, installing new or improved systems, engineering indicators of performance assessment, on-farm water distribution by the applied irrigation system, and example of practices problems by case studies at different sites. Over the long term, irrigation must be adequate but not excessive to prevent harmful accumulation of salt in the root zone and to prevent a high water table that may contribute to salt accumulation at the soil surface [3].

Infiltrated depths of water must be relatively uniform to meet the crop's need and leach salt adequately, without excessive surface runoff or deep percolation. To meet such depth and uniformity requirements, irrigation systems must be suited to the site conditions, well-designed, and well-managed [4].

The chapter provides an in-depth comparison of design restrictions, characterization, and approaches to each situation. As such, after reading this chapter, an interested reader should be able to identify both successful and problematic approaches used to cope with various aspects of the surface irrigation. Such an outcome should

H. A.-D. M. Hiekal (✉)
Irrigation and Drainage Research Unit, Desert Research Center, Cairo 11753, Egypt
e-mail: hmhekal@drc.gov.eg

E.-S. E. Omran and A. M. Negm (eds.), *Technological and Modern Irrigation Environment in Egypt*, Springer Water,
https://doi.org/10.1007/978-3-030-30375-4_11

prove useful to researchers, practitioners, water managers, and policymakers who are looking to improve their baseline understanding of surface irrigation development from different disciplines and levels of management.

By evaluating the level of the existing irrigation system, it possible to understand the farmer's practices in their traditional farms and enhancement it by improving the common irrigation system.

Thorough understanding of the effects of surface irrigation design quality is needed. This chapter summarizes the previous work on the effects of irrigation water distribution on water consumed, soil, and crop production.

2 Primary Theories of Water Flow by Pipes

By gravity that the water stored in the tank goes down by its own weight inside the pipes and run out. The water pressure is the force which water exerts in the walls of the container it is contained (pipe's walls, reservoir's wall).

The pressure in a considered point corresponds (or its equivalent) to the weight of the water column above this point. Knowing that the density of water is 1 g cm^{-3}, we can easily calculate the water column weight above a given point:

$$\text{Water column weight} = \text{water density} \times \text{water column height} = 1\ \text{g cm}^{-3}$$

2.1 Water Pressure—Static and Dynamic Head

- Static water pressure:

The pressure exerted by static water depends only upon the depth of the water, the density of the water, and the acceleration of gravity. The pressure in static water arises from the weight of the water and is given by the expression:

$$P_{\text{static water}} = \rho g h \tag{1}$$

where

$\rho = m/V$ water density
g acceleration of gravity
h depth of water.

The pressure from the weight of a column of liquid of area A and height h is shown in Fig. 1.

Because of the ease of visualizing a column height of a known liquid, it has become common practice to state all kinds of pressures in column height units, like

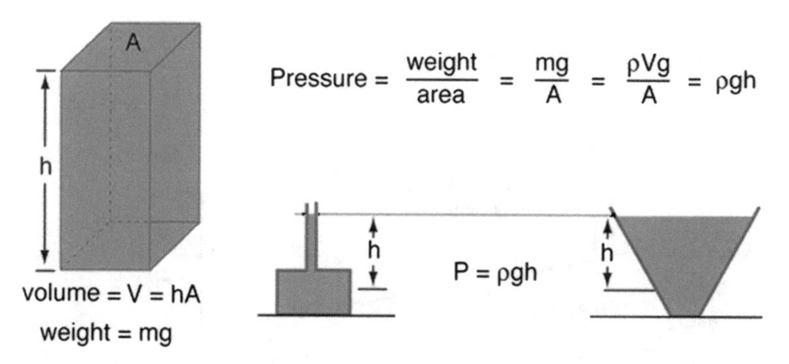

Fig. 1 Schematic form of static pressure concept

mmHg or cm H_2O, etc. Pressures are often measured by manometers in terms of a liquid column height and do not depend on the shape, total mass, or surface area of the liquid.

Pressures in column height units are mmHg or cm H_2O and often measured by manometers in terms of a liquid column height and do not depend on the shape, total mass, or surface area of the liquid.

- **Hydraulic and energy grade line for pipe flow**

Hydraulic calculations are required to design irrigation pipes. A hydraulic grade line analysis is required for all designs to ensure that water flows through the pipes in the manner intended.

The total energy of flow in a pipe section (with respect to a reference datum) is the sum of the elevation of the pipe center (elevation head). The pressure exerted by the water in the pipe expressed or shown by the velocity head and the height of a column of water (pressure head, or piezometric head, if a piezometer is provided in the pipe).

The total energy of flowing water, when represented in figure, is termed as energy grade line or energy gradient. The pressure of water in the pipe represented by elevation when drawn in line is termed as hydraulic grade line or hydraulic gradient as shown in Fig. 2.

- **Types of flow in pipe—Reynolds number**

The flow of water in pipes is of two types: laminar and turbulent. In laminar flow, the fluid moves in layers called luminous. In turbulent flow, secondary random motions are superimposed on the principal flow, and mixing occurs between adjacent sectors. In 1883, Reynolds introduced a dimensionless parameter (which has since been known as Reynolds number) that gives a quantitative indication of the laminar to turbulent transition. Reynolds number R_N according to [5] is

$$R_N = \frac{\rho V d}{\mu} \tag{2}$$

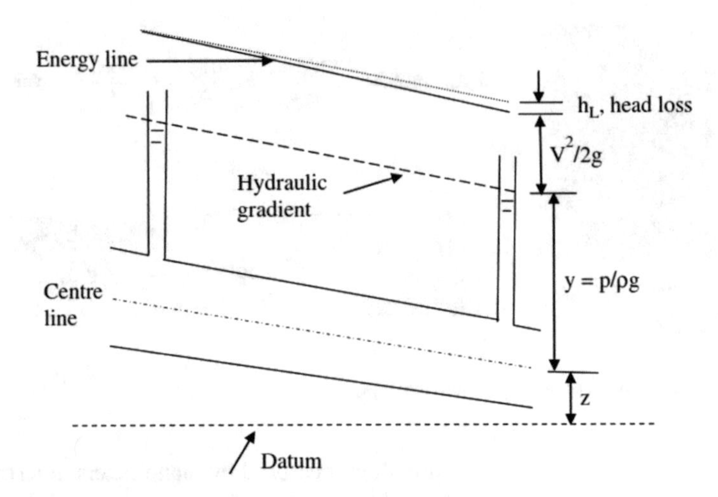

Fig. 2 Schematic of hydraulic and energy line in pipe flow

where

ρ density of fluid (kg m^{-3})
V mean fluid velocity (m s^{-1})
d diameter of the pipe (m)
μ coefficient of viscosity of the fluid (kg m^{-1} s^{-1}).

Generally, a flow is laminar if $R_N \leq 2100$. A transition between laminar and turbulent flow occurs for R_N between 2100 and 4000 (transition flow). Above 4000, the flow is turbulent. At turbulence range, the flow becomes unstable, and there is increased mixing that result in viscous losses which are generally much higher than those of laminar flow.

The Reynolds number can be considered in another way, as

$$R_N = \frac{\text{Inertia forces}}{\text{Viscous forces}} \quad (3)$$

The inertia forces represent the fluid's natural resistance to acceleration. The viscous forces arise because of the internal friction of the fluid. In a low Reynolds number flow, the inertia forces are small and negligible compared to the viscous forces, whereas in a high Reynolds number flow, the viscous forces are small compared to the inertia forces.

- **Velocity Profile of Pipe Flow**

Typical velocity profile of a pipe flow is shown in Fig. 3. The velocity is zero at the surface, increases after that, and reaches its maximum at the center of the pipe.

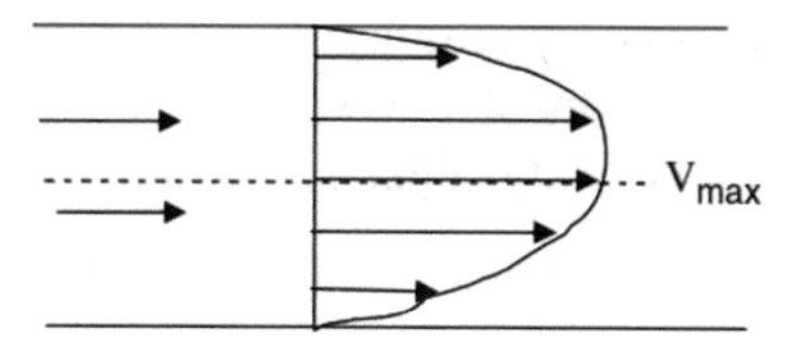

Fig. 3 Diagram showing velocity distribution in pipe flow

2.2 *Calculation of Head Losses in Pipe Flow*

- **Causes and components of head loss**

When fluid flows through the pipe, the internal roughness of the pipe wall can create local eddy currents vicinity to the surface, thus adding a resistance to the flow of the pipe. Pipes with smooth walls have only a small resistance to flow (frictional resistance). Smooth glass, copper, and polyethylene have small frictional resistance, whereas cast iron, concrete, steel pipe, etc., create larger eddy currents and effect on frictional resistance.

- **Importance of designing pipe sizes for irrigation water flow**

Determination of head loss (meaning loss of energy) in the pipe is necessary because the pump and motor (power) combination should be matched to flow and pressure requirement [6]. Oversizing makes for inefficiencies that waste energy and cost money.

- **Components of head loss**

Head loss in the pipe may be divided into the following:

- Major head loss
- Minor head loss.

Major head loss consists of loss due to friction in the pipe. Minor loss consists of loss due to change in diameter, change of velocity in bends, joints, valves, and similar items.

- **Factors affecting head loss**

Frictional head loss (h_{L}) in the pipe can be functionally expressed as follows:

$$h_{\mathrm{L}} = f(L, V, D, n, \rho, v) \tag{4}$$

where

L length of pipe
V velocity of flow
D pipe diameter

n roughness of the pipe surface (internal surface, over which flow occurs)
ρ density of flowing fluid
ν viscosity of the flowing fluid.

The mode of action of the factors affecting head loss is as follows:

- head loss varies directly as the length of the pipe
- it varies almost as the square of the velocity
- it varies almost inversely as the diameter
- it depends on the surface roughness of the pipe wall
- it is independent of pressure.

- **Significant head loss equations** according to [5]:

- *Darcy–Weisbach formula for head loss*:

The Darcy–Weisbach formula for head loss in a pipe due to friction in turbulent flow can be expressed as

$$h_f = f\frac{LV^2}{D2g} \tag{5}$$

where

h_f head loss due to friction (m)
f friction factor (or Darcy's friction coefficient)
L length of pipe (m)
V velocity of flow (m s^{-1})
g acceleration due to gravity (m s^{-2}) = 9.81 m s^{-2}
D inner diameter of the pipe (m).

Darcy introduced the concept of relative roughness, where the ratio of the internal roughness of a pipe to the internal diameter of the pipe affects friction factor for **turbulent flow**.

- *Head loss under laminar flow—Hagen–Poiseuille equation*:

$$h_f = \frac{32\mu VL}{wD^2} \tag{6}$$

where

V velocity of flow (m s^{-1})
L length of pipe (m)
D inner diameter of the pipe (m)
w specific wt. of the fluid (kg m^{-3})
μ viscosity of the flowing fluid (kg s^{-1} m^{-2}).

Table 1 Minor loss coefficient for different fittings, after [4]

Fittings	Minor loss coefficient (c)
Fully open ball valve	0.05
Threaded union	0.08
Fully open gate valve	0.15
½ closed gate valve	2.1
Fully open angle valve	2
Threaded long radius 90° elbows	0.2
Flanged 180° return bends	0.2
Flanged tees, line flow	0.2
Threaded tees, line flow	0.9
Threaded tees, branch flow	2.0
Fully opened globe valve	10

- **Calculation of minor loss**

Minor loss can be expressed as

$$h_{\text{minor}} = c\frac{V^2}{2g} \tag{7}$$

where c is the minor loss coefficient. Thus, the total minor loss can be calculated by summing the minor loss coefficients and multiplying the sum with the dynamic pressure head. Minor loss coefficients of different components/fittings are given in Table 1.

- **How to minimizing head loss in pipes?**

One of the main aims of pipe design is to minimize the head losses associated with pipe length (frictional loss), bends, diameter change, and transitions. Minimization of head loss will keep the diameter of the pipeline to the minimum (necessary to achieve the design flow capacity), and therefore, its cost will be reduced. Head losses in the pipe can be minimized by:

– Using large diameter pipe in the mainline
– Minimizing bends or turns
– Making/selecting internal surface of the pipe smoother.

2.3 *Designing Pipe Sizes for Irrigation Water Flow*

Selection of pipe size should be based on the following:

– hydraulic capacity (discharge) requirement

- head loss, and
- economy.

In the short run, a small diameter pipe may require lower initial cost, but due to excessive head loss, it may require a higher cost in the long run. Pipe size based on hydraulic capacity can be found as

$$A = \frac{Q}{V} \tag{8}$$

where

Q required discharge ($m^3\ s^{-1}$)
A cross-sectional area of the pipe (m^2)
V permissible velocity of flow ($m\ s^{-1}$).

The diameter of the pipe can be found from the relation,

$$A = \pi D^2/4 \tag{9}$$

$$D = \sqrt{\frac{4A}{\pi}} \tag{10}$$

where π represents a constant, approximately equal to 3.14159.

The pipe must have the capacity to supply peak demand (Q). After calculating the maximum size required, the second step is to calculate the head loss per unit length (say 100 m) and for the whole irrigation farm. Extra power and cost necessary for the head loss should be calculated for the entire useful life of the pipe.

3 General Considerations for Designing Surface Irrigation System

Irrigated agriculture faces a number of difficult problems in the future. One of the major concerns is the generally poor efficiency with which water resources have been used for irrigation. A relatively safe estimate is that 40% or more of the water diverted for irrigation is wasted at the farm level through either deep percolation or surface runoff. Agricultural irrigation future has many challenges, such as global warming, the low efficiency with which water resources have been used for irrigation, and 40% or more of the water diverted for irrigation is wasted through either deep percolation or surface runoff. These losses may not be lost when one views water use in the regional context since return flows become part of the valuable resource elsewhere. However, these losses often represent certain opportunities for water because they delay the arrival of water at downstream diversions and because they almost universally produce poorer quality water. One of the more evident problems in the future is the growth of alternative demands for water such as urban and industrial needs. These use to place a higher value on water resources and therefore tend to focus attention on wasteful practices. Irrigation science in the future will undoubtedly face the problem of maximizing efficiency. These losses often represent certain foregone opportunities for water because they delay the arrival of water at downstream diversions, and this produce low-quality water. The big problem in the future is the growth of alternative demands for water such as urban and industrial needs. In the future, the agriculture irrigation will undoubtedly face the problem of maximizing efficiency.

Irrigation in arid areas of the world provides two essential agricultural requirements:

- a moisture supply for plant growth which also transports essential nutrients and
- a flow of water to leach or dilute salt in the soil. Irrigation also benefits croplands through cooling the soil and the atmosphere to create a more favorable environment for plant growth [7].

- **Many decisions must be made before installing an irrigation system**

Some determinations are technical in nature, some economic, and others involve a close scrutiny of the operation and crop to be irrigated.

- Location, quantity, and quality of water should be determined before any type of irrigation system is selected. No assumptions should be made about the water supply. The challenge is technical, economical, and others like the operation and crop. And also location, quantity, and quality of water should be determined before selecting any type of irrigation system.
- Make sure that the water source is significant enough to meet the irrigation system's demand by test pumping groundwater sources or measuring flow rate of streams.
- Determination of the water advance or infiltration advance is an analysis problem, whereas computation of the inflow rate or system layout (e.g., length, width, and slope) is a design problem. The analysis of flow in surface irrigation is complex due to the interactions of several variables such as infiltration characteristics, inflow rate, and hydraulic resistance.

The design is more complex due to interactions of input variables and the target output parameters such as irrigation efficiency, uniformity, runoff, and deep percolation. In most cases, the aim of the surface irrigation system design is to determine the appropriate inflow rates and cutoff times so that maximum or desired performance is obtained for a given field condition.

- **The surface irrigation method** (border, basin, and furrow) should be able to apply an equal depth of water all over the field without causing any erosion as shown in Fig. 4.

To minimize the water percolation losses, the opportunity time (the difference between advance and recession periods—will discuss later) should be uniform throughout the plot and equal to the time required to put the required depth of water into the soil. Runoff from the field can be eliminated by controlling the inflow rate

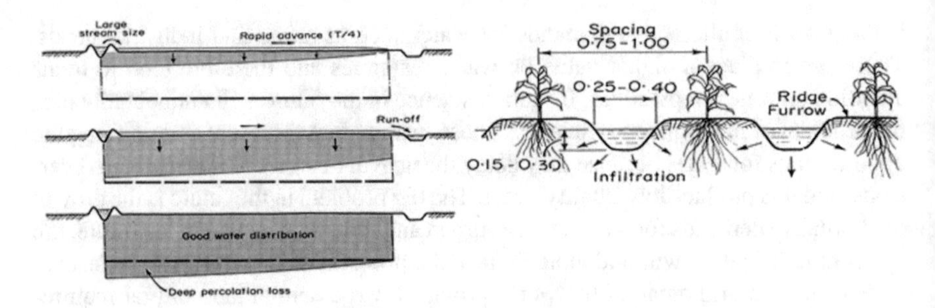

Fig. 4 Furrow irrigation: infiltration approaches

at which inflow decays with a time exactly correspond with the decay of the average infiltration rate with time for the entire length of the field. Inflow is usually cut back in discrete steps.

3.1 Variables in Surface Irrigation System

Important variables in surface irrigation system include the following: (i) infiltration rate, (ii) surface roughness, (iii) size of stream, (iv) slope of land surface, (v) erosion hazard, (vi) rate of advance, (vii) length of run, (viii) depth of flow, (ix) depth of water to be applied, (x) infiltration depth. These are schematically presented in Fig. 5a–c.

3.2 Hydraulics of Surface Irrigation System

The surface irrigation system and some of its features may be divided into the following four component systems: (1) water supply, (2) water conveyance or delivery, (3) water use, and (4) drainage. For the complete system to work well, each must work conjunctively toward the common goal of promoting maximum on-farm production. Historically, the elements of an irrigation system have not functioned well as a system, and the result has too often been very low project irrigation efficiency.

There are the following three phases of waterfront in a surface irrigation system:

- advance phase
- wetting phase (or ponding), and
- recession phase.

The advance phase starts when water first enters the field plot and continues up to the time when it has advanced to the end of the plot as shown in Fig. 6. The period between the time of advance completion and the time when the inflow is cut off or shut off is referred to as wetting or ponding or storage phase. After termination of

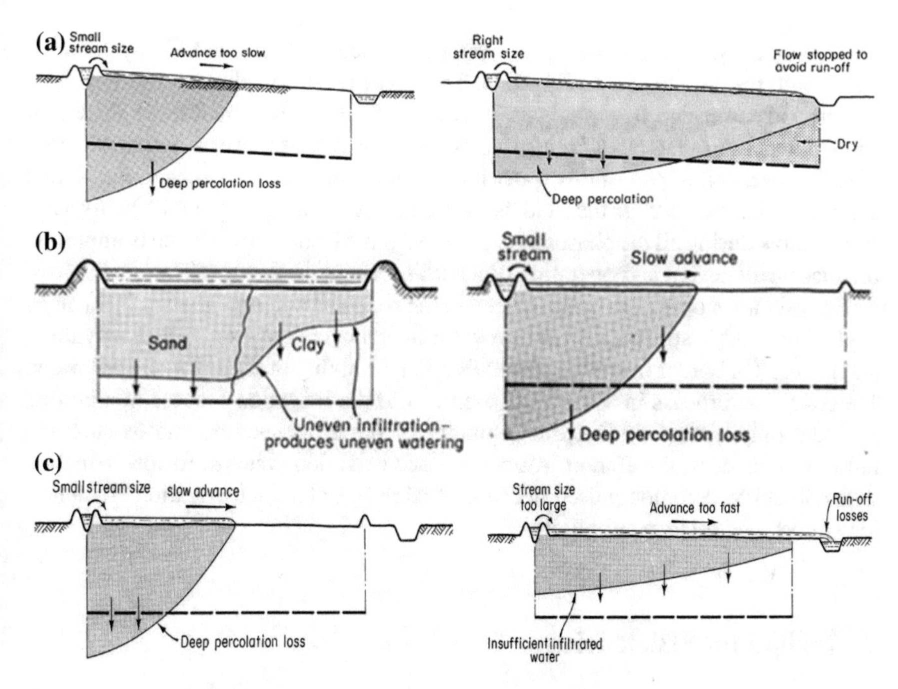

Fig. 5 **a** Furrow irrigation: problems of moisture distribution. **b** Basin irrigation: problems of moisture distribution. **c** Border irrigation: problems of moisture distribution

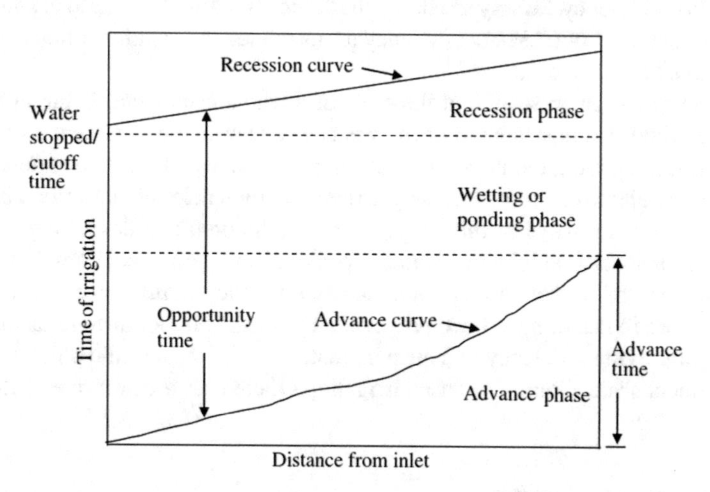

Fig. 6 Schematic presentations of phases of waterfront in surface irrigation system

the inflow, the ponding water or the waterfront recedes from the field by draining and/or into the next field by infiltration. This is the recession phase.

Unsteady overland flow analysis is required for the design and management of surface irrigation systems. When sufficient water is released over a porous medium in surface irrigation, part of this water infiltrates into the soil as shown in Fig. 4, and the remainder moves over the field as overland flow (runoff). Hydraulic analysis of surface flow during all the phases of irrigation from advance to recession is important for successful design and operation of a surface irrigation system [8].

Furrows are sloping channels formed in the soil. The amount of water that can be applied in a single application via furrow (or in other conventional surface irrigation, that is, flood or border irrigation) depends upon the ability of the soil to absorb water. The irrigation process in a furrow is identical to the irrigation process in a border, with the only difference that the geometry of the cross section, and as such the infiltration process, is different. Among surface irrigation systems, furrow irrigation with cutback is commonly used because of its potential higher irrigation efficiency, lower cost, and relative simplicity.

4 Irrigation Efficiencies

There are many ways of thinking about, determining, and describing concepts relating to irrigation efficiency. Simply speaking, the "efficiency" implies a ratio of something "in" to something "out." Many efficiency terms related to irrigation efficiency are in use or have been proposed.

Efficiency can be measured at the scale of a whole catchment, at the individual plant scale, and at almost any level in between. The scale of measurement is of critical importance in tackling the issue of improving efficiency and must be matched with the specific objective. Commonly used irrigation efficiencies are described below.

There are many ways of thinking about, determining, and describing concepts relating to irrigation efficiency. Simply speaking, the "efficiency" implies a ratio of something "in" to something "out." Many efficiency terms related to irrigation efficiency are in use or have been proposed. The critical importance in tackling the issue of improving efficiency and must be matched with the specific objective is the measurement scale. Commonly used irrigation efficiencies are described below.

4.1 Application Efficiency

Water application efficiency (Ea) expresses the percentage of irrigation water contributing to root zone requirement. It indicates how well the irrigation system can deliver and apply water to the crop root zone. Hence, the application efficiency takes into account losses such as runoff, evaporation, spray drift, deep drainage, and application of water outside the target crop areas. Of these factors, deep drainage

Table 2 Attainable application efficiencies under different surface irrigation systems

Type of surface irrigation system	Attainable efficiency range (%)
Border	75–85
Basin	80–90
Furrow	65–80

and runoff are probably the major causes of inefficiency and are generally due to overwatering.

Application efficiency defined by different researchers and varies slightly in the expression [8–14]. In broad term, application efficiency is the percentage of water delivered to the field that is ready for crop use; this parameter relates the total volume of water applied by the irrigation system to the volume of water that has been added to the root zone and is available for use by the crop. Thus, the application efficiency (Ea) is calculated according to [15] as

$$\mathrm{Ea} = (W_\mathrm{s}/W_\mathrm{f}) * 100 \quad (11)$$

where

Ea water application efficiency, %,
W_s irrigation water available to the crop (amount of water stored in the root zone), m^3, and
W_f water delivered to the field (amount of water added), m^3.

In simple words, irrigation water available to the crop = root zone soil moisture after irrigation − root zone soil moisture before irrigation.

Water delivered to the field = flow meter reading.

Application efficiency is primarily affected by the management of irrigation and may vary significantly between irrigation events.

Table 2 presents attainable application efficiencies under different irrigation systems (adapted from [13]).

Where the target water depth is the intended application amount, typically the soil moisture deficit amount. Thus, Potential Application Efficiency (PAE) is defined as the application efficiency when the target water depth is just satisfied (i.e., the target depth is equal to the minimum depth in the water distribution). Thus, PAE is the ratio of the target depth to the depth applied. PAE is typically used for design to assure full irrigation everywhere [10]. One of the limitations of surface irrigation is the difficulty in uniformly applying small depths of water. Surface systems are typically designed for application depths of 100 mm or more, while pressurized systems can typically be designed to apply as10 mm. When designer attempts to apply smaller depths with surface irrigation systems, the distribution of infiltrated water typically deteriorates. Therefore, one important aspect of the design is to be able to apply small depths of water uniformly.

4.2 Storage Efficiency/Water Requirement Efficiency

Storage efficiency indicates how well the irrigation satisfies the requirement to completely fill the target root zone soil moisture. Thus, storage efficiency (ES) is represented as

$$\mathrm{ES} = (\text{change in root zone soil moisture}) \times 100/(\text{target change in root zone soil moisture}) \tag{12}$$

where the change in the root zone soil moisture content is not measured directly, the storage efficiency can be approximated by relating the average depth of water applied over the field to the target root zone deficit. The root zone deficit is calculated using soil type, crop root zone, and soil moisture content data. In this case, the storage efficiency is calculated as

$$\mathrm{ES} = (\text{average depth applied}) \times 100/(\text{root zone deficit}) \tag{13}$$

The maximum storage efficiency is 100%. Calculations with a result above 100% indicate losses due to runoff or deep drainage [10].

4.3 Irrigation Distribution Uniformity

Irrigation uniformity is a measure of how uniform the application of water is to the surface of the field. That is, an expression describes the evenness of water applied to a crop over a specified area, usually a field, a block, or an irrigation district. The value of this parameter decreases as the variation increases.

Distribution uniformity is primarily influenced by the system design criteria. Poor uniformity of application is often easily identified by differences in crop response and/or evidence of surface waterlogging or dryness. The part of the field receiving more than the average depth may suffer from inefficiencies due to waterlogging and/or runoff, while the other part receiving less than the average may suffer from undue water stress. Thus, uniform irrigation is important to ensure maximum production and minimum cost.

An important component of the evaluation of in-field irrigation performance is the assessment of irrigation uniformity. If the volume of water applied to a field is known, then the average applied in depth over the whole field can be calculated. In most cases, one half of the field receives less than the average depth and one half more than the average depth applied. Hence, if the average volume applied is the target application required to meet the crop requirements, one half of the field has been over-irrigated (reducing the efficiency of application), while the other half of the field has been under-irrigated (potentially reducing yield). Thus, a major aim of irrigation management is to apply water with a high degree of uniformity while keeping wastage to a minimum.

4.4 *Uniformity Coefficient*

Uniformity coefficient, introduced by Christiansen [16], it is defined as the ratio of the difference between the average infiltrated amount and the average deviation from the infiltrated amount, to the average infiltrated amount. That is

$$\mathrm{UCC} = \left[\frac{\sum_{i=1}^{i=N} Z_i - Z_{\mathrm{av}}}{Z_{\mathrm{av}} * N}\right] * 100 \tag{14}$$

where

UCC Christiansen uniformity coefficient (or simply uniformity coefficient)
Z_i infiltrated amount at point i
Z_{av} average infiltrated amount
N number of points used in the computation of UCC.

Christiansen developed uniformity coefficient to measure the uniformity of sprinkler systems, and it is most often applied in sprinkler irrigation situation. It is seldom used in other types of irrigation. Values of UCC typically range from 0.6 to 0.9.

4.5 *Low-Quarter Distribution Uniformity*

Low-quarter distribution uniformity ($\mathrm{DU_{lq}}$) is defined as the percentage of the average low-quarter infiltrated depth to the average infiltrated depth as [10]:

$$\mathrm{DU_{lq}} = 100 * \frac{\mathrm{LQ}}{M} \tag{15}$$

where

$\mathrm{DU_{lq}}$ distribution uniformity at low quarter (or simply distribution uniformity, DU)
LQ average low-quarter depth infiltrated (mm)
M average depth infiltrated (mm).

The "average low-quarter depth infiltrated" is the average of the lowest one-quarter of the measured values where each value represents an equal area.

For calculation of DU of low one half, substitute "low quarter" by "average low half depth received or infiltrated." has been applied to all types of irrigation systems.

In trickle irrigation, it is also known as "emission uniformity." In sprinkler situation, it is termed "pattern efficiency."

The DUlq, the relationship between DU and UCC, can be approximated as follows [17]

$$\text{UCC} = 100 - 0.63(100 - \text{DU}) \tag{16}$$

$$\text{DU} = 100 - 1.59(100 - \text{UCC}) \tag{17}$$

5 Performance Evaluation

Describe how to determine the performance of basin/furrow irrigation. It is assumed that the net irrigation water need of the crop is known (i.e., the net irrigation depth). This is compared with what happens during the actual irrigation practice. Field application efficiency thus obtained is a good measure for the evaluation of the performance.

5.1 Concept, Objective, and Purpose of Performance Evaluation

- **Concept**

Performance terms measure how close an irrigation event is an ideal one. An ideal or reference irrigation is one that can apply the right amount of water over the entire area of interest without loss. Evaluation is a process of establishing a worth of something. The "worth" means the value, merit, or excellence of the thing.

Performance evaluation is the systematic analysis of an irrigation system and/or management based on measurements taken under field conditions and practices are normally used and comparing the same with an ideal one. Traditionally, irrigation audits are conducted to evaluate the performance of existing irrigation systems. A full irrigation audit involves an assessment of the water source characteristics, pumping, distribution system, storage, and in-field application systems. However, audits are also conducted on several components of the on-farm irrigation system [18].

- **Objectives**

The modernization of an irrigated area must start with a diagnosis of its current situation. Following this procedure, the specific problem affecting water use can be addressed and that may lead to a feasible solution. The specific objectives of performance evaluation are as follows:

- To identify the causes of irrigation inefficiencies
- To identify the problem/weak point of irrigation management
- To diagnose the water management standard of the irrigation project
- To determine the main principles leading to an improvement in irrigation performance.

- **Purpose of performance evaluation**

The purpose of performance assessment is to measure, through consistently applied standards, various factors that indicate either by comparison across systems whether a system is performing "well" or "badly" in a relative sense or by a system-specific analysis to see how the system is operating in relation to its own objectives [19]. The specific purposes are as follows:

- to improve irrigation performance
- to improve management process
- to improve the sustainability of irrigated agriculture.

- **Benefits of evaluation**

Evaluation leads to the following benefits:

- Improved quality of activities
- Improved ability of the managers to manage the system
- Savings of water and energy
- Ensure maximum production/benefit and minimum cost.

- **Factors affecting irrigation performance**

The performance of an irrigation system at field scale depends on several design variables, management variables, and system variables *or* factors. These factors characterize an irrigation event. Mathematically, it can be expressed as

$$P_{\text{ir}} = f(q_{\text{in}}, A, L, W, N, S_0, I_n, t_{\text{cutoff}}, S_{\text{w}}, D_{\text{ru}}, P, R_{\text{d}}, \text{ET}, \ldots) \quad (18)$$

where

P_{ir} performance of an irrigation event
f function
q_{in} inflow rate or application rate (to the furrow or per unit width of border or basin)
A sectional form of the unit plot to be irrigated (especially for furrow)
L length of run of the flow
W width of the section *or* unit plot
N roughness coefficient of flow for the plot (Manning's N)
S_0 longitudinal slope of the plot
I_{n} infiltration characteristics of the soil
t_{cutoff} time cutoff

S_w soil water status at the time of irrigation (i.e., condition of deficit)
D_{ru} reuse of drainage runoff (if applicable)
P pressure of the flow system (especially for gated/perforated distribution pipes)
R_d root zone depth of the crop during the irrigation event
ET atmospheric water demand or evapotranspiration demand.

5.2 *Performance Indicators*

Activities of irrigation systems start at the point of water supply headwork or pump. Impacts of irrigation are not limited to the field but also extend to the socioeconomic conditions of the target audience. In general, a set of indices or indicators are used for evaluating the performance of an irrigation scheme. Indicators are termed as performance indicators. No single indicator is satisfactory for all descriptive purpose. Moreover, there are uncertainties about the exact values of some indicators. Several indicators can give an overall picture of the irrigation project. Typically, high engineering efficiency implies a reduction of losses. Beneficial uses include crop water use, salt leaching, frost protection, crop cooling, and pesticide and fertilizer applications [20]. For convenience in understanding and application, the indicators can be grouped as

- Engineering
- on-farm water use indicators
- Crop and water productivity
- Socioeconomic.

5.3 *Engineering Indicators*

Engineering indicators are those which are related to pump, water headwork, water supply, water conveyance system, and energy use [4, 21]. Indices under this category include the following:

– Pumping plant efficiency	– Headwork's efficiency
– Water conveyance efficiency	– Water delivery performance
– Irrigation system efficiency (or overall efficiency)	– Equity of water delivery
– Channel density	– Water supply—requirement ratio
– Water availability and shortage	– Energy use efficiency

5.4 On-Farm Water Use Indicators

These indicators concern the efficiency of on-farm water application and the uniformity of water distribution along the irrigated field. Indicators under this category are as follows:

– On-farm water loss	– Deep percolation fraction/deep percolation ratio
– Runoff fraction/tail-water ratio	– Water application efficiency
– Storage efficiency/water requirement efficiency	– Application efficiency of low quarter
– Distribution efficiency or uniformity	– Low-quarter distribution uniformity

5.5 Crop and Water Productivity

Indicators under this category are as follows:

– Area irrigated	– Irrigation intensity
– The duty of discharge/supply water	– Crop productivity (yield rate)
– Water productivity	– Irrigation water productivity

5.6 Socioeconomic Indicators

In some cases, cost–benefit or social uplift and social acceptance aspects are measured. These are called socioeconomic indicators. Indicators under this category include the following:

– Irrigation benefit–cost ratio	– Cost per unit production
– Irrigation cost per unit area	– Farmers income ratio

6 Ideal Situation for Estimation of Irrigation System

The field (soil) and crop condition should represent the ideal/normal field condition during the evaluation of an irrigation system. The conditions can be summarized as follows:

(a) The field soil should be stable, not new, refilled, or a developed one.
(b) The crop condition should be representative, not just after emergence or at ripening stage but in between (good coverage).
(c) The soil should be dry enough—appropriate time for irrigation.
(d) Water supply/water pressure should be sufficient enough to apply inflow the designed rate.

7 Performance Assessment of Surface Irrigation System

– **Pumping plant evaluation**

Pumping system efficiency can contribute substantially to energy saving. Pumping plant evaluation requires a pump test, which checks the flow rate capacity, lift, discharge pressure and/or velocity, rated discharge capacity, and input horsepower.

Pump discharge can be measured by flow meter (in the vicinity of the pump outlet), flume, or by the coordinate method [22]. The rated discharge capacity of the pump can be read from the manufacturer's manual *or* the pump rating written on the pump body. If a mechanical engine is used to power the pump, its capacity can be read from its rating seal or manual. If an electrical motor is used to operate the pump, power consumption by the motor can be measured by "Clip-On meter" or "Multi-meter" or from the change in power reading in the "electric meter" for a certain period. Rated capacity of the motor can be read from its body.

Knowing the above information, overall pumping plant efficiency and efficiency of each component (such as motor or engine efficiency, pump efficiency) can be calculated.

7.1 Border Irrigation Evaluation

Field observations and measurements required for conducting a border irrigation system evaluation include the following:

– Border dimension	– Advance phases and time
– Slope of the border	– Recession time
– Inflow rate	– Topography of the field
– Runoff rate and volume (if any)	– Crop type and stage of the crop
– Irrigation time (duration)	

The measurement steps and procedures are as follows:

- The border dimensions can be measured using a "measuring tape."
- Soil surface elevations (at different points, 10–30 m intervals along the borders) can be determined using a "total topographic station" or "level instrument." Slope and standard deviation of soil surface elevations can be determined from the measured data.
- The "inflow" or "irrigation discharge" can be measured using suitable flow measuring devices such as mini-propeller meter and flume.
- The advance phase can be determined from the recording of the advance time to reference points located along the border (every 10–30 m).
- A number of flow depth measurements are to be performed across the border, every 5–10 m. The average of all measurements is used to represent the flow depth at this point and time.
- The flow depth at the upstream end of the border is to be measured shortly before the cutoff.
- In the open border, surface runoff (if conditions permit) is to be monitored. The runoff can be measured using the mini-propeller meter or a flume.
- A hydrograph is to be established from discharge measurements, and it is time for integration will yield the runoff volume.
- Infiltration in "ring infiltrometer" and border infiltration can be correlated, and a relationship can be established. Then, the infiltration parameters can be determined.
- To estimate the infiltrated depths of water (required for computing uniformity and efficiency indices), field data from the evaluation can be utilized to derive the infiltration parameters of a Kostiakov type infiltration equation. The infiltration parameters (K, a) and roughness coefficient (N) can be determined through the solution of inverse surface irrigation problem [23].
- For that, a hydrodynamic one-dimensional surface irrigation model (e.g., SIR-MOD) can be used. Such a model is to be executed using tentative values of the coefficient "K" and the exponential "a" from the Kostiakov infiltration equation and the Manning's N. The parameters should be adjusted until the model satisfactorily reproduces the experimental values of flow depth and irrigation advance for each evaluation [24].
- Performance indices—application efficiency and the low-quarter distribution uniformity—should be determined using the formula described in an earlier.

7.2 Basin Irrigation Evaluation

Basins have no global slope, but the undulations of the soil surface can have an important effect on the advance and recession process of an irrigation event.

For evaluation of basin irrigation, measurements should be made during representative irrigation events. The required measurements are as follows:

- advance, water depths at selected locations
- surface drainage or recession commonly measured performance indices for basin irrigation are as follows:
- application efficiency
- distribution uniformity
- deep percolation ratios
- requirement efficiency or storage efficiency.

For basin irrigation, tail-water ratio is zero. [25] defined the distribution uniformity (DU) for basin irrigation.

7.3 Furrow Irrigation Evaluation

Generally, the evaluation of furrow irrigation system is restricted to a single or small number of adjacent furrows due to the intensive measurement process. Complete inflow, advance, and runoff measurements are used to accurately determine soil infiltration rate for a small number of furrows.

The working step and procedures for the evaluation of furrow irrigation system are as follows:

- Measure the length and spacing of furrow.
- Measure soil moisture (before irrigation).
- Install the equipment (e.g., flume, scale, moisture measuring equipment).
- Start irrigation.
- Record the flow rate (at 5–10 min intervals, until the constant flow rate is achieved).
- Record the advance data after 6, 12, and 24 h from the starting of irrigation.
- Record the water depth at different points (10, 20, 50 m) at several time intervals.
- Record the cutoff time.
- Record recession data (water depth) at several distances (10, 20, 50 m) from the starting point at several time intervals.
- Record the depth of ponding at lower ¼ the part of the furrow.
- Record the runoff volume (if the process permits).
- Measure soil moisture up to the desired depth (root zone) at different points throughout the furrow after reaching field capacity.
- Determine the wetted cross section of the furrow at several sections and average them.

Volume balance approach can be applied to find out different components of water balance (e.g., infiltration, deep percolation). Volume balance approach is based on the principle of mass conservation. At any time, the total volume of water that has entered the furrow must be equal to the sum of the surface storage, subsurface storage (infiltrated), deep percolation (if any), and runoff (if any).

8 Improving Performance of Surface Irrigation System

Improving the performance of an irrigation system is to take remedial measures for correcting the fault/deficiency, which has been identified during evaluation/diagnosis process. Besides, a number of techniques can be used in the design of a system to increase its uniformity and efficiency [26] Mentioned that for surface irrigation systems, the inflow rate could be matched with the soil intake rate, slope, and length of the run; the cutoff time can be matched thereby. Another technique is that water use is more efficient with afternoon irrigation as the evaporative loss is minimal. Some common problems/faults and suggestive measures for improving the performances are summarized according to [27]:

- Pumping plant efficiency is low:
  - Renovate the moving parts
  - In case of deep well, wash out the well screen
- Water conveyance efficiency is low:
  - Renovate/perform lining the conveyance channel
  - Reduce the field channel density
- Water delivery performance is not satisfactory:
  - Perform efficient/economic channel design
  - Recast/ensure delivery system
- Channel density is high:
  - Reduce the channel length by straightening through the command area
- On-farm water loss is high:
  - Compact the borders of each plot
  - Improve the water-holding capacity of the soil by adding organic manures
  - Reduce relative percentage of sand by adding silt or clay soil
- Water supply—requirement ratio is not good:
  - Recast the supply amount, or
  - Change the cropping pattern (if possible), altering high water-demanding crops, or

  - Search for new source of supply

- Deep percolation:
  - Line the channels soil
- Runoff fraction is high:
  - Maintain correct slope of land
  - Apply correct flow rate and time for flow (cutoff time)
  - Take care of the borders; construct high barriers
- Water application efficiency is low:
  - Minimize on-farm water loss
  - Estimate correct amount of water demand
  - Apply correct flow rate based on infiltration characteristics
  - Level the land with appropriate slope
  - Maintain correct slope of the water run considering infiltration rate and flow rate
  - Improve water-holding capacity of the soil
- Water storage efficiency is not satisfactory:
  - Correctly estimate the crop root zone depth before irrigation
  - Estimate correct amount of water demand
- Distribution uniformity is low (poor distribution of infiltrated water over the field):
  - Apply correct flow rate based on infiltration rate and
  - the slope of the run
  - Design the length of run based on infiltration rate, slope,
  - flow rate
  - Cut off the flow at proper time
- Low-quarter distribution uniformity is low:
  - Apply correct flow rate based on infiltration rate
  - Design the length of run based on infiltration rate, slope, and flow rate
  - Cut off the flow at proper time (after reaching the waterfront at tail end)
- Area irrigated per unit flow (Duty) is not satisfactory:
  - Reduce conveyance, seepage, and percolation loss
  - Schedule irrigation properly (apply correct amount of water based on need)
  - Improve water-holding capacity of the soil
- The intensity of irrigation is low:
  - Reduce all possible losses
  - Increase irrigation efficiency
  - Schedule crops and crop rotations

- Crop productivity is low:
  - Ensure proper irrigation
  - Ensure proper management of other inputs (like balance fertilizer)
  - Ensure other cultural management (proper population, weeding, pesticide, and insecticide application, if needed)
- Water productivity is below the normal range:
  - Schedule irrigation properly
  - Reduce tail-water runoff
  - Minimize on-farm water loss
  - Maximize utilization of stored soil moisture
  - Ensure other crop management aspects
- Irrigation water productivity is below the desired limit:
  - Schedule irrigation properly
  - Reduce tail-water runoff
  - Minimize on-farm water loss
  - Maximize utilization of stored soil moisture
  - Ensure other crop management aspects
- Irrigation benefit–cost ratio (B–C ratio) is low:
  - Minimize irrigation cost by proper scheduling and reducing all sorts of water loss
  - Maximize production by proper management of other inputs and selecting appropriate crop type and variety
  - Maximize utilization of stored soil water and rainwater, if available
- Cost per unit production is high:
  - Similar to that of B–C ratio
- Irrigation cost per unit area is high:
  - Similar to that of B–C ratio
- Farmer's income ratio is not satisfactory:
  - Similar to that of B–C ratio.

9 Case Studies from Egypt

9.1 Improving Irrigation Efficiency

Because crop irrigation is practiced in areas with dry climates under climate changes, much of the water use in those areas is for agriculture. Most of the irrigation systems are surface or gravity systems, which typically have efficiencies of 50–60%. This means that 40–50% of the water applied to the field is used for evapotranspiration by the crop, while 40–50% is "lost" from the conveyance system, by surface runoff from the lower end of the field, and by deep percolation of water that moves downward through the root zone as shown in Fig. 7.

Increased irrigation efficiencies allow farmers to irrigate fields with less water, which is an economical benefit. Also, increased irrigation efficiencies generally mean better water management practices, which, in turn, often give higher crop yields [28]. Thus, increasing field irrigation efficiencies also saves water by increasing the crop production, thus allowing more crops to be produced with less water [29].

Field irrigation efficiencies of gravity systems can be increased by better management of surface irrigation systems (changing rate and/or duration of water application), modifying surface irrigation systems (changing the length or slope of the field, including using zero slope or level basins). Surface irrigation systems often can be designed and managed to obtain irrigation efficiencies of 80–90%.

Thus, it is not always necessary to use a sprinkler or drip irrigation systems when high irrigation efficiencies are desired.

First applied of the method and evaluated at different sites in Egypt to enhance water application efficiency (Ea), storage efficiency (Es), and water distribution uniformity as shown in Figs. 8, 9, and 10. Through controlled PVC, spill pipes with 1 m length and 63 mm diameter installed in the ditch of irrigation canal against the upper ridge of the field, which convey the water according to the required flow rate (one spill pipe for each furrow). The temporary dam (barrier) was used (if needed) to keep a constant hydraulic head above the inlets of a group of spill pipes to realize inflow rate adequately for each spill pipe (equal inflow rate each furrow) during irrigation events.

Fig. 7 Irrigation with traditional borders system using an open excavation channel

Fig. 8 Preparation and installation of spill pipes

Fig. 9 After operation with long furrows irrigation system

The number of spill pipes (each group of furrows) determined to depend on the gross water discharge pass in irrigation channel by gravity.

The operating technique of the developed system starts with the water spill pipes being closed until the water height reaches in the channel above the level of the spill pipes at least 12 cm (through the temporary dam in the irrigation channel.) At that time, the farmer can remove the plastic caps to allow water to pass and start irrigation, by ensuring that outflow will be equally behind a number of spill pipes. Before opening the second group of spill pipes which it was closed previously, the farmer firstly must be waiting to re-rise the water level in irrigation channel as shown before and so on during the third group of spill pipes until finishing the irrigation event. By this method of operating technique the water will distribute in equal flow rates to irrigated furrows and/or borders with shorter advance times. [30] showed

Fig. 10 After installation and evaluated of spill pipes with excavated channel

that more water losses by deep percolation in the soil especially occurred in first and second watering events as shown in Figs. 11 and 12.

Second applied of the method and evaluated at different sites in Egypt to enhance irrigation performance as shown in Fig. 13 by the simple fabricated way.

When irrigation water directed from the pumping unit (the control head) to the farm by network pipelines and control valves. Thus, more PVC pipe diameters (110–160 mm) can use and fabricate as shown in Fig. 13 using outflow orifices each 72 cm for fixation one PVC Nibble each orifice with controlled by PVC cap easy take off during watering, this method of controlled perforated pipes (CPP) was used and evaluated instead of metal gated pipes.

Fig. 11 During irrigation with long borders system

Fig. 12 After operation with long border irrigation system

Fig. 13 Preparation and collection of controlled perforated pipes (CPP)

The group of controlled orifices number opened depending on the general flow directed to the irrigation site as shown in Figs. 14 and 15.

The results of long evaluation periods through different research studies at different sites in Egypt as [13, 31, 32] proved that the positive effects on different field crops yields cultivated under various conditions with improved management practices compared with traditional management were significant increases.

9.2 Some of the Obstacles and Constraints

(a) **Financial/economic**

- Certain **water conservation/demand management (WC/DM)** measures depend on financial outlay by end users, who may not have adequate resources
- Water is allocated to consumers irrespective of economic value or efficiency of use
- Water institutions own water supply infrastructures

Fig. 14 During irrigation with long borders system

Fig. 15 During irrigation with two long borders system (CPP) instead of excavation open channel

– Lack of funding or disproportionate funding for supply-side measures at the expense of WC/DM.

(b) **Technical/institutional**

- lack of adequate knowledge of the cause of growth in demand
- current planning practices choose the cheapest solution without regard to operating costs
- lack of understanding of the consumer and water usage patterns
- lack of cooperation among local authorities
- lack of cooperation among water services institutions
- officials and industry sectors protect their interests.

10 Conclusions

Egypt is highly vulnerable to climate change, which increases the water demand and causes a loss of crops. Thus, one of the main challenges facing the sequential government during the previous decades was to enhance the agriculture sector by increasing the efficiency of water use.

Irrigation efficiency is greatly dependent on the type and design of water conveyance and distribution systems. Designing of economic pipe diameter is important to minimize cost, water loss, and land requirement.

To achieve high performance in surface irrigation system, it must be designed to irrigate uniformly, with the ability to apply the right depth at the right time. Properly designed, installed, maintained, and managed irrigation system greatly reduce the volume of irrigation water and hence save energy and money. Besides, it improves the crop yield and quality.

Developed water distribution systems are more effective, efficient, and far better than the conventional system, and developed water distribution systems at different locations in the country are running in good condition without any major constraint.

This developed distribution system could be widely adopted as a model in the field for increasing agricultural production in Egypt. Among all the water distribution systems, the PVC pipe line system is the most suitable. Improved and efficient water management practices can help to maintain farm profitability in an era of increasingly limited and more costly water supplies.

11 Recommendations

Evaluation helps to identify the problems resulted from mismanagement and the measures required to correct them. No single indicator is satisfactory for all descriptive purposes.

In general, a set of indices are used for evaluating the performance of surface irrigation system. Most commonly used indices are described in this chapter. Of course, adequate monitoring and evaluation of performance are needed to improve water management practices in order to achieve an increase in overall efficiency.

References

1. El-Fellaly S, Saleh EM (2004) Egypt's experience with regard to water demand management in agriculture. In: Eight international water technology conference, Alexandria, Egypt, p 5–25
2. Abdin AE, Gaafar I (2009) Rational water use in Egypt. In: El Moujabber et al (eds) Technological perspectives for rational use of water resources in the Mediterranean Region, CIHEAM, Bari, pp 11–27 (Options Méditerranéennes: Série A. Séminaires Méditerranéens; n. 88)
3. Abdel Ghaffar E, Shaban M (2014) Investigating the challenges facing drainage water reuse strategy in Egypt using empirical modeling and sensitivity analysis. Irrig Drain 63:123–131
4. Ali MH (2001) Technical performance evaluation of Boyra deep tube-well—a case study. J Inst Eng Bangladesh 28(1):33–37
5. Chilton RA, Stains R (1998) Pressure loss equations for laminar and turbulent non-Newtonian pipe flow. J Hydraulic Eng 124:522–529
6. Burt CM, Clemmens AJ, Strelkoff TS, Solomon KH, Hardy L, Howell T, Eisenhauer D, Bleisner R (1997) Irrigation performance measures-efficiency and uniformity. J Irrig Drain Eng 123(6):423–442
7. Walker WR (1989) Guidelines for designing and evaluating surface irrigation systems. FAO Irr. and Drain. Paper No 45, 137 p, Rome, Italy
8. James LG (1988) Principles of farm irrigation system design. Wiley (Ed), New York, 543 pp
9. Bos MG, Nugteren J (1974) On irrigation efficiencies. International Institute for Land Reclamation and Improvement, Wageningen, Netherlands
10. Merriam JL, Shearer MN, Burt CM (1983) Evaluating irrigation systems and practices. In: Jensen ME (ed) Design and operation of farm irrigation systems. ASAE Monograph No 3, USA
11. USDA (1997) Irrigation systems evaluation procedures. National Eng. Handbook, Chapter 9, Part 652. Natural Resources Conservation Service USDA, Washington, DC
12. Bos MG, Murray-Rust DH, Murray DJ, Johnson HG, Snellen WB (1994) Methodologies for assessing performance of irrigation and drainage management. J Irrig Drain Syst 7:231–261
13. Sharkawy SFT, Khafaga HS, Hiekal HAM, Mousa AA (2017) Increasing salt tolerance of Egyptian clover by using integrated management system under marginal conditions in El-Tina Plain- North Sinai—Egypt. Egypt J Appl Sci 32(9):355–378
14. Heermann DF, Wallender WW, Bos GM (1990) Irrigation efficiency and uniformity. In: Hoffman GJ, Howell TA, Solomon KH (eds) Management of farm irrigation systems. ASAE, St. Joseph, Michigan, pp 125–149
15. Wigginton DW, Raine SR (2001) Measuring irrigation system performance in the Queensland Dairy Industry. National Centre for Eng. in Agric. Publication 179729/5, Toowoomba, Australia
16. Christiansen JE (1941) The uniformity of application of water by sprinkler systems. Agric Eng 22:89–92
17. Strelkoff T, Clemmens AJ (2007) Hydraulics of surface systems. In: Hoffman et al (eds) Design and operation of farm irrigation systems, ASABE, St. Joseph, Michigan, pp 436–498
18. Pereira LS (1999) Higher performance through combined improvements in irrigation methods and scheduling: a discussion. Agric Water Manage 40(1):153–169
19. Solomon KH (1988) Irrigation systems and water application efficiencies. Center for Irrigation Technology Research, CAIT Pub # 880104. California State University, California
20. Hoffman GJ, Evans RG, Jensen ME, Martin DL, Elliot RL (2007) Design and operation of farm irrigation systems, 2nd edn. American Society of Agric. and Biological

21. Sarma PBS, Rao VV (1997) Evaluation of an irrigation water management scheme—a case study. Agric Water Manage 32:181–195
22. Kruse EG (1978) Describing irrigation efficiency and uniformity. J Irrig Drain 104(IR 1):34–41
23. Katapodes ND, Tang JH, Clemmens AJ (1990) Estimation of surface irrigation parameters. J Irrig Drain Eng ASCE 116(5):676–696
24. Walker WR (2003) SIRMOD III surface irrigation simulation, evaluation and design, guide and technical documentation. Biological and Irrig Eng, Utah State University, Logan
25. Clemmens AJ, Dedrick AR (1982) Limits for practical level-basin design. J Irrig Drain Div ASCE 108(2):127–141
26. Hornbuckle JW, Christen EW, Faulkner RD (2003) Improving the efficiency and performance of furrow irrigation using modeling in South-Eastern Australia. In: Workshop on improved irrigation technologies and methods: R&D and testing, Proceedings of 54th executive council of ICID and 20th European regional conference, Montpellier, France, 14–19 Sept
27. Ali MH (2011) Practices of irrigation and on-farm water management, vol 2. Springer, New York, p 546
28. Playa E, Mateos L (2006) Modernization and optimization of irrigation systems to increase water productivity. Agric Water Manage 80:100–116
29. Abu-hashim M., Negm A (2018) Deficit irrigation management as strategy under conditions of water scarcity; potential application in North Sinai, Egypt. In: Negm A, Abu-hashim M (eds) Sustainability of agricultural environment in Egypt: Part I soil-water-food Nexus, Hdb Env Chem. Springer International Publishing AG 2018. https://doi.org/10.1007/698_2018_292
30. Clemmens AJ, El-Haddad Z, Strelkoff TS (1999) Assessing the potential for modern surface irrigation in Egypt. Trans ASAE 42(4):995–1008
31. Hiekal HAM, Khafaga HS, Sharkawy SFT, Ali AA, Al-Dakheel A (2016) Enhancing irrigation system and improving soil and crop management for forage sorghum production in marginal environment. Egypt. J Appl Sci 31(11):259–292
32. Khafaga HS, Heikal HAM, Abdel-Nabi AS, Sharkawy SFT, Al-Shaer HM (2017) Acclimatization fodder beet plants under two irrigation systems in saline soils at Sahl El Tina—North Sinai. Menoufia J Plant Prod 2(12):479–494

Micro-sprinkler Irrigation of Orchard

S. A. Abd El-Hafez, M. A. Mahmoud and A. Z. El-Bably

1 Introduction

Evaluation of the irrigation system and suggested improvements are proper steps and full understanding of micro-irrigation system, citrus and soil and water requirements. This chapter provides vital information for successfully growing citrus in Nubaria, as well as detailed advice for micro-irrigation systems on citrus.

A sprinkler irrigation system generally includes sprinklers, laterals, sub-mains, main pipelines, pumping plants and boosters, operational control equipment and other accessories required for efficient water application. In some cases, sprinkler systems may be pressurized by gravity and therefore pumping plants may not be required.

The planning and design of irrigation systems should aim at maximizing the returns and minimizing both the initial capital outlay and the costs per unit volume of water used, thus contributing both directly and indirectly to the overall reduction of the production costs and the increase of returns. In other words, planning and design are a process of optimizing resources [1].

The orchard sprinkler is a small spinner or impact sprinkler designed to cover the interspace between adjacent trees; there is little or no overlap between sprinklers. Orchard sprinklers are designed to be operated at pressures between 10 and 30 psi, and typically, the diameter of coverage is between 5 and 10 m. They are located under the tree canopies to provide approximately uniform volumes of water for each

S. A. Abd El-Hafez · M. A. Mahmoud · A. Z. El-Bably (✉)
Water Requirements and Field Irrigation Research Department, Agriculture Research Center (ARC), Soil, Water and Environment Research Institute (SWERI), 9th Cairo University Street, Giza, Egypt

E.-S. E. Omran and A. M. Negm (eds.), *Technological and Modern Irrigation Environment in Egypt*, Springer Water,
https://doi.org/10.1007/978-3-030-30375-4_12

tree [2]. Water should be applied fairly even to areas to be wetted even though some soil around each tree may receive little or no irrigation. The individual sprinklers can be supplied by hoses and periodically for each position [3, 4].

Improvements in irrigation water management and irrigation systems save water, nutrients, chemicals, time and money. For high yield and quality, management of irrigation system should establish to provide the required amount of water at the optimum timing. The main issue for the field crops growing under orchard trees is how to improve the distribution uniformity [5, 6].

For better irrigation system management, farmers should sophisticate their systems with available technologies for scheduling irrigation according to crops requirements, applications frequencies and controlling volumes [7, 8]. Installing soil moisture tensiometers in each section of different spacing will enable the farmer to monitor how much water should be added to the tree, and thus, the farmer can save time (operating hours) and labors when he identifies how much to irrigate [9, 10]. At the same time, high distributing water and controlling the irrigation system are very significant to the orchard. The automated irrigation and fertilization system will collaborate with the soil moisture sensors to provide an efficient means of irrigation scheduling [11].

The objective of this chapter is to recommend an irrigation system that would enable the farmers to improve the distribution uniformity in the orchard, to improve micro-sprinkler irrigation system and all the evaluation processes based on accurate data which provide reliable information to the farmers and to apply enough water to satisfy the needs of the soil, trees and evapotranspiration rates during the hottest time of the year.

2 Types and Irrigation System

Micro-spry irrigation system, also known as micro-jet or spray emitter irrigation system which is very similar to a drip irrigation system except that each emitter (sprayer) can wet larger areas compare to emitters of drip irrigation systems. Spray emitters are normally used on coarse-textured soils (sandy soil), where wetting sufficiently large areas would require a large number of drippers. This is because coarse-textured soils have inherently vertical wetting compared to do fine textures soil (clayey soil). They are also used for trees where one to two sprayers per tree can go over the wetted area requirements [12, 13].

The crop water requirements of a citrus tree and their irrigation leaching requirements were calculated.

2.1 Percentage Wetted Area

The desirable percentage wetted (P_w) for our example would be 50%. It is, therefore, necessary to identify a sprayer, which can cover 50% of the surface area commanded by one tree, which is 18 m^2 (36 × 0.5). Another consideration is the soil infiltration rate. In order to avoid a runoff, the application rate of the sprayer should not exceed the soil infiltration rate. The area directly wetted by the sprayer (A_w) can be calculated using the following equation [14].

$$A_w = \frac{\pi \times d^2}{4} \times \frac{(\omega)}{360} \tag{1}$$

where

d is the diameter of a circle approximated by the wetting pattern
ω is the approximate angle of the pie-wedged shape cut out of the circle.

The soil infiltration rate is 6 mm/h, a prayer from manufacturer catalog should be selected, and the wetted area is calculated using the above equation.

Referring to Table 1, obtained from a manufacturer, the performance characteristics of a sprayer are provided for different pressures and nozzle sizes. This table is for 360° water distribution.

Assuming that a blue nozzle is selected operating at 14 m head, the wetted diameter is 6.0 m. The wetted area is calculated using the above equation

$$A_w = \frac{3.14 \times 6^2}{4} \times \frac{(360)}{360} = 28.3\,\text{m}^2$$

Because this is far above the minimum required area of 18 m^2 ($P_w = 50\%$), another option could be considered. Looking at the wetted diameter of a sky blue nozzle operating at 14 m head, which is 5.0 m, the same calculation can be made [10, 15]:

$$A_w = \frac{3.14 \times 5^2}{4} \times \frac{(360)}{360} = 19.6\,\text{m}^2 \text{ which is close to } 18\,\text{m}^2.$$

The highest application rate occurring near the emitter is 3.5 mm/h, which is well within 6 mm/h soil infiltration rate. If the blue nozzle operating at 14 m head is chosen, the application rate near the emitter would be 4.5 mm/h, which is also below the soil infiltration rate.

Table 1 Performance sprayers, being 20 m above the ground level, for 360° water distribution

Type (model)	Color code	Pressure (m)	Flow rate (lph)	Wetted diameter (m)	Constant, k	Exponent, x
025	Sky blue	14	19	5.0	5.15	0.5
		20	23	5.5		
		25	26	6.0		
		30	28	6.0		
040	Blue	14	33	6.0	9	0.5
		20	40	7.0		
		25	45	7.0		
		30	49	7.0		
050	Green	14	39	6.5	10.5	0.5
		20	47	6.5		
		25	53	6.5		
		30	58	6.5		
060	Gray	14	49	6.5	13	0.5
		20	58	7.0		
		25	65	7.0		
		30	71	7.0		
070	Black	14	56	7.5	15	0.5
		20	67	8.0		
		25	75	8.0		
		30	82	8.0		
090	Orange	14	74	7.5	19.7	0.5
		20	88	8.0		
		25	99	8.0		
		30	108	8.0		
120	Red	14	100	7.5	26.6	0.5
		20	119	8.0		
		25	113	9.0		
		30	146	9.0		
160	Brown	14	130	8.5	34.7	0.5
		20	155	8.5		
		25	173	8.5		
		30	190	9.0		
200	Yellow	14	167	8.5	44.75	0.5
		20	200	9.0		
		25	224	9.0		
		30	245	9.0		

2.2 Irrigation Frequency and Duration

$IR_n = 6.04$ mm d^{-1} or $(6.04 \times 6 \times 6 \times 0.5) = 217$ (l day per tree)
Tree spacing = 6 m × 6 m
Effective root zone depth (ERD) = 1 m
Desirable soil wetted area = $S_p \times S_r \times P_w = 6 \times 6 \times 0.5 = 18$ m^2
Available moisture = 120 mm/m
Moisture depletion = 20%.

The solution could be as shown in Table 2.

Procedures for evaluating under-tree sprinklers having non-overlapping (or slightly overlapping) patterns of application are described and discussed in this chapter.

The uniformity of the watering pattern produced by over-tree sprinklers, useful for frost protection and climate control as well as for irrigation, can be evaluated only at the top of the tree canopy level. Interference of the catch pattern by the trees makes soil surface measurements meaningless. However, ground-level distribution is of most importance to irrigation. Observations indicate how much soil is dry, and probing can indicate uniformity of application. Under-tree systems requiring overlap from adjacent' sprinklers to obtain uniformity can be evaluated by the standard technique for open-field evaluation [16].

The *orchard sprinkler* is a small spinner or impact sprinkler designed to cover the interspace between adjacent trees; there is little or no overlap between sprinklers. Orchard sprinklers are designed to be operated at pressures between 10 and 30 psi, and typically, the diameter of coverage is between 5 and 10 m. They are located under the tree canopies to provide approximately uniform volumes of water for each tree. Water should be applied fairly even to areas to be wetted even though some soil around each tree may receive little or no irrigation (see Fig. 1).

The individual sprinklers can be supplied by hoses and periodically moved to cover several positions, or there can be a sprinkler provided for each position.

Table 2 Solution

Parameters types	Sky blue nozzle	Blue nozzle
Wetted area, A_w	• 19.6 m^2	• 28.3 m^2
Percentage wetted area, P_w	• 100 × (19.6/(6 × 6)) = 54%	• 100 × (28.3/(6 × 6)) = 79%
Available soil moisture/tree	• (120/1000) × 19.6 × 1000 = 2352 l/tree	• (120/1000) × 28.3 × 1000 = 3396 l/tree
Readily available moisture	• 2352 × 0.2 = 470 l/tree	• 3396 × 0.2 679 l/tree
Irrigation frequency at peak	• 470/217 = 2.2 days, say 2 days	• 679/217 = 3.1 days, say 3 days
IR_n to be applied per irrigation	• 217 × 2 = 434 l/tree	• 217 × 3 = 651 l/tree

Fig. 1 Orchard sprinkler operating from a hose line

The following questions relative to the use of orchard sprinklers should be considered before selecting equipment [17, 18].

1. Is an under-tree sprinkler system the most practical irrigation system for the orchard?
2. Does wetting the soil around the tree trunk induce diseases and would a shield give the trunk sufficient protection?
3. Will irrigation spray damage the fruit?
4. Do low branches and props seriously interfere with the pattern's uniformity?
5. Would salinity of the irrigation water damage leaves, which are wetted?
6. Is the water supply sometimes inadequate making it?
7. Desire to use sprinklers that can be adjusted to wet a smaller area when necessary?
8. Is a crop going to be raised between tree rows while trees are small? If so, what is the expected crop height?

3 Evaluation

The irrigation objectives must be known before the operation of the system can be evaluated intelligently. *Uniformity of application* and the *efficiency of storing* water for plant use are the two most important points to be considered. For evaluating orchard sprinkler systems, uniformity and efficiency must be qualified, for often it is not practical to try to have complete coverage [15, 19, 20]. Fortunately, mature trees have such extensive root systems that they can extract soil moisture wherever it is available. Therefore, the roots may absorb any available stored water.

The data needed for evaluating an existing under-tree non-overlapping system are:

1. Depth (or volume) of water caught in a radial row (or rows) of catch containers.

2. Duration of test.
3. Duration and frequency of normal irrigations.
4. Flow rate from tested sprinkler.
5. Pressures throughout the system.
6. *MAD* and *SMD*.
7. Sprinkler locations relative to trees.
8. Spacing and arrangement of trees.
9. Interference of sprinkler jets by branches.
10. Sequence of operation.
11. Percent of ground area wetted.
12. Additional data indicated on Form 1 (Annex 1).

3.1 Equipment Needed

The equipment needed is essentially the same as for the full evaluation of sprinkler–lateral systems.

1. A pressure gauge (0–50 psi) with pilot attachment is useful but not essential.
2. A stopwatch or watch with an easily visible second hand.
3. A large (at least 4.0 l) containers with volume marked.
4. A bucket, funnel, 1.2 m lengths of hose and a tin sheet or other means for deflecting the sprinkler jets and any leakage into the container.
5. Approximately twenty catch containers such as 1-quart oil cans.
6. A measuring stick (or ruler) to measure the depth or 500-ml graduated cylinder to measure the volume of water caught in containers.
7. A soil probe or auger.
8. A tape for measuring distances in laying out the radial rows of catch containers.
9. A shovel for smoothing areas where containers are to be set and for checking profiles of soil, root and water penetration.
10. Manufacturer's sprinkler performance charts.
11. Form 1 for recording data.

3.2 Field Procedure

Information obtained from the following field procedure should be recorded on a data sheet similar to Form 1 (Annex 1).

1. Choose radial row locations where water will be caught from only one sprinkler. It is best to test several sprinklers at several locations to check for system variations and improperly adjusted sprinklers. To save time, it is practical to test the sprinklers simultaneously with different adjustments and pressures.

2. Fill in parts 1 and 2 of Form 1 concerning the crop, field, root depth and *MAD*.
3. Check and record in part 3 the *SMD* in the area of the pattern that will receive full irrigation. This area should represent half or more of the sprinkler pattern and should not be affected by overlap or tree drip. Also, determine and record the soil texture and estimate the available soil moisture capacity in the root zone.
4. Trees and spacing between trees 7 m × 7 m (see Form 1, part 4).
5. Check and record in part 5 the sprinkler makes and model, size of the nozzles, the normal sprinkler spacing and the location of sprinklers relative to the trees.
6. Obtain the normal duration and frequency of irrigation from the operator and record them.
7. Obtain and record the rated sprinkler discharge and pressure from the design data and the manufacturer's catalog.
8. Observe sprinkler operation at a pressure higher and lower than normal, and then set the pressure back to "normal" for the evaluation test. Note arid record the height of the jet trajectory, tree and wind interference and characteristics of sprinkler rotation.
9. Measure and record the sprinkler pressure, wetted diameter and total discharge including any leakage from the test sprinkler and with two or three other sprinklers spaced throughout the system. Where the jet is too diffuse or small to use a pitot tube, the pressure gauge may need to be connected into the sprinkler riser. The overall uniformity of the system can be evaluated better by determination of flow rate than by pressure checks; however, knowledge of pressure is useful.
10. Set out a radial row of catch containers along a radius of the sprinkler's wetted circle (as in Fig. 2). If unusual conditions such as strong wind or a steep slope exist, four rows of containers should be used; however, if the wind is negligible, as it often is in orchards, one row is adequate. Remove any potential interference of catchment caused by weed, branches, props or other objects. Be sure that all containers are empty. Space the first container 0.3 m from the sprinkler and align the rest 0.6 m apart to cover the full range of the jet.

    Note and record in part 10 the starting time of each test and continue the test until at least 2.5 cm is caught in some containers and note the time the test is stopped. If practical, continue each test for full-length irrigation to obtain data that are represented by normal irrigation practice. Be careful that containers do not overflow.

    Measure the depth or volume of water caught in each container. Record each measurement in the space above the corresponding radial distance of the container from the sprinkler.
11. Check the sprinkler pressure at 20–40 systematically selected locations throughout the system (e.g., at the two ends and midpoints of each manifold) and record the maximum, minimum and average pressure in part 11.
12. Note in part 12 the type of system operation and such operating conditions as the speed of the wind, impact on trees and resulting drip, overlap on adjacent sprinkler patterns if any and uniformity of sprinkler rotation.

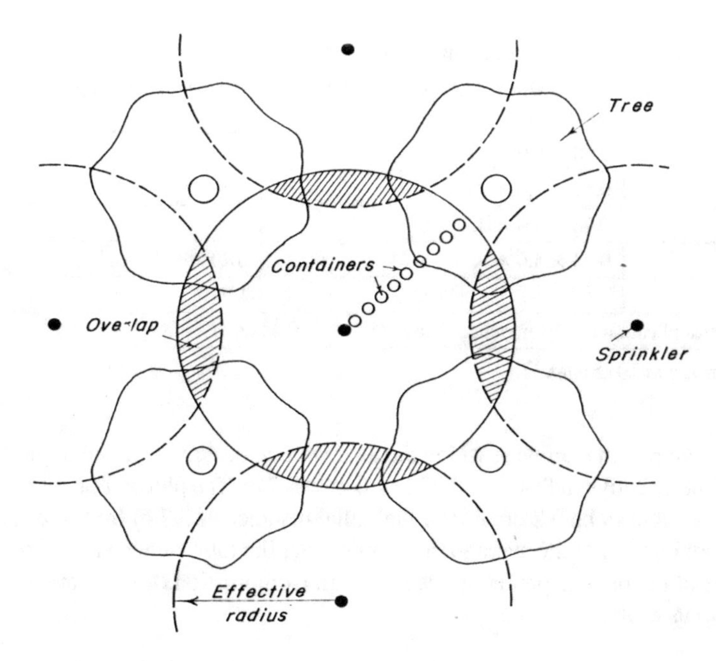

Fig. 2 Layout for a typical test of orchard sprinkler system in an orchard having a square pattern of trees

Check the general uniformity and the depth of wetting with the soil probe immediately following a normal irrigation. After one or two days, check the depth again to determine whether the irrigation was adequate.

3.3 Utilization of Field Data

Information recorded in the field should be reduced to a form that can be conveniently studied and used. It is usually assumed that the water caught is equivalent to the water infiltrated. The depths or volumes of water caught should be converted to rates in cm per hour, cm h^{-1}; the rate profile should be plotted as shown on Form 1, part 10; and the effective radius, R_e, noted. R_e of the sprinkler in the reported test was 4.0 m, which is the radius at which the rate profile plot crosses the zero line.

3.4 Average Application Rate

From the R_e of 4.0 m, the radius at which the approximate average application rate occurs for each concentric quarter of the area can be computed by multiplying R_e by 0.40 for the inner quarter, 0.60 for the second quarter, 0.78 for the third quarter and 0.93 for the outer quarter.

Table 3 Compuation of the average rate application rate

Quarter of area	Radius where average rate occur occurs (m)	Average rate from graph (cm h^{-1})
Inner	$0.40 \times 4.0 = 1.6$	0.50
Second	$0.60 \times 4.0 = 2.4$	0.56
Third	$0.78 \times 4.0 = 3.1$	0.46
Outer	$0.93 \times 4.0 = 3.7$	0.20
Total		1.7
Average application rate over wetted area = (1.7/4) = 0.43 cm h^{-1}		

See Form 1, part 10 (Annex 1)

For example, the radius at which the average rate occurs in the outer quarter is at 93% of the effective radius, i.e., $0.93 \times 4.0\,\text{m} = 3.7$ m. The plot on Form 1 shows the application rate to be 0.2 cm h^{-1} at the radial distance of 3.7 m from the sprinkler. An approximation of the average rate caught over the total wetted area is the sum of the rates at the quarter points divided by four. Computation of the average rate can be set up in Table 3.

3.5 Distribution Characteristic

Since only part of the surface area may be wetted, the uniformity of irrigation should be evaluated by the distribution characteristics DC instead of DU [17]. Since only part of the area is left dry, the remaining smaller wetted area should be irrigated proportionally more often to supply the total water needed to balance evapotranspiration. For example, if only half of the area is wetted, the frequency of irrigation must be doubled.

For a single non-overlapping sprinkler, DC is the percent of the total wetted area that has received and infiltrated more than the average depth [21].

$$\text{DC} = \frac{\text{Area that has received more than average depth}}{\text{Total wetted area}} \times 100 \qquad (2)$$

The DC can be determined (see Form 14.1, part 10) by first drawing a line (see dotted line part 10) representing the average rate of 0.43 cm h^{-1} across the rate profile line and noting the radius of 3.3 m where the two lines cross. Then, calculating the ratio of this radius to the total radius and multiplying the square of the ratio by 100 gives

$$\text{Radius ratio} = (3.3\,\text{m}/4.0\,\text{m}) = 0.82$$

and

$$DC = (\text{Radius Ratio})^2 \times 100$$
$$= (0.82)^2 \times 100 = 68\%.$$

The DC relates to the uniformity of that portion of the central wetted area that may contribute to deep percolation losses even under good management. High DC values indicate that the adequately irrigated area may be relatively large while the potential losses from deep percolation are low. The DC can approach 100%; this would indicate an extremely uniform application provided there be very little overlap or tree interference. A DC greater than 50% is considered satisfactory, and the computed value of 68% for the example problem indicates a very good pattern.

3.6 Storage Efficiency

The most important objective of the field evaluation is to determine how effectively the water is being applied. Since orchard irrigation almost always leaves some areas and depths under-irrigated but still results in a very satisfactory irrigation program, the term *storage efficiency, SE,* is used instead of *AELQ* [17, 22, 23]. In the area wetted, the *SE* should be determined so that the effectiveness of the irrigation can be evaluated. Neither *PELQ* nor *AELQ* can be used to evaluate orchard systems, which wet only part of the area, since the average low quarter depth could be near zero.

$$SE = \frac{\text{average depth stored under circular wetted area}}{\text{average depth applied to circular wetted area}} \times 100 \quad (3)$$

In computing the *average depth stored* in the circular wetted area under each sprinkler, it is assumed that all the water that falls on each spot within the wetted area up to the *SMD* is stored. Water more than the *SMD* is lost by deep percolation. The following procedure will aid in calculating the average depth stored.

First, determine what depth would be applied at each catch point by multiplying the rate values calculated in part 10 by the duration of normal irrigation, which this example was 24 h. Then, plot the depths of the application at various radial distances from the sprinkler as shown in Fig. 3 and draw a line across the depth profile representing the *SMD*. For this illustration, the *SMD* was 10 cm and was assumed uniform (although it seldom is). All moisture above the *SMD* line would be stored in the soil. Overlap and/or distortions caused by the trees are not considered.

The *average depth of moisture stored* under the circular area represented by the area above the *SMD* line may be estimated by dividing the wetted area into subareas. The average depths applied to and stored in the various portions of the area can be multiplied by the percent of the area receiving that depth, and the sum of these products will equal the average depth stored. The entire area inside the radius at the intersection between the *SMD* line and the depth profile will store the *SMD*. If the profile is fairly uniform, one average value is adequate for the area beyond the *SMD*

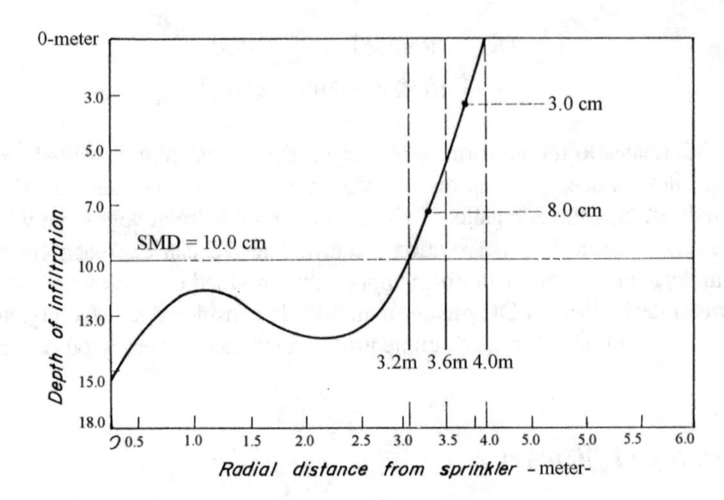

Fig. 3 Profile of water application along the sprinkler radius for a 24-hour set

line intersection. However, if profiles are curved, computations of depth from two areas will give slightly more precise results. For Fig. 3, one outer section would be adequate, but two were used for demonstration.

The steps used to calculate the average depth and the numerical values based on Fig. 3 are:

1. Find the radius at the intersection of *SMD* with the depth profile (3.2 m) and one other radius (3.6 m); this divides the under-watered profile into two convenient subareas.
2. Determine the ratio between these radii and the effective radius of 3.99 m, (3.3/4.0) = 0.82, (3.6/4.0) = 0.90).
3. Square the radius ratios to find the corresponding portion of the area included inside each radius, $[(0.82)^2 = 0.67, (0.90)^2\ 0.81]$.
4. Determine the portion of the total area included in each of the three subareas defined by the two intermediate radii. For this example, they are: 0.67, 0.82 – 0.67 = 0.15 and 100 – 0.81 = 0.19.
5. Estimate the average depth in each subarea from the depth profile (these can be taken at the middle of each subarea with adequate accuracy). From Fig. 3, these are the SMD of 10, 8 and 3 cm.
6. Multiply each subarea portion by the corresponding average depth. The sum of the products equals the average depth of water stored in the root zone under the circular wetted area.

$$0.67 \times 10.0 = 6.7\,\text{cm}$$
$$0.15 \times 8.0 = 1.2\,\text{cm}$$
$$0.19 \times 3.0 = 0.6\,\text{cm}$$

Average depth = 8.5 cm stored under wetted circular area.

The average depth of water applied to the circular wetted area is computed by using the sprinkler discharge rate of 0.244 $m^3\ h^{-1}$ (see Form 1, part 9, test column) and the wetted radius, R, 4.0 m, to obtain:

$$\begin{aligned}\text{Application rate} &= \frac{1}{\pi} \times \frac{\text{sprinkler discharge } (\text{m}^3\,\text{h})}{R_e\ (\text{m}) \times R_e\ (\text{m})} \\ &= \frac{1}{3.14} \times \frac{0.244}{4.0 \times 4.0} = 0.49\,\text{cm}\,\text{h}^{-1} \end{aligned} \quad (4)$$

and for a 24-hour set

Average depth applied to wetted circular area = 0.49 × 24 = 11.8 cm

The SE can be computed (assuming negligible overlap and drip, which could cause some water to go too deep) by

$$\text{SE} = (8.5/11.6) \times 100 = 73\%$$

4 Analysis and Recommendations

Several observations and recommendations can be based on the information recorded in Form 1 and the preceding computations.

Uniformity on the tested area was good as indicated by the DC of 68%. If this percent had been much higher, it would have indicated that a greater depth had been infiltrated near the perimeter; this would result in a little water going too deep because of overlap unless the effective radius of 4 m was reduced. If this were the condition, the wetted diameter should be reduced from 7.95 to nearly 7.0 m, which is the tree spacing (see shaded areas in Fig. 3).

The pressures, discharges and wetted diameters of the sprinkler tested and other sprinklers checked were all reasonably close (see sample Form 14.1, parts 9 and 11). The efficiency reduction (ER) is caused by the variations in pressure throughout the system in accordance with the presented formula

$$\begin{aligned}\text{ER} &= 0.2\frac{\text{maximum presure} - \text{minimum presure}}{\text{average system presure}} \\ \text{ER} &= 0.2 \times (21 - 18)/18 = 0.03 \text{ (or 3\%)}.\end{aligned} \quad (5)$$

This indicates that the general system uniformity was very good.

Water losses from causes other than deep percolation, such as loss from evaporation, are equal to the difference between the average application rate (0.48 cm h^{-1}) and the average catch rate (0.43 cm h^{-1}). This is equal to $[(0.48 - 0.43)/0.48] \times 100 = 10\%$ of the water applied a percentage that is too high for evaporation only. However, it is a reasonable figure because it includes any errors in measurement. These losses cannot be controlled by management practices.

Losses by deep percolation can be identified by the differences between the average depths infiltrated (0.43 cm $h^{-1} \times 24$ h = 10.3 cm) and average depth stored, (8.4 cm). Thus, 2.0 cm or 18% of the applied water goes too deep; this is a large amount for a partial area irrigation program. Observing the depth profile and the 10 cm SMD line on Fig. 3 shows that deep percolation is appreciable in the central portion of the pattern even though it is a nearly uniform pattern. A depth of 12.7 cm infiltrates near the sprinkler while only 10 cm can be stored. This excess depth occurred because the 24-hour set time is too long.

4.1 Improvements

A major improvement would be the reduction of losses due to deep percolation. This could be accomplished by

1. Reducing the duration of irrigation to less than 24 h.
2. Lengthening the interval between irrigations by 1 or 2 days and increasing the MAD to near 5 inches.
3. Reducing the pressure or nozzle size to reduce the flow rate so the 24-hour duration could be continued.

The result of any of these changes would need to be re-evaluated to see whether it was better than the results achieved under the present system. The pattern could become worse or improve, as will be shown.

Alternate side irrigation is generally a good management practice. It is especially good when only a portion of the total area is wetted because it provides additional safety by reducing the average crop stress between irrigations.

Adjusting the duration of irrigation: The optimum duration of irrigation Ti, to replace the SMD, can be found by trial. Figure 3 shows that 12.7 cm represents the approximate maximum infiltrated depth for a 24-hour set and that SMD is only 10 cm. Ti can be estimated from

$$\text{Ti} = 10.0/12.7 \times 24 = 19\,\text{h}$$

Storage efficiency (72%) is a fairly low value particularly in view of the DC value of 68%. SB is low because the 24-hour irrigations being used are too long and cause excess deep percolation. Instead of using the original 24-hour set duration, 19 h can be used and a new value of SB can be determined. This will require plotting a new profile of depth infiltrated similar to Fig. 3 and proceeding with the evaluation outlined earlier to obtain the following:

$$SE = 8.1/9.1 \times 100 = 89\%$$

The analysis indicated the unmeasured losses remained at about 10%, but the losses to deep percolation were reduced to approximately 1%. Average depth stored in the wetted circular area was reduced from the initial 8.3 to 8.1 cm because less of the area received the full SMD of 10.1 cm. This will require reducing the irrigation interval to 8.1/8.4 = 97% of the initial interval, which is not very significant. However, the application time will be considerably reduced to 19/24 = 79% of the original. Nineteen-hour irrigation may be inconvenient, but it would be most efficient.

Average depth applied: The ratio of the wetted area to the actual tree-covered area must be determined before the average depth (or volume) of water to be applied to a field and the proper frequency of irrigation, based on anticipated evapotranspiration rates, can be computed. The circular wetted area provided by each sprinkler for each tree is as follows:

$$\text{Wetted area} = \pi r^2 = 3.14 \times 4.0 = 50\,\text{m}^2$$

and the total area serviced by each sprinkler on a 7.0- by 7.0-m spacing is 49 m^2.

Evapotranspiration and water applied are computed by assuming the entire soil area of the field is functioning. Therefore, for the 24-hour set where the average depth stored in the actual circular wetted area is 8.4 cm, the average depth of water stored over the whole orchard is as follows:

$$\text{Average depth stored} = (50/49) \times 8.4\,\text{cm} = 8.5\,\text{cm}$$

This value is to be used to compute the amount of water to be replaced and the irrigation interval.

5 Conclusion

Good management of micro-sprinkler irrigation system provides great water and soil conservation and reduces applied water requirement. Using micro-sprinkler irrigation system improves the existing water productivity; water use; efficiency, economic returns, increase yield and great control of applied water through improving DC and SE for the field crops growing under orchard area.

6 Recommendation

Increase production while decreasing water use and costs is considered one of the most important contemporary issues today. In order for agriculture to continue to feed the Egyptian people, farmers have to learn to utilize from existing water supplies as far as possible in new lands. A lot of farmers are converting to micro-irrigation for their trees and field crops because of the potential to save water and increase efficiencies. Micro-irrigation system can save water by applying less water exactly where it is needed. In order to maximize yield and profits, farmers have to apply excellent irrigation practices as well as consistent maintenance of their micro-irrigation system. Micro-sprinkler irrigation system is a very common method for irrigating tree crops in Egypt as explained throughout this chapter.

Annex 1

Form 1 Orchard Sprinkler Irrigation Evaluation

1. Location *Nubaria* Observer *Alaa El-Bably* Date *6/17/2015*
2. Crop *citrus*, root zone depth *1.5 m*, MAD *50%*, MAD *10.0 cm*
3. Soil: texture *sandy loam*, available moisture *4.1 cm/30 cm*, SMD *10 cm*
4. Tree: pattern *square* spacing *7.2 m* by *7.2 m*
5. Sprinkler: make *BE*, model *B-21*, nozzles #1 by cm spacing *7.2* by *7.2 m*, location to trees *center*
6. Irrigation: duration 24 h, frequency *21 days*
7. Rated sprinkler discharge *0.222 m³ h* at *20 psi* and diameter *8.0 m*
8. Sprinkler jet: height *1.0*, interference *negligible*
9. Actual sprinkler pressure and discharge (see back for location):

Sprinkler locations	Test	2	3	4
Pressure (psi)	19	21	18	19
Catch volume (l)	4	4	4	4
Catch time (s)	54	52	55	54
Discharge (m^3 h)	0.244	0.245	0.244	0.243
Wetted diameter (m)	7.8	8.1	7.8	7.8

Comments: *sprinkler performance good with smooth rotation*

10. Container row test data in units of *cm*, volume/depth ml/cm
    Test: start *7:20 pm*, stop *8:00 am*, duration *12 h 40 min = 1 2.67 h*

Catch (*cm*)	*7.1*	*6.1*	*6.4*	*7.1*	*7.1*	*5.31*	*1.3*
Rate (cm h^{-1})	*0.56*	*0.48*	*0.51*	*0.56*	*0.56*	*0.41*	*0.10*

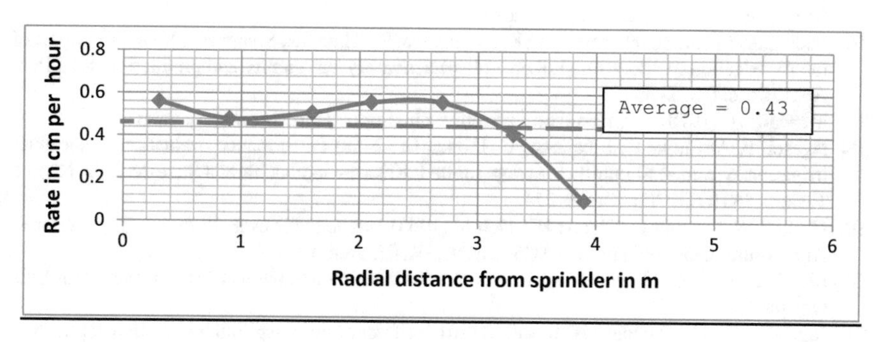

11. Discharge pressures: max 21 psi, min 18 psi, ave *19 psi*
12. Comments: *The citrus tree branches did not obstruct the sprinkler jets and the sprinklers rotated smoothly and uniformly. The system is the portable hose-pull type.*

References

1. FAO (2001) Sprinkler irrigation systems, planning, design, operation and maintenance. Irrigation manual 8, Harare, Zimbabwe
2. Department of Agriculture and Natural Resources Conservation Service (USDA) (2013) Microirrigation. Part 623: National engineering handbook, chapter 7, pp 19–191
3. Michael M (2014) Micro-irrigation design for avocado orchard in California. Agricultural Systems Management, Bio Resource and Agricultural Engineering Department, California Polytechnic State University, San Luis Obispo, CA
4. Yuan S, Ransford D, Zhu X, Junping L, Kun T (2017) Optimization of movable irrigation system and performance assessment of distribution uniformity under varying conditions. Int J Agric Biol Eng 10:72–79. Open Access at https://www.ijabe.org
5. Ayars J, Phene C (2007) Automation. In: Lamm FR, Ayars JE, Nakayama S (eds) Micro-irrigation for crop production, chap. 7, pp 259–284
6. Boswell MJ (1984) Micro-irrigation design manual. James Hardie Irrigation Co., El Cajon, CA
7. Burt C, Styles S (1994) Drip and microirrigation for trees, vines, and row crops. Irrig. Training and Research Center, Calif. Polytechnic State Univ., San Luis Obispo, CA 292 pp.
8. Hassan F (1998) Microirrigation management and maintenance. Agro Industrial, Fresno, CA
9. Howell TA, Meron M (2006) Irrigation scheduling. In: Lamm FR, Ayars JE, Nakayama FS (eds) Microirrigation for crop production. Design, operation, and management, 2nd edn. Elsevier, The Netherlands, pp 61–130
10. Keller J, Bliesner R (1990) Sprinkler and trickle irrigation. Chapman and Hall, New York
11. Godin R, Broner I (2013) Micro-sprinkler irrigation for orchards fact sheet no. 4.703. Crop Series|Irrigation. www.ext.colostate.edu

12. Christiansen J (1942) Irrigation by sprinkling. Bulletin 670, Agricultural Experiment Station, University of California, Berkeley, CA
13. FAO (2002) Localized irrigation systems, planning, design, operation and maintenance. Irrigation manual 9, Harare, Zimbabwe
14. ASAE (1990) ASAE EP405.1. Design and installation of micro-irrigation systems
15. FAO (2002) Monitoring the technical and financial performance of an irrigation scheme. Irrigation manual 14, Harare, Zimbabwe
16. Sudhakar P, Hanumantharayappa SK, Swamy GMR, Jalaja SK, Sivaprasad V (2018) Impact of micro irrigation methods on mulberry (*Morus alba* L.) leaf quality and production. Int J Pure Appl Biosci 6:332–339
17. Merriam JL (1978) Farm irrigation system evaluation: a guide for management
18. Ngasoh F, Anyadike C, Mbajiorgu C, Usman M (2018) Performance evaluation of sprinkler irrigation system at Mambilla beverage limited, Kakara-Gembu, Taraba State-Nigeria. Niger J Technol (NIJOTECH) 37:268–274
19. Griddle W, Sterling D, Claude HP, Dell G (1965) Methods for evaluating irrigation systems. Agricultural handbook, no. 82. SCS, USDA, Washington, DC
20. Merriam JL (1968) Irrigation system evaluation and improvement. Blake Printery, San Luis Obispo, CA
21. James D, Awu D, Pamdaya N, Kasali M (2018) Effect of micro-sprinkler irrigation application efficiency for Okra (*Abelmoschus esculentus*) farming. Int J Basic Appl Sci 6:98–103
22. Pair C (ed) (1975) Sprinkler irrigation, 4th edn. Sprinkler Irrigation Association, Silver Spring, MD
23. SCS national engineering handbook (1967) Planning farm irrigation systems, chap 3, sect 15. USDA, Washington, DC

Drip Irrigation Technology: Principles, Design, and Evaluation

Mostafa Fayed

1 Introduction

The steady increase in population is an urgent factor for increasing agricultural production to provide food in the world to cope with this increase in the population. Increasing agricultural production requires horizontal expansion by adding new lands, which will require additional water to meet the irrigation needs of different crops. This expansion requires the search for new sources of water and rationalizing the use of available water, which has become a priority for planners and officials in the agricultural and water sectors in all countries of the world [1].

Traditional and non-conventional water sources in Egypt are limited and are growing scarcer year after year. The limited share of the Nile water that Egypt receives is not expected to increase in the future. The lack of these resources and sometimes their scarcity, coupled with the continuous decline in the quality of water used in irrigation, has created an urgent need for research and development in irrigation technology to provide water and energy, improve field management of water, and optimize the use of water resources. That is because the agriculture consumes more than 85% of the total demand for water. Drip irrigation is one of the modern innovations of irrigation—which is undoubtedly a clear advance in irrigation technology [1, 2].

Perhaps one of the most important methods of modern irrigation technology represented in the system of drip irrigation achieved some of these demands and avoided some of the disadvantages that emerged with other irrigation methods. Drip irrigation is one of the modern innovations of irrigation—which is undoubtedly a clear advance in irrigation technology [1].

M. Fayed (✉)
Water and Irrigation Systems Engineering, Faculty of Agricultural Engineering, Al-Azhar University, Nasr City, Cairo, Egypt
e-mail: mostafa.fayed@azhar.edu.eg

E.-S. E. Omran and A. M. Negm (eds.), *Technological and Modern Irrigation Environment in Egypt*, Springer Water,
https://doi.org/10.1007/978-3-030-30375-4_13

The design and evaluation of drip irrigation system are described briefly in this chapter. Sufficient information is presented to familiarize the reader with:

1. Definition of drip irrigation system.
2. The benefits and problems of the drip irrigation system.
3. The latest types of drip irrigation system in Egypt.
4. The major components of drip irrigation systems.
5. Design and evaluation steps of drip irrigation system.

2 Drip Irrigation

Drip, trickle, localized, or pressurized irrigation, by definition, is an irrigation technology. Irrigation water and chemical solutions are connected to drip irrigation systems in the quantities required and calculated accurately and at slow rates in the form of separate or continuous points using mechanical tools known as pointers placed at specific points along the water supply lines. The rate of disposition of these filters ranges from 1 to 12 L/h. In this method, the root zone is wetted only with the rest of the soil surface. This is done for both vegetable plants and fruit trees, but with varying water rates. The method of drip irrigation is characterized by slowly adding water to the plant at intervals between twice a day and once every three days, depending on the type of plant, soil type, and climatic conditions. Fertilizers and some chemicals can also be delivered through a high-efficiency drip irrigation method for each plant. Drip irrigation can be applied in the field above the soil surface or under the soil surface. The water distributed through the network drip irrigation pipes to pressure ranging from 15 to 200 kPa [1, 3].

Drip irrigation is superior to other irrigation methods (surface irrigation and sprinkler irrigation), which can be used very efficiently in sandy soil and fields with different terrains. It can also be used in the arid and wetland. Well-designed drip irrigation provides up to 50% water content compared to surface irrigation and 30% compared to spray irrigation by reducing water losses that are lost by deep leakage, surface runoff, or evaporation in other irrigation methods. Drip irrigation provides energy because it works at low operating pressure compared to sprinkler irrigation systems [1, 3].

The method of drip irrigation has spread in many countries such as America, Australia, Mexico, and England, and recently spread in the reclamation lands in Egypt. One of the first countries to consider this system was England in 1945, where it began to be used only in the field of experiments and then expanded to be used in irrigation. Then came Australia, America, and Israel, where its use in Israel increased widely in the late 1960 and early 1970 [2].

In Egypt, a drip irrigation system was designed, constructed, and tested for the first time in 1975 by El-Awady et al. [4]. The system used micro-tube emitters under the low head, thus reducing amenability to clogging among other advantages.

Previously, the term "micro-irrigation" has recently emerged. It is more comprehensive than drip irrigation and is used to describe the irrigation method which is characterized by the following [3, 5]:

1. The application of water at low rates.
2. The application of water over long irrigation time.
3. The application of water at frequent intervals.
4. The application of water directly near or into the plant root zone.
5. The application of water by low operating pressure delivery system.
6. Water is used to transport fertilizers and other agricultural chemicals to plants.

The term "micro-irrigation" describes a family of irrigation systems that apply water through small devices. These devices deliver water onto the soil surface very near the plant or below the soil surface directly into the plant root zone. Growers, producers, and landscapers have adapted micro-irrigation systems to suit their needs for precision water application. Micro-irrigation systems are immensely popular not only in arid regions and urban settings but also in sub-humid and humid zones where water supplies are limited or water is expensive. In irrigated agriculture, micro-irrigation is used extensively for row crops, mulched crops, orchards, gardens, greenhouses, and nurseries. In urban landscapes, micro-irrigation is widely used with ornamental plantings, http://www.durhampump.com/irrigation/micro-system/.

Micro-irrigation can be defined as the slow application of water on, above, or below the soil by surface drip, subsurface drip, bubbler, and micro-sprinkler systems. Water is applied as discrete or continuous drips, tiny streams, or miniature spray through emitters or applicators placed along a water delivery line adjacent to the plant row [3].

3 Drip Irrigation Types

3.1 Surface Drip Irrigation System

The surface drip irrigation system is a type of micro-irrigation. This system uses emitters and lateral lines laid on the soil surface or attached above ground on a trellis or tree (Fig. 1). Surface drip irrigation is widely used to irrigate perennial crops (tree and vines) and annual row crops [3]. Generally, discharge rates are less than 12 L/h for point source emitters and less than 12 L/h m for line source emitters. The advantages of surface drip irrigation system are easy installation, inspection, maintenance, cleaning, and changing emitters, in addition to the possibility of observing the wetting patterns on soil surface and measuring the discharge rates of individual emitters. On the other hand, the lateral lines of this system can interfere with some agricultural processes such as plowing, harvesting, and others [6].

Fig. 1 Surface drip irrigation

3.2 Subsurface Drip Irrigation System

This technology has gathered momentum during the last two decades. Subsurface drip irrigation system is a type of micro-irrigation. It is defined as the application of water below the soil surface by buried emitters (Fig. 2). The discharge rate of emitters for subsurface irrigation system is generally in the same range as those for surface drip irrigation rates and is usually less than 7.5 L/h. This system is suitable for a wide variety of horticultural and agronomic crops and, in many respects, is applicable to those crops presently under surface drip irrigation. The advantages of the subsurface drip irrigation system are negligible interference with farming activity, elimination of mechanical damage to laterals, decreased weed infestation, elimination of runoff, and evaporation from the soil surface may be zero, and improved uptake of nutrition elements by the roots, notably phosphorous. The disadvantages are high costs for burying the laterals into the soil, plugging hazard by intruding roots and sucked-in soil particles, inconvenience in monitoring the performance of drippers and laterals, and strict maintenance is mandatory [3, 6, 7].

3.3 Bubbler Irrigation System

In bubbler irrigation, water is applied to the soil surface as a small stream or fountain to flood small basins or the soil adjacent to individual trees, typically from a small diameter tube 1–13 mm or commercially available emitters, where the discharge rates of these emitters are less than 250 L/h (Fig. 3). Because the emitter discharge

Fig. 2 Subsurface drip irrigation

Fig. 3 Bubbler irrigation system

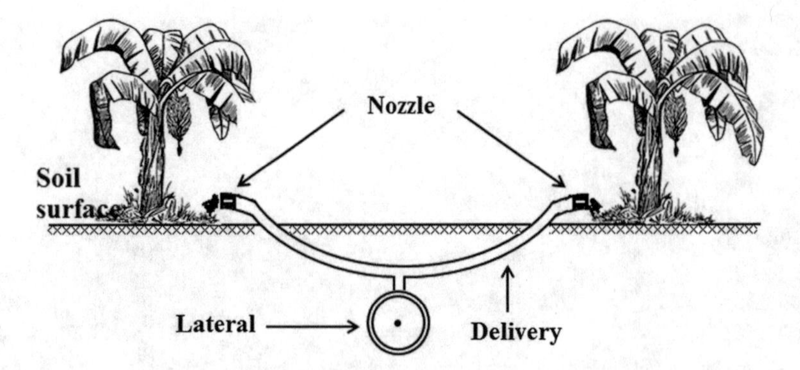

Fig. 4 Typical installation of new bubbler irrigation system [8]

rate normally exceeds the infiltration rate of soil, a small basin or furrows are usually required to contain the water or control the runoff and erosion. Because of the large diameter tubes, bubbler systems are not as prone to clog and normally have higher discharge rates than surface and subsurface drip systems. Operating pressure for bubbler irrigation system is ranging from 100 to 300 kPa. This system is well suited for perennial crops, particularly on orchards and vineyards with level typography [3, 6].

The advantages of bubbler irrigation are reduced filtration, maintenance or repair, and energy requirements compared with other types of drip irrigation systems. However, larger size lateral lines are usually required with the bubbler systems to reduce the pressure loss associated with the higher discharge rates [3].

In recent decades in Egypt, a bubbler irrigation system has been used successfully in Faiyoum Governorate for the irrigation of fruit trees. Also, in 1999 at Valoga Village, South Tahrir, Behera Governorate, Egypt, Elmesery [8] studied and evaluated a new design and installation of low head bubbler irrigation, which can use drip irrigation networks with available materials to irrigate banana trees (Fig. 4). The method uses nozzles which can be conveniently adjusted for uniformity. The development of a bubbler system can easily be adjusted for uniformity by changing of the nozzle, sizes along laterals, according to the following equations [8]:

In the case of single delivery:

$$\mathrm{dn_n} = \left[\frac{q_\mathrm{d}}{6.31 g^{0.5} (\mathrm{HE_n}/L)^{0.5} d^{1.14}}\right]^{0.735} \tag{1}$$

In the case of two deliveries:

$$\mathrm{dn_n} = \frac{q_\mathrm{d} L^{0.5}}{4.7 g^{0.5} d^{1.5} (\mathrm{HE_n})^{0.5}} \tag{2}$$

where dn_n is the nozzle diameter (cm), q_d is the delivery discharge (cm^3/s), g is the gravitational acceleration (cm/s^2), HE_n is the effective head (cm), L is the delivery length (cm), and d is the delivery diameter (cm).

3.4 Micro-sprinkler Irrigation System

Micro-sprinkler irrigation is also called spray irrigation. In micro-sprinkler irrigation, water is sprayed on the soil surface near individual trees as a small spray, jet, fog, or mist, where the water travels through the air becoming instrumental in the distribution of water (Fig. 5). Discharge rates for point source spray emitters (micro-sprinklers) are generally less than 175 L/h. Micro-sprinkler irrigation system is primarily used to irrigate trees and vine crops. Micro-sprinklers can be spaced to cover the entire land surface as with conventional sprinkler systems or a portion of the land surface like other trickle systems. The primary advantage of spray over bubbler irrigation is lower application rates which decrease the potential for runoff and erosion. Losses due to wind drift and evaporation are, however, greater with spray irrigation than with other drip systems [3, 6].

Fig. 5 Micro-sprinkler irrigation system

3.5 Mobile Drip Irrigation System (MDIS)

The idea of a mobile drip irrigation system (MDIS) is a combination of several advantages of the stationary drip irrigation with the lateral move irrigation system as center pivot, linear move, or boom trailer irrigation machines. Generally, MDIS combines the efficiency of surface drip irrigation (95%) with the flexibility and economics of center pivot irrigation. Crops can be effectively irrigated at very low litters per minute well deliveries. MDIS is combining the labor-saving advantages of the mobile and semi-mobile irrigation methods with the water- and energy-saving effects of drip irrigation system [1, 9–11].

The use of drip tubes with a moving irrigation system appears firstly introduced in 1974 by Rawlins et al. [12] which combined several advantages of drip irrigation system and lateral move systems. Efforts were made to commercialize the technology as early as 1992 under the name of Drag-N-Drip by Sherman Fox of Trickle Irrigation Specialties Co. of Salt Lake City, Utah. Newer efforts at commercializing the technology are being made by T-L Irrigation of Hastings, Nebraska (www.tlirr.com), under the trade name of Precision Mobile Drip Irrigation (PMDI), which utilizes in-line drip hoses to distribute water directly to the ground [10, 13, 14].

In Egypt, a linear mobile drip irrigation system (MDIS) was designed, built by using classic dripping irrigation materials and tested on pea crop for the first time in 2016 by Fayed [15] in the research field of Water and Irrigation Systems Engineering Department, Faculty of Agricultural Engineering, Al-Azhar University, Nasr City, Cairo, Egypt.

The linear mobile drip irrigation system (MDIS) is designed generally and assembled as shown in Figs. 6 and 7 [15]. It includes two towers with four driven wheels for moving the system in a linear desired direction. The towers carrying a water supply pipe were closed from one end, and the other end was connected to water source by two polyethylene (PE) tubes with a diameter of 18 mm by thread adapter (18 mm × 19 mm male "V"). The water source is generally a water pump, fertilizer apparatus and hydraulic equipment as a pressure gauge, flow meter, and a pressure regulator. The supply pipe includes eight saddles (63 mm × 12.7 mm female) at spacing of 50 cm, each saddle connected to one end of a vertically oriented drop line assemblies by tavlit push fit elbow male thread (16 mm × 12.7 mm male). The other end of the drop line was connected to manual flush valve of 16 mm. Each valve was connected to the upper end of the drip tube of 16 mm, and the lower end was closed by line end of 16 mm. PCJ online drippers were installed on the lower end of the drip tube, which was spaced above the ground.

Under this design, the wetting front advance in three directions, horizontal (*H*), vertical (*V*), and diagonal (*D*), is determined by drip tube discharge rate or system speed. The following equation can be used to compute the wetting front advance in three directions in loamy sand soil [15].

$$\alpha = 76.146\, q_{\mathrm{dt}}^{0.127} \tag{3}$$

Fig. 6 First use of linear mobile drip irrigation on a pea field in Egypt in 2016. Photograph was taken in December 2015 in the research field of Irrigation Systems Engineering Department, Faculty of Agricultural Engineering, Al-Azhar University, Nasr City, Cairo, Egypt [15]

$$\alpha = 42.859\, S^{0.127} \tag{4}$$

where α is the wetting front advance in any direction H, V, and D (m), q_{dt} is the discharge rate of drip tube (m^3/h), and S is the linear mobile drip irrigation system speed (m/h).

Generally, the mobile drip irrigation system (MDIS) consists of polyethylene (PE) drip line pulled through the field by the linear move irrigation system. As the drip lines are pulled behind the system, the integrated emitters deliver a uniform water pattern across the full length of the irrigated area. Because the drip lines deliver water directly to the soil surface, evaporation and wind drift are virtually eliminated, allowing more water to reach the soil surface and deeper percolation to the plant's root zone. https://www.businesswire.com/news/-home/20150324006412/en/Netafim-USA-Names-Teete.

3.5.1 Advantages of Mobile Drip Irrigation System

Mobile drip irrigation system provides certain advantages over other move systems as follows [11, 13, 15]:

- Increases yield.
- Saving water 10–20% compared with center pivot sprinklers.
- Used on either tall or short crops.
- No water losses due to wind drift and evaporation.
- Eliminates wheel track issues.

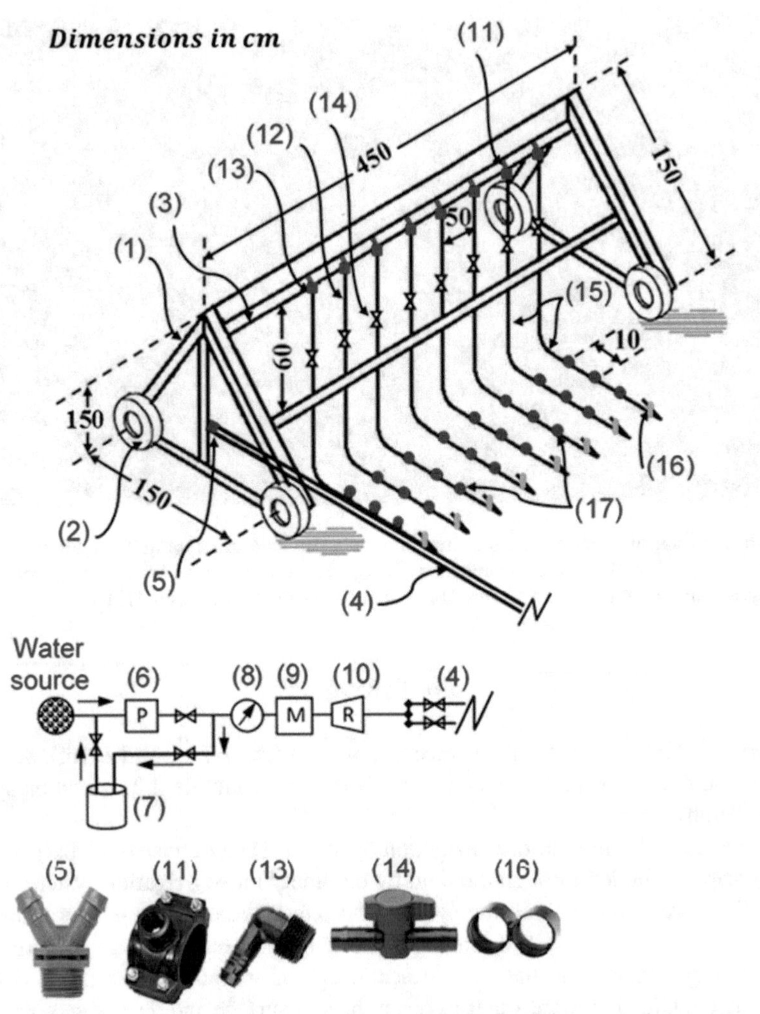

(1) Tower. (2) Driven wheel (175/70R13 82H). (3) Water supply pipe (UPVC 63 *mm*). (4) Two PE tubes 18 *mm*. (5) Thread adapter (18 *mm* ×19 *mm* male "V"). (6) Water pump. (7) Fertilizer apparatus. (8) Pressure gauge. (9) Flow meter. (10) Pressure regulator. (11) Saddles (63*mm* ×12.7 *mm* female). (12) Vertically oriented drop line assemblies. (13) Tavlit push fit elbow male thread (16 *mm* ×12.7 *mm* male). (14) Manual flush valve (16 *mm*). (15) Drip tube (16 *mm*). (16) Line end (16 *mm*). (17) PCJ on-line drippers.

Fig. 7 Components of linear mobile drip irrigation system (MDIS) as designed and constructed in Egypt 2016 [15]

- High irrigation efficiency.
- Flexibility of movement.
- Low operating pressure.
- Low capital requirements.
- Low labor requirements.

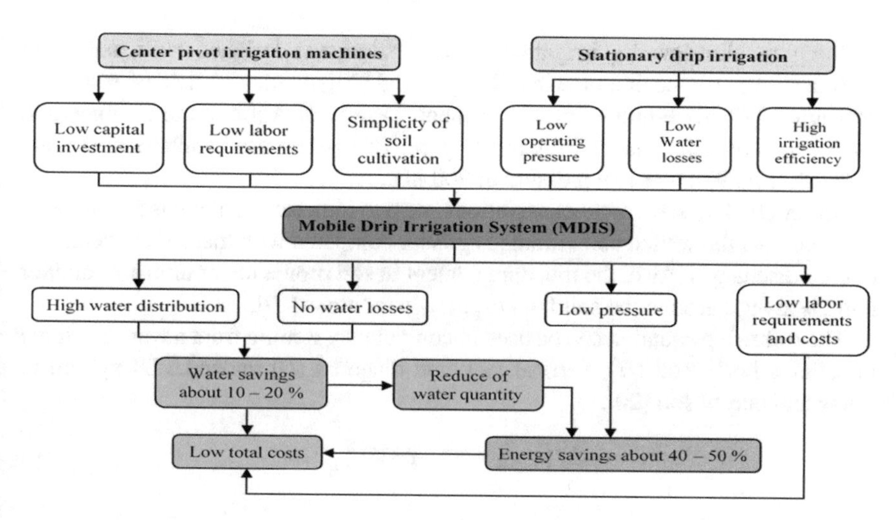

Fig. 8 Water and energy savings and the advantages of the MDIS [11]

- Soil cultivation under MDIS machines is easy.
- Soil stays moist without crusting, while soil compaction is reduced or eliminated.

Figure 8 illustrates the possibility of saving water and energy as well as the advantages of the mobile drip irrigation system [11].

The disadvantages of MDIS are [13]:

- Complex mechanical and electrical system.
- The MDIS (or other similar irrigation systems) represents a sizable capital investment.
- The system follows a straight line and does not correct for unequally spaced rows.

3.6 Ultra-low Drip Irrigation (ULDI) System

Ultra-low drip irrigation (ULDI) is a new aspect of micro-irrigation that has been researched in the past half-decade. Known as minute or ultra-low rate irrigation system or micro-drip irrigation, this new idea involves applying water at a rate close to that of plant water uptake, even lower than the natural soil infiltration rate. This process is accomplished by using spatters or pulsating drippers [16–19].

ULDI system has been developed that provides emitter discharge rates lower than 0.5 L/h [17]. ULDI system is able to apply water similar to or even lower than the soil water intake capacity. The ULDI system does apply 0.5 mm/h or even only a quarter of a millimeter per hour. Such an extremely low application rate will fit most soil water intake capacity. This low water application rate makes it possible to irrigate continuously during peak days around the clock [17–19].

In Egypt, ultra-low drip irrigation (ULDI) system was designed, built, and tested on squash crop for the first time in 2012 by Fayed [20], in research field of Water and Irrigation Systems Engineering Department, Faculty of Agricultural Engineering, Al-Azhar University, Nasr City, Cairo, Egypt. The system used emitters of 0.5 and 0.25 L/h (Fig. 9) under low pressure of 200 kPa.

Under ULDI system, field observations seem to indicate that there is no saturated zone and that the wetted soil volume is greater compared with that for conventional emitter discharges. Also, the moisture content in soil profile under ultra-low emitter point is always around the soil field capacity conditions [19].

The following equation can be used to compute the wetting front advance in three directions, horizontal (H), vertical (V), and diagonal (D) under ULDI system in sandy and clayey soil [20].

$$H = 11.54\, t^{0.369} \cdot K^{0.053} \cdot q^{0.316} \tag{5}$$

$$V = 13.63\, t^{0.410} \cdot K^{0.115} \cdot q^{0.295} \tag{6}$$

$$D = 13.45\, t^{0.396} \cdot K^{0.095} \cdot q^{0.302} \tag{7}$$

where H, V, and D are the wetting front advances in horizontal, vertical, and diagonal, respectively (cm), t is the time of water application (h), K is the hydraulic conductivity of soil (cm/h), and q is the ultra-low emitter discharge rate (L/h).

3.6.1 Advantages of Ultra-low Drip Irrigation (ULDI) System

The advantages of ULDI system can be summarized as follows [16]:

1. Optimum growth conditions due to the ability to maintain optimum balance of air, water, and nutrients in the soil.
2. Better utilization of available space. Plant density can be increased.
3. Quicker turnaround of plant materials reducing growth cycles.
4. Higher yields.
5. Minimize leaching of nutrients that occur with excess water flow.
6. The ultra-low rate system is much cheaper than the common micro-irrigation systems, and smaller polyvinyl chloride (PVC) tube size reduced horsepower requirements.
7. No runoff on heavy soil.
8. No water loss through the root zone on very sandy soil.
9. Water and fertilizer saving up to 40–50%.
10. Better quality.
11. Water could be applied efficiently on shallow soils in hilly areas.

(1) Feeder lateral 16 mm. (2) On-line, pressure-compensated dripper 0.5 L/h (use with flat 2 or 4-way manifold). (3) Micro-tube 6 mm. (4) Angle arrow dripper.

Fig. 9 First use of ultra-low drip irrigation on a squash field in Egypt in 2012. Photographs were taken in June 2011 in the research field of Irrigation Systems Engineering Department, Faculty of Agricultural Engineering, Al-Azhar University, Nasr City, Cairo, Egypt [15]

4 Components of Drip Irrigation System

Drip irrigation system network has several components as shown in Fig. 10 [21]. Drip irrigation system is made up of many additional components. These components have to be compatible with each other, with the crop demands, and with the characteristics of the field to be irrigated.

The components are classified into six principal categories [1]:

1. Water source and pumping unit: A pumping system from an on-surface or underground source or a connection to public, commercial, or cooperative supply network.
2. Delivery system: Mainline, sub-mains, and manifolds (feeder pipelines).
3. Drip laterals (irrigating pipelines): Before it is discharged into the field through point source emitters, bubblers, or micro-sprinklers.
4. Control accessories: Valves, water meters, pressure gauges, flow regulators, automation devices, backflow preventers, vacuum and air release valves, etc.
5. Filtration system: Filters are sometimes omitted in the bubbler and spray sprinkler systems.
6. Chemical equipment: Drip systems may or may not include chemical injection equipment for the injection of plant nutrients and water treatment agents.

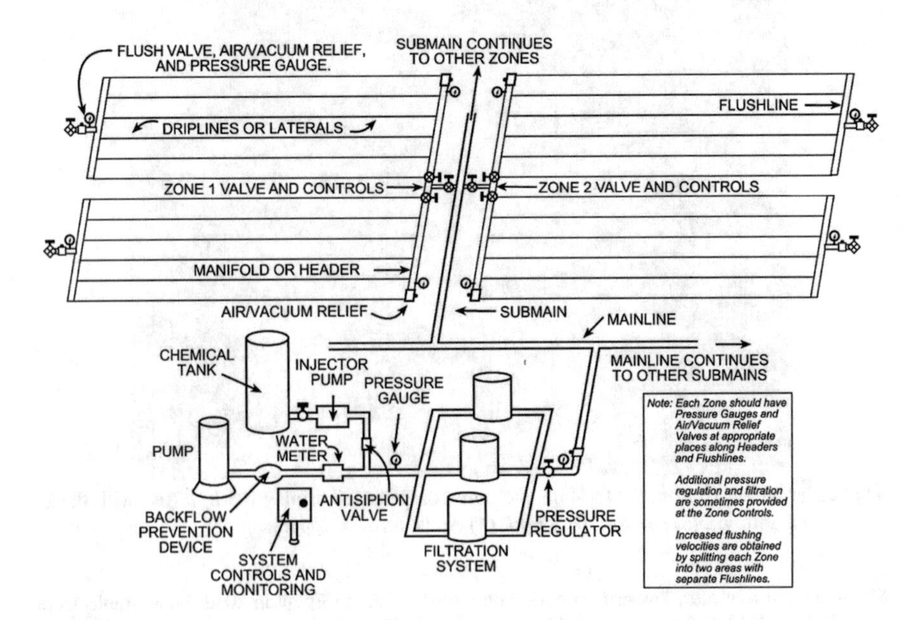

Fig. 10 Components of a drip irrigation system [21]. *Credit* Kansas State University

5 Design of Drip Irrigation System

5.1 *General Steps for Design*

When designing a drip irrigation system for a land area, it is necessary to obtain some important information that must be known to the designer. The information is required to design a drip irrigation system that can be classified as follows:

- **Climate data**: such as temperature, wind speed, rainfall distribution and its quantity. These data are required to calculate the evaporation of the plant.
- **Soil properties**: such as soil type, construction, porosity, soil density, field capacity, permanent wilting point, salinity ratio, leaching properties, and nutrients.
- **Plant properties**: type of yield, root depth, crop coefficient, and vegetative area.
- **Field data**: field form, area, drainage channels, and obstacles, etc.
- **Water source**: location, discharge rate, and water quality.
- **Economic information**: It includes the cost of equipment, energy cost, operation, labor, maintenance, and expected return.
- **Hardware and materials needed for design**: (These devices include pipes, valves, controllers, filters, fertilization vessels, meters, etc.) the necessary information about these devices such as performance curves, manufacturing different coefficients, etc.

5.2 *Elements that Are Calculated When Designing Drip Irrigation System*

The complete design of the drip irrigation system includes the following:

5.2.1 Calculate Maximum Net Depth of Each Irrigation Application (d_n)

It can be calculated from the following equation [22]:

$$d_{\mathrm{n}} = \mathrm{AW} \times D_{\mathrm{s}} \times P_{\mathrm{w}} \times \mathrm{dep} \tag{8}$$

The available water (AW) is given by the following equation [6]:

$$\mathrm{AW} = (\theta_{\mathrm{F.C}} - \theta_{\mathrm{P.W.P}}) \times \gamma_{\mathrm{b}} \times 1000 \tag{9}$$

where d_n is the maximum net depth of each irrigation application in (mm), AW is available water in (mm/m), D_s is the soil depth to be considered in (m), P_w is the wetted area as a percent of the total area as a fraction, dep is the portion of available moisture depletion allowed or desired, $\theta_{F.C}$ is soil moisture content on a weight basis at field capacity as a fraction, $\theta_{P.W.P}$ is soil moisture content on a weight basis at permanent wilting point as a fraction, and γ_b is the relative density of soil in (g/cm^3).

The ratio P_w depends on the type of plant and the distance between plants, either the distance between emitters on the line (S_e) or the distance between lines (S_l), depending on the type of planning used in the drip irrigation network.

In case of single straight lateral pattern, values of P_w are obtained from tables directly at the planting distances. In case of double straight lateral pattern, value of P_w is calculated from the following equation [22]:

$$P_w = \frac{(P_{w1}S_1) + (P_{w2}S_2)}{S_r} \tag{10}$$

where P_{w1} is taken from Table 1 for distance S_1, S_1 is the narrow spacing between pairs of laterals (m), P_{w2} is taken from Table 1 for distance S_2, S_2 is the wider distance between the laterals which is equal to ($S_r - S_1$) (m), and S_r is the distance between tree rows (m).

In case of zigzag or pigtail pattern, value of P_w is calculated from the following equation [22]:

$$P_w = \frac{100 \times n \times S_{ep} \times S_w}{S_t \times S_r} \tag{11}$$

where n is the number of emission points per tree, S_{ep} is the distance between emission points (m), S_w is the width of the wetted strip which is the S_1 value taken from Table 1 which would give $P = 100\%$ for the given emitter discharge and soil (m), S_t is the distance between trees in the rows (m), and S_r is the distance between tree rows (m).

5.2.2 Calculate the Number of Emitters per Tree or Plant (N_e)

The number of emitters per tree or plant varies according to patterns of lateral as follows:

In case of single straight lateral pattern for each row of trees, the number of emitters per tree or plant can be calculated from the following equation [6, 22]:

$$N_e = \frac{k \times P_w \times S_t \times S_r}{D_w \times S_l} \tag{12}$$

where N_e is the number of emitters per tree or plant, k is constant equal to 100, P_w is the wetted area as a percent of the total area (%), S_t is the distance between trees or plants in the row (m), S_r is the distance between tree rows or plant rows (m), D_w is the maximum diameter of wetted circle formed by a single point source emission device (cm), and S_l is the distance between emitter lines (cm).

Table 1 Percentage of soil wetted various discharge rates and spacing for a single row of uniformly spaced emission points in straight line applying 40 mm of water per cycle over the wetted area [22]

Effective spacing between laterals, S_l (m)[a]	Emission point discharge (L/h)[b]														
	Less than 1.5			2			4			8			More than 12		
	Recommended spacing of emission points along the lateral for coarse, medium, and fine textured soils—S_e (m)[c]														
	C	M	F	C	M	F	C	M	F	C	M	F	C	M	F
	0.2	0.5	0.9	0.3	0.7	1.0	0.6	1.0	1.3	1.0	1.3	1.7	1.3	1.6	2.0
	Percentage of soil wetted[d]														
0.8	38	88	100	50	100	100	100	100	100	100	100	100	100	100	100
1.0	33	70	100	40	80	100	80	100	100	100	100	100	100	100	100
1.2	25	58	92	33	67	100	67	100	100	100	100	100	100	100	100
1.5	20	47	73	26	53	80	53	80	100	80	100	100	100	100	100
2.0	15	35	55	20	40	60	40	60	80	60	80	100	80	100	100
2.5	12	28	44	16	32	48	32	48	64	48	64	80	64	80	100
3.0	10	23	37	13	26	40	26	40	53	40	53	67	53	67	80
3.5	9	20	31	11	23	34	23	34	46	34	46	57	46	57	68
4.0	8	18	28	10	20	30	20	30	40	30	40	50	40	50	60
4.5	7	16	24	9	18	26	18	26	36	26	36	44	36	44	53

(continued)

Table 1 (continued)

Effective spacing between laterals, S_l (m)[a]	Emission point discharge (L/h)[b]														
	Less than 1.5			2			4			8			More than 12		
	Recommended spacing of emission points along the lateral for coarse, medium, and fine textured soils—S_e (m)[c]														
	C	M	F	C	M	F	C	M	F	C	M	F	C	M	F
	0.2	0.5	0.9	0.3	0.7	1.0	0.6	1.0	1.3	1.0	1.3	1.7	1.3	1.6	2.0
	Percentage of soil wetted[d]														
5.0	6	14	22	8	16	24	16	24	32	24	32	40	32	40	48
6.0	5	12	18	7	14	20	14	20	27	20	27	34	27	34	40

Courtesy Rain Bird Corporation

[a]Where double laterals (or laterals having multi-exit emitters) are used in orchards and enter with the average lateral spacing as S_l, provided the spacing between any two laterals (or rows of emission points) is equal to or greater than an S_l value which would give P %

[b]When irrigation is applied in relatively short pulses, the horizontal spread of the wetted zone is less than for heavier applications. Therefore, the table should be entered with an effective emission point discharge of approximately half the instantaneous discharge. In soils with hardpans, clay or sand lenses, or other stratifications which enhance the horizontal spread of moisture, the table may be entered with up to double the installation rate for regular application (or for the instantaneous rate for short pauses)

[c]The texture of the soil is designated by C, coarse; M, medium; and F, fine. The emission point spacing is equal to approximately 80% of the largest diameter of the wetted area of the soil underlying the point (closer spacings on the lateral do not affect the percentage area wetted)

[d]The percentage of soil wetted is based on the area of the horizontal section approximately 0.30 m beneath the soil surface. In wide-spaced tree crops, caution should be exercised where less than 1/3 of the soil volume will be wetted in low rainfall areas or 1/5 in high rainfall areas. In close-spaced crops, most of the soil volume may need to be wetted to assure a sufficient water supply to each plant

In the other patterns, double laterals for each row of trees, multi-exit emitter hose to distribute water, zigzag single lateral for each row of trees and pigtail with emitters connected to single lateral, and the number of emitters per tree or plant can be calculated from the following equation [2]:

$$N_e = \frac{2k \times P_w \times S_t \times S_r}{S_l(S_l + D_w)} \tag{13}$$

In the double lateral pattern for each row of trees, the distance between double laterals should be equal to the value of D_w where the largest wet area can be obtained. In the case of zigzag, pigtail, and multi-exit patterns, the distance between emitters must be equal to D_w value in each direction.

In case of micro-sprinklers, the number of it can be calculated from the following equation [2]:

$$N_e = \frac{P_w \times S_t \times S_r}{100\left(A_s + \frac{D_w + P_s}{2k}\right)} \tag{14}$$

where A_s is the irrigated area by micro-sprinkler (m^2) and P_s is circumference of the wetted circle (cm). In the previous equation, D_w can be calculated as follows [2]:

$$S_e = D_T + \frac{D_w}{2k} \tag{15}$$

where D_T is the ejaculation distance of the micro-sprinkler (m) and S_e is the distance between emitters on the line (m).

5.2.3 Calculate the Number of Operating Stations (or Units) into Which a Field Is Divided (*N*)

Naturally, it is difficult to irrigate the total area at the same time, so the total area should be divided into a number of operating stations that will be irrigated alternately during the irrigation period. The number of operating stations (or units) into which a field is divided is calculated from the following equation [2, 6, 22]:

$$N \leq \frac{H_a \times q_t}{S_e \times S_l} \times \frac{0.90\,\mathrm{EU}}{\mathrm{ET_O}} \tag{16}$$

or

$$N \leq \frac{H_a \times F}{h} \tag{17}$$

where N is the number of operating stations (or units) into which a field is divided, H_a is actual daily operating hours for pump and motor (it is ranging from 12 to

15 h/day), and q_t is total discharge rate for single tree (L/h). If each tree is irrigated by three emitters, q_t is equal to 3 × discharge rate of single emitter. S_l is the distance between lines, EU is the emission uniformity as a fraction, ET_O is the transpiration rate of the crop (mm/day), F is the irrigation interval (day), and h is total time of operation for each operational unit during each irrigation cycle (h).

5.2.4 Calculate the Total Discharge Rate of Drip Irrigation Network (Q)

The total discharge rate (system capacity) of the drip irrigation system network can be calculated from the following equation [2, 22]:

$$Q\ (\mathrm{m^3/h}) = \frac{A \times F \times \mathrm{ET_O}}{N \times h \times 0.90\,\mathrm{EU}} \times 10^{-3} \tag{18}$$

or

$$Q\ (\mathrm{m^3/h}) = \frac{A \times q_t}{N \times S_e \times S_l} \times 10^{-3} \tag{19}$$

or

$$Q\ (\mathrm{m^3/h}) = \frac{A \times d_g}{N \times h} \times 10^{-3} \tag{20}$$

Also, the total discharge rate of drip irrigation network can be calculated by another method in m^3/day as follows [2]:

$$Q\ (\mathrm{m^3/day}) = \frac{A \times F \times H_a \times \mathrm{ET_O}}{N \times h \times 0.90\,\mathrm{EU}} \times 10^{-3} \tag{21}$$

where Q is total discharge rate of drip irrigation network (m^3/h), A is the total area to be irrigated (m^2), and d_g is the gross depth of each irrigation application (mm).

The gross depth of each irrigation application can be calculated directly by the following equation [2, 6, 22]:

$$d_g = \frac{d_n}{E_a} = \frac{F \times \mathrm{ET_O}}{E_a} \tag{22}$$

where E_a is the efficiency of irrigation (efficiency of applying water) by a drip irrigation system as a fraction.

The irrigation efficiency (E_a) is expressed by the following equation [2, 22]:

$$E_a = K_s \times \mathrm{EU} \tag{23}$$

where K_s is a coefficient less than one and expresses deep percolation losses in which [2, 6]:

$$K_s = \frac{\text{Average water stored in root volume}}{\text{Average water applied}} \quad (24)$$

Since it is difficult to calculate the value of K_s, it is estimated based on soil type as follows [23]:

Coarse sand or light topsoil with	
Gravel subsoil	$K_s = 0.87$
Sand	$K_s = 0.91$
Silt	$K_s = 0.95$
Loam and clay	$K_s = 1.00$

5.2.5 Calculate the Time of Operation for Each Operational Unit During Each Irrigation Cycle (*h*)

The time of operation for one section of total area sections can be calculated from the following equation [2, 6, 22]:

$$h = \frac{S_e \times S_l \times d_n}{q_t} \quad (25)$$

5.2.6 Calculate the Irrigation Requirement Under Drip Irrigation System (I.R)

The irrigation requirement for trees or vegetable crops under drip irrigation system can be calculated from the following equation [2]:

$$\text{I.R} = \frac{\text{ET}_o \times K_c \times K_r \times S_e \times S_l}{E_a \times \text{EU}\,(1 - \text{LR})} \quad (26)$$

where I.R is the irrigation requirement per one tree or one plant (L/day), K_c is crop coefficient, K_r is the appropriate ground cover reduction factor, and LR is the amount of water required for leaching of salt (a number less than one).

The crop coefficient (K_c) is taken from FAO's tables [24]. The amount of water required for the leaching of salts (LR) can be calculated from the following equation [2]:

$$LR = \frac{EC_i}{EC_d} \tag{27}$$

where EC_i is an electrical conductivity of irrigation water (dS/m or mmhos/cm) and EC_d is an electrical conductivity of the saturated soil extract (dS/m or mmhos/cm).

The appropriate ground cover (GC) reduction factor is a number less than one and can be calculated from the following equations [2, 23]:

- **Keller and Karmeli equation**:

$$K_r = \frac{GC}{0.85} \tag{28}$$

- **Freeman and Garzoli equation**:

$$K_r = 0.50\,(GC + 1) \tag{29}$$

- **Decroix and CTG REF equation**:

$$K_r = 0.10 + GC \tag{30}$$

FAO 1984 provides the reduction factors suggested by various researchers in order to account for the reduction in evapotranspiration as shown in Table 2 [23].

Table 2 Values of K_r suggested by different authors [23]

GC (%)	Crop factor K_r according to		
	Keller and Karmeli	Freeman and Garzoli	Decroix CTG REF
10	0.12	0.10	0.20
20	0.24	0.20	0.30
30	0.35	0.30	0.40
40	0.47	0.40	0.50
50	0.49	0.75	0.60
60	0.70	0.80	0.70
70	0.82	0.85	0.80
80	0.94	0.90	0.90
90	1.00	0.95	1.00
100	1.00	1.00	1.00

6 Evaluation of Drip Irrigation System

Evaluation of drip irrigation system includes evaluation of the following elements:

1. The gross depth of each irrigation application (mm).
2. The irrigation requirement per tree or plant (L/day).
3. Are the hours of daily operation given the necessary irrigation requirements? Especially when the maximum daily water requirement under Egyptian conditions at the age of the tree that needs the maximum water requirement.
4. Distribution of moisture in the soil profile in three directions as follows:
   - In horizontal direction between trees or plants along one row.
   - In vertical direction (horizontal) between rows.
   - In the depth of both previous directions.
5. Distribution of salts in the soil profile at the same previous directions.
6. Efficiency of the filters.
7. Efficiency of the emitters and their suitability for soil type and plant type.
8. Water losses.
9. Network efficiency in general.
10. Economics and costs of drip irrigation network.

6.1 Experimental Evaluation of Drip Irrigation Network

The drip irrigation network to be evaluated for a specific period of 3–4 h is operated, and the following elements are measured [2].

6.1.1 Watching Water Pressure (on Pressure Meter) Before and After the Filter Unit

It is normal to have pressure before the filters greater than after, but at 50 kPa, and if the difference is more, it means there are deposits in the filters and therefore filters should be washed if the reverse washing is done manually. If reverse washing is done automatically, it is not necessary to note filters, only to see efficiency of the filters.

6.1.2 Determination of the Net Depth of Each Irrigation Application (d_n)

Previously, the net depth of each irrigation application (d_n) under drip irrigation system has already been calculated in Sect. 5.2.1.

6.1.3 Determination of the Irrigation Requirement (I.R)

Previously, irrigation requirement per tree or plant (I.R) under drip irrigation system has already been calculated as follows [2, 23]:

$$\text{I.R (L/day)} = \frac{\text{ET}_\text{O} \times K_\text{c} \times K_\text{r} \times S_\text{e} \times S_\text{l}}{E_\text{a} \times \text{EU}(1 - \text{LR})} \tag{31}$$

By operating the required hours according to the irrigation program, the amount of water added per tree or plant can be calculated as follows [2]:

$$\text{Water amount added per tree or plant (L/day)} = q_\text{t} \times H \tag{32}$$

6.1.4 Distribution of Moisture and Salts in Soil Profile

This is done by taking samples of different depths [(0–20), (20–30), (30–40), (40–60), and (60–80) cm] and may increase depending on soil type, depth of irrigation water, and age of plant. This is done for all the depths between the plants or trees on the row and between the rows, and then the results obtained are drawn in contour lines.

6.1.5 Emitter Evaluation

There are some very important parameters to evaluate the performance of emitters. These parameters are of emitter discharge (q), the emitter discharge exponent (x), the coefficient of variation of the discharge (C_v), emission uniformity (EU), emitter flow variation (q_var), and application efficiency (E_a). These parameters are described below:

1. **Pressure–Discharge Relationship**

In the design of drip irrigation systems, the relationship between emitter discharge and operating pressure is calculated based on the emitter flow function given by Keller and Karmeli [22, 25] as follows:

$$q_\text{e} = K_\text{d} \cdot h_\text{d}^x \tag{33}$$

where q_e is emitter discharge rate (L/h), K_d is constant of proportionality that characterizes each emitter, h_d is the operating pressure head at the emitter (m), and x is emitter discharge exponent that is characterized by the flow regime.

The discharge rate of several emitters (more than 20 emitters) must be measured along the drip line at different operating pressures. Then, the results are drawn to know the difference in the discharge rates. The average discharge rate of emitters for each line is calculated to predict the moisture distribution which can result from this and the extent of differences in this moisture or salt distribution along the drip line.

2. **Manufacturer's Coefficient of Variation (C_v)**

Manufacturer's coefficient of variation of the discharge (C_v) is one of the significant parameters related to the uniformity and efficiency of the system. It could be obtained by taking a random sample of emitters and measuring the discharge rates at the same temperature and pressure. It can be calculated by using the following equation [23]:

$$C_v = \frac{\mathrm{Sd}}{q_{\mathrm{avg}}} \tag{34}$$

where C_v is manufacturer's coefficient of variation, Sd is estimated standard deviation of the discharge rates of sampled set of emitters (L/h), and q_{avg} is the average discharge rates of sampled set of emitters (L/h).

The standard deviation values can be calculated in the same manner using the following equation [23, 26]:

$$\mathrm{Sd} = \left(\frac{(q_1^2 + q_2^2 + \cdots + q_n^2) - n(q_{\mathrm{avg}})^2}{n-1} \right)^{0.5} \tag{35}$$

where $q_1, q_2, \ldots, q_n$ is discharge rate of emitters tested (L/h) and n is the number of emission devices tested.

The coefficient of discharge variation values can be classified according to ASAE Standards [26] as shown in Table 3.

3. **Emission Uniformity (EU)**

In order to determine if the system is operating at acceptable efficiency, the uniformity of emission can be evaluated by the following equation [6, 22, 23, 26]:

$$\mathrm{EU} = 100 \times \left(1.0 - \left[\frac{1.27}{\sqrt{n}} C_v\right]\right) \times \left(\frac{q_{\min}}{q_{\mathrm{avg}}}\right) \tag{36}$$

Table 3 Recommended classification of manufacturer's coefficient variation (C_v) [26]

Emitter type	C_v range		Classification
Point source		<0.05	Excellent
	0.05	0.07	Average
	0.07	0.11	Marginal
	0.11	0.15	Poor
		>0.15	Unacceptable
Line source		<0.10	Good
	0.10	0.20	Average
		>0.20	Marginal to unacceptable

Table 4 Recommended ranges of design emission uniformity (EU) [26]

Emitter type	Spacing (m)	Topography	Slope (%)	EU range (%)
Point source on perennial crops	>4	Uniform	<2	90–95
		steep or undulating	>2	85–90
Point source on perennial or semipermanent crops	<4	Uniform	<2	85–90
		steep or undulating	>2	80–90
Line source on annual or perennial crops	All	Uniform	<2	80–90
		steep or undulating	>2	70–85

where EU is the design emission uniformity (%), n is the number of emitters per plant, and q_{min} is the minimum emitter discharge for minimum pressure in the sub-unit (L/h).

General criteria for EU values for systems which have been in operation for one or more seasons are: greater than 90%, excellent; between 80% and 90%, good; 70–80%, fair; and less than 70%, poor [27].

Table 4 shows range for EU values recommended by ASAE for use in Eq. 36 [26]. Economic considerations may dictate a higher or lower uniformity than those given in Table 4 [6].

4. **Variation from Hydraulic Design (q_{var})**

Flow variation within laterals occurs due to pressure head variations and manufacturing variations of the individual emitters. Both of these processes will be discussed and related as to how they can be used in the design process. Lateral emitter flow variation (q_{var}) was determined as the following equation given by Wu and Gitlin [28]:

$$q_{var} = \frac{q_{max} - q_{min}}{q_{max}} \times 100 \tag{37}$$

where q_{var} is lateral emitter flow variation (%), q_{max} is maximum emitter flow rate (L/h), and q_{min} is minimum emitter flow rate (L/h).

Wu and Gitlin [28] recommended the general criteria for emitter flow variation (q_{var}) values which are 10% or less is generally desirable, acceptable when between 10 and 20%, and unacceptable when greater than 20%.

5. **Application Efficiency (E_a)**

The efficiency of applying water by a drip irrigation system depends on the uniformity of application (EU). Application efficiency (E_a) can be calculated by using the following equation [2, 22, 24]:

$$E_a = K_s \times EU \tag{38}$$

6.1.6 Water Losses and Network Efficiency

The drip irrigation network should be continuously observed during operation, especially the pressure meters and the emitters to avoid any decrease in the operating pressure or the discharge rate of the emitters. If happened, it means a leakage in one of the main or sub-main lines exist. The leak position should be detected and repaired immediately [2].

The various components of a drip irrigation system network require different preventative and operational maintenance procedures to ensure proper system performance and extend the life of the component. These procedures that can be implemented to roughly evaluate performance are as follows [29]:

The first step is to check the hourly flow rate at the main flow meter and compare it with the designed flow rate [design flow rate (L/h) = number of emitters × emitter nominal flow rate (L/h)].

Second step is the pressure gauges that are installed in the plot have to be checked. The measured values have to be compared to the designed pressure for each set. The pressure difference between inlet and outlet and dirt accumulation in filters have to be checked as well.

Also, a periodic visual inspection of the status of the media could help assess drip irrigation network effectiveness and indicate if the backlashing schedule needs to be adjusted, or perhaps chlorination is needed; due to excess biomass growth regularly scheduled maintenance operations, such as the monthly flushing of all laterals, can prevent accumulation of fines and the potential for irreversible emitter plugging. It is advisable to follow the instructions provided by the suppliers of all the components of the drip irrigation system with care [30].

6.1.7 Economics and Costs of Drip Irrigation Network

The cost of the drip irrigation network is calculated after the appropriate design has been made and then the total costs are calculated. Annualized operating costs can be estimated by calculating fixed and variable annualized costs [2].

7 Summary

Drip or trickle irrigation is the most efficient method of irrigating. The drip irrigation system works by applying water slowly, directly, and at frequent intervals to the soil through mechanical devices called emitters or drippers, localized at selected points along the water delivery line. Also, the drip irrigation can apply the fertilizers and pesticides with irrigation water. The different types of drip irrigation comprise: surface, subsurface, bubbler, micro-sprinkler, mobile drip irrigation, and ultra-low drip irrigation. All drip irrigation systems consist of a pumping unit, fertigation unit, filtration unit, a control head, main and sub-main pipes, laterals, and emitters.

The complete design of the drip irrigation system includes calculation of: the maximum net depth of each irrigation application (d_n), the number of emitters per tree or plant (N_e), the number of operating stations (or units) into which a field is divided (N), the total discharge rate of drip irrigation network (Q), the time of operation for each operational unit during each irrigation cycle (h), and the irrigation requirement under drip irrigation system (I.R).

The evaluation of drip irrigation system includes evaluation of: watching water pressure (on pressure meter) before and after the filter unit, determination of the net depth of each irrigation application (d_n), determination of the irrigation requirement (I.R), distribution of moisture and salts in soil profile, emitter evaluation, water losses and network efficiency and economics and costs of drip irrigation network.

References

1. Al-Amoud AI (1998) Trickle irrigation systems. King Saud University, Kingdom of Saudi Arabia (in Arabic)
2. Badr AA (1990) Irrigation engineering and drainage. Cairo University, Egypt (in Arabic)
3. Lamm FR, Ayars JE, Nakayama FS (2007) Microirrigation for crop production: design, operation, and management, vol 13. Elsevier, Amsterdam
4. El-Awady MN, Amerhom GW, Zaki MS (1975) Trickle irrigation trial on pea in conditions typical of Qalubia. Ann Agric Sci Moshtohor 4:235–244
5. Aung KH, Thomas FS (2003) Introduction to micro-irrigation. North Dakota State University, NDSU Extension Service, AE-1243
6. James LG (1988) Principles of farm irrigation systems design. Wiley, New York, pp 260–298
7. Waller P, Yitayew M (2016) Irrigation and drainage engineering. Springer, Cham, Heidelberg, New York, Dordrecht, London, pp 1–18
8. Elmesery AEA (1999) A study on design and evaluation of bubbler irrigation system. Ph.D. thesis, Faculty of Agriculture, Al-Azhar University, Cairo, Egypt
9. Sourell H (2000) Mobile drip irrigation—an alternative to irrigation with nozzles. In: 6th international microirrigation congress (Micro 2000), Cape Town, South Africa, 22–27 Oct 2000. International Commission on Irrigation and Drainage (ICID), pp 1–9
10. Lamm F (2003) Drip irrigation laterals on center pivot irrigation. Irrigation at K-State Research and Extension. https://www.ksre.k-state.edu/irrigate/photos/mdi.html
11. Derbala AA (2003) Development and evaluation of mobile drip irrigation with center pivot irrigation machines. Ph.D. thesis, Landbauforschung Völkenrode, FAL Agricultural Research, Special Issue 250. https://d-nb.info/996797408/34
12. Rawlins SL, Hoffman GW, Merrill SD (1974) Traveling trickle system. In: Proceedings of the international drip irrigation congress, 2nd, San Diego, pp 184–187 [cited from Phene CJ et al (1985)]
13. Phene CJ, Howell TA, Sikorski MD (1985) A traveling trickle irrigation system. Adv Irrig 3:1–49 (online). https://books.google.com.eg/books?id=RQ-XgBAAAQBAJ&printsec=frontcover&hl=ar&source=gbs_ge_summary_r&cad=0#v=onepage&q&f=false
14. Hezarjaribi A (2008) Site-specific irrigation: improvement of application map and a dynamic steering of modified centre pivot irrigation system. Doctoral dissertation, Universitätsbibliothek Giessen. http://geb.uni-giessen.de/geb/volltexte/2008/5759/
15. Fayed MH (2016) Design and evaluation of mobile drip irrigation system. Ph.D. thesis, Faculty of Agriculture, Al-Azhar University, Cairo, Egypt
16. Lubars P, Richard M (2008) Minute or ultra-low micro-irrigation. http://www.scribd.com/doc/8145273/p13

17. Koenig E (1997) Methods of micro-irrigation with very small discharges and particularly low application rates. Water Irrig 365:32–38 [cited from Assouline S (2002), in Hebrew]
18. Assouline S, Cohen S, Meerbach D, Harodi T, Rosner M (2002) Microdrip irrigation of field crops: effect on yield, water uptake, and drainage in sweet corn. Soil Sci Soc Am J 66:228–235
19. Assouline S (2002) The effects of microdrip and conventional drip irrigation on water distribution and uptake. Soil Sci Soc Am J 66:1630–1636
20. Fayed MH (2012) A study on some factors affecting ultra low drip irrigation. MS thesis, Faculty of Agriculture, Al-Azhar University, Cairo, Egypt
21. Lamm F (2015) SDI in the Great Plains. Drawings of microirrigation component. K-State Research and Extension. https://www.ksre.k-state.edu/sdi/images/photos-/dmc/MIS3.jpg
22. Keller J, Karmeli D (1975) Trickle irrigation design (No. 04; TC805, K3). Rain Bird Sprinkler Manufacturing Corporation, Glendora, CA
23. Savva AP, Frenken K (2002) Irrigation manual: planning, development, monitoring and evaluation of irrigated agriculture with farmer participation, vol IV. Module 9: localized irrigation: planning, design, operation and maintenance. FAO. http://www.fao.org/3/a-ai598e.pdf
24. Savva AP, Frenken K (2002) Irrigation manual: crop water requirements and irrigation scheduling, vol ii. Module 4. FAO. http://www.fao.org/3/a-ai593e.pdf
25. Keller J, Karmeli D (1974) Trickle irrigation design parameters. Trans ASAE 17(4):678–684
26. ASAE Standard (1994) Design and installation of microirrigation systems. EP405.1, pp 724–727
27. Merriam JL, Keller J (1978) Farm irrigation system evaluation: a guide for management, 3rd edn. Utah State University, Logan, UT, USA
28. Wu IP, Gitlin HM (1974) Drip irrigation design based on uniformity. Trans ASAE 17:429–432
29. Sne M (2009) Micro irrigation: technology and application, 2nd edn. CINADCO, pp 117–122. https://www.scribd.com/doc/20157961/Micro-Irrigation-Technology-and-Applications
30. Dasberg S, Or D (1999) Drip irrigation (applied agriculture), 1st edn. Springer, Berlin, Heidelberg, New York, pp 100–126. ISSN 1433-7576

Water Reuse and Treatment

Irrigation with Magnetically Treated Water Enhances Growth and Defense Mechanisms of Broad Bean (*Vicia Faba* L.) and Rehabilitates the Toxicity of Nickel and Lead

Eman R. Abuslima, Amal H. Saleh and Ahmed I. Mohamed

1 Introduction

The different industries such as smelters, tanneries, metal refineries, and mining operations are considered the major sources of heavy metal release to the soil which threaten the agricultural sustainability [1, 2]. Proper amounts of heavy metals are required for plant growth; however, the ability of plants to accumulate essential metals equally enables them to acquire other nonessential metals [3].

Nickel (Ni) is known to be an essential plant nutrient which is required for the proper functioning of many metalloenzymes such as urease [4–6]. At high concentration, however, Ni has been found reduced shoot and root growth [7], and excess Ni levels also affect nutrient absorption via roots [7, 8] and inhibit antioxidant enzyme activity [9]. Consequently, such impairments lead to crop yield shortage, when excessive Ni concentrations are present in soils.

Although lead (Pb) is one of the nonessential plant nutrients, studies from different parts of the world though "have revealed that vegetables grown in contaminated soils can accumulate unhealthy amounts of lead in the edible parts of the plant" [10]. It often leads to diminished growth, deformation of cellular structures, ion homeostasis, and reductions in chlorophyll biosynthesis, and induces over-production of reactive oxygen species (ROS) which consequently lead to lipid peroxidation and alteration of total proteins and antioxidant level in plants [2, 11–13]. Therefore, it severely affects crop growth and productivity [14, 15].

Reactive oxygen species (ROS) once produced readily attacks biological structures and biomolecules and results in metabolic dysfunction [16]. Hence, activation

E. R. Abuslima (✉) · A. H. Saleh
Department of Botany, Faculty of Science, Suez Canal University, Ismailia 41522, Egypt
e-mail: eman_ramadan@science.suez.edu.eg

A. I. Mohamed
Department of Soil and Water, Faculty of Agriculture, Suez Canal University, Ismailia 41522, Egypt

E.-S. E. Omran and A. M. Negm (eds.), *Technological and Modern Irrigation Environment in Egypt*, Springer Water,
https://doi.org/10.1007/978-3-030-30375-4_14

of the anti-oxidative defense system is one of the mandatory strategies to scavenge ROS that plants provoke [15, 17]. Antioxidants, both enzymatic such as superoxide dismutase (SOD), ascorbate peroxidase (APX), peroxidase (POX), and catalase (CAT), and non-enzymatic such as glutathione (GSH), are involved in direct and/or indirect detoxification of ROS in plants [18, 19].

SOD can sequester O_2^- primarily in chloroplasts, reducing lipid membrane peroxidation. Hence, CAT which is located in the peroxisomes of plant cells, its main role is to lessen H_2O_2, which is produced by the SOD reaction, while APX can remove H_2O_2; however, it is distributed in the peroxisomes as well as chloroplasts, cytosol, and mitochondrion [20]. POX also can reduce H_2O_2 accumulation and eliminate MDA, leading to the integrity of cell membrane lipids and maintaining cell membrane stability [21, 22]. In addition, glutathione has a role in the detoxification of ROS, but induces the regulation of sulfur transport and upregulating stress defense genes as well [20, 23].

Anthocyanins and carotenoids also play a dual role as pigments distributed in various fruits and vegetables and as non-enzymatic antioxidants which can protect plants from biotic and abiotic stresses [24]. Moreover, plants also synthesize organic compounds or osmolytes such as proline and soluble sugars to protect essential proteins and enzymes affecting plant growth and development [25] and to maintain cell osmotic potential [26–28].

Modern agricultural efforts are now in search of an efficient and eco-friendly production technology for improving the crop productivity without harming the environment, and magnetic water treatment is one such area [29, 30]. Water in liquid form can be affected by magnetic fields [31, 32] which assist its purification [33]. Lee et al. [34], Iwasaka and Ueno [35], and McMahon [36] have declared that the size of the water clusters changes when exposed to a magnetic field. Su and Wu [37] discussed how MTW can break up water clusters into single or smaller clusters; hence, the activity of water is improved, and the hydration can be done more efficiently.

Morejon et al. [38] observed an increase in germination of *Pinus tropicalis* seeds from 43% in control to 81% with MTW and a marked improvement in the seedling growth after germination as well. Irrigation of common bean plants with magnetic water increased the growth characteristics, nucleic acids (RNA and DNA), photosynthetic pigments, photosynthetic activity, and translocation productivity of photoassimilates compared to control plants [39] and total protein content in broad bean compared with control plants [40].

Another interesting feature of MTW is that they appeared to enhance tolerance to abiotic stresses. Bagherifard and Ghasemnezhad [41] observed that under water stress conditions the magnetized water treatment caused more activity of antioxidant enzymes and free proline in mung bean plant compared to ordinary water. In artichoke leaves, a significant difference at $p \leq 0.001$, in phenols and total antioxidant activity, was recorded from irrigated plants with magnetized salty water as compared to irrigated plants with tap water [42].

Broad bean is considered a main ingredient of Egyptian people's diet on a daily basis. Therefore, increasing production of broad bean and improving yield quality are the major targets to meet the demand of the increasing Egyptian population

[43]. Although the heavy metal tolerance of this crop has been previously explored, however, so far as we know, there is no study dealing with the effect of magnetically treated water in alleviating the toxicity of heavy metals in *Vicia faba* L. that has been conducted before. Therefore, the present study was carried out to perform further investigation on the impact of magnetically treated water on increasing tolerance toward more abiotic stresses such as heavy metals on *Vicia faba* L.

2 Materials and Methods

2.1 Soil Preparation, Plant Material, and Growth Conditions

Sandy soil samples were collected from the surface horizon (0–30 cm) of Research Farm of the Faculty of Agriculture, Suez Canal University. Soil samples were artificially contaminated with two doses of Ni and Pb provided as aqueous solutions of $NiSO_4$ and $Pb(NO_3)_2$, at the beginning of the experiment. The concentrations of Ni and Pb selected for the treatments were (250 and 500), and (500 and 1000) mg kg^{-1} dry weight soils, respectively, which are higher than the upper metal concentration level in soil were considered toxic to plants [44]. The zero level of heavy metal treatment was considered as the control (C). The choice of Ni and Pb concentrations was based on previous studies [45]. The treatments were mixed uniformly with soil and incubated in laboratory at the temperature range of 25–28 °C for a month. After incubation, 20 kg of soil from each group was placed in each plastic pot (50 cm in height and 30 cm in diameter) and fertilized with 50.0 mg N kg^{-1} dry soil as urea, 50.0 mg P kg^{-1} as diammonium phosphate, and 50.0 mg K kg^{-1} as potassium sulfate.

Seeds of broad bean (*Vicia faba* L.) were obtained from Agricultural Research Center (ARC), Ismailia, Egypt. Broad bean (*Vicia faba* L.) seeds were surface sterilized with 2.5% sodium hypochlorite for 15 min and then washed thoroughly with distilled water. Six seeds of broad bean were grown in each pot. Each group from the above five groups was divided into two subgroups: One was irrigated with tap water, and the other was irrigated with magnetically treated water (MTW). Irrigation water was passed through a magnetic field (1000 G) magnetron unit of 0.5-inch diameter to obtain magnetically treated water. Soil moisture was maintained at 80% of field capacity based on temperature changes and relative moisture to avoid leakage of water. Broad bean samples were collected after six weeks post-sowing. For biomass determination, after measurement of fresh weight, shoots and roots were oven-dried at 80 °C until reaching a constant weight.

2.2 *Biochemical Measurements*

2.2.1 Estimation of Photosynthetic Pigments

The photosynthetic pigments (chlorophyll a + b and carotenoids) were determined in *Vicia faba* L. according to the spectrophotometric method recommended by Metzner et al. [46]. The pigment contents were extracted from a known fresh weight of leaves in 85% (v/v) aqueous acetone. The extract was centrifuged at 4000 × *g* for 10 min. The absorbance was measured against a blank of pure 85% aqueous acetone at three wavelengths 663, 644, and 452.5 nm taking into consideration the dilution made. Then, pigment contents (Chl a, Chl b, and carotenoids) were calculated as $\mu g\ g^{-1}$ DW.

2.2.2 Estimation of Total Carbohydrates

The total carbohydrates were measured by acid hydrolysis of polysaccharides into simple sugars and estimated by the spectrophotometric method recommended by Hedge and Hofretier [47]; "100 mg of dried sample is hydrolyzed in a boiling tube containing 5 ml of 2.5 N hydrochloric acid in a boiling water bath for 3 h. Then, 'cool to room temperature. Neutralize it with solid sodium carbonate until the effervescence ceases; make up the volume to 100 ml and centrifuge'" (https://www.omicsonline.org/proximate-and-phytochemical-analysis-of-seed-coat-fr). The supernatant (1 ml) was mixed with 1 ml of 5% and 5 ml of 96% H_2SO_4. The reaction mix was boiled in a water bath at 25–30 °C for 20 min. Absorbance was recorded at 490 nm after cooling. The amount of sugar was calculated against reference standards prepared from glucose and calculated as mg/g DW of leaves.

2.2.3 Estimation of Total Proline

Proline was determined following the method of [48]. Fresh leaf tissue (500 mg) was homogenized in 10 ml of 3% sulfosalicylic acid. The homogenate was centrifuged at 10,000 × *g* for 10 min, and the supernatant was transferred to a fresh 1.5 ml tube. The extracted solution was reacted with an equal volume of glacial acetic acid and ninhydrin reagent (1.25 g ninhydrin in 30 mL glacial acetic acid and 20 mL 6 M H_3PO_4) and incubated at 95 °C for 1 h. The reaction was terminated by placing the tube in an ice bath. The reaction mixture was mixed with 4 ml of toluene by vortexing for 5 min with a test tube mixer. Absorbance of the chromophore-containing toluene was determined at 520 nm against the toluene blank. The proline content was calculated by using standard curve of pure proline.

2.2.4 Determination of Oxidative Stress (H_2O_2 and MDA)

Leaf tissues (200 mg) were homogenized in 2 ml of 3% trichloroacetic for hydrogen peroxide (H_2O_2) estimation, according to the modified method of [49]. The absorbance was measured at 390 nm. For malondialdehyde (MDA) determination, leaf tissues (300 mg) were homogenized in 2 ml of 0.1% trichloroacetic acid and MDA content was estimated according to [50]. The MDA concentration was determined by dividing the difference into absorbance (A532–A600) by its molar extinction coefficient (155 mM^{-1} cm^{-1}).

2.2.5 Estimation of Total Protein and Antioxidants

Total protein content was determined according to [51]. Fresh leaf samples (500 mg) were extracted in 5 mL phosphate buffer (0.1 M, pH 7.5). The extracts were centrifuged, and the supernatants were collected; 100 µl of extract was completed to 1 ml phosphate buffer (pH 7.5) added to 5 ml of Bradford reagent. Absorbance was recorded photometrically at 595 nm using bovine serum albumin as a standard. Protein contents were expressed as mg g^{-1} DW. The remaining supernatant was used for the following assays of antioxidant enzyme activities.

SOD (EC 1.15.1.1) was assayed by measuring its ability to inhibit the photochemical reduction of nitro blue tetrazolium (NBT) [52]. The reaction mixture contained enzyme extract added to 0.25 ml of 13 mM methionine, 80 µM NBT, and 0.1 mM EDTA, and the total volume of 3.0 mL was made with buffer in each set. Then, 0.25 mL of 50 µM riboflavin was added to each set in the last. The tubes were shaken and placed 30 cm away from light source for 20 min. The reaction was stopped by switching off the light. The absorbance was recorded at 560 nm. One unit of enzyme activity represents the amount of enzyme required for 50% inhibition of NBT reduction at 560 nm.

APX (EC 1.11.1.11) was assayed following [53]. The assay mix (2 mL) contained 25 mM potassium phosphate buffer (pH 7.0), 0.1 mM EDTA, 0.25 m Mascorbate, 1.0 mM H_2O_2 and 0.2 mL enzyme extract, and H_2O_2. The reaction was initiated by addition of hydrogen peroxide, and oxidation of ascorbate was followed by the decrease in absorbance at 290 nm at 30 s interval for 5 min. One unit of APX activity is defined as the amount of enzyme that oxidizes 1 µM of ascorbate per min at room temperature.

POX (EC 1.11.1.7) was assayed following [54]. The reaction mix consisted of 3 ml of the buffer solution (phosphate buffer 0.1 M, pH 7), 0.05 mL guaiacol solution, 0.1 mL enzyme extract, and 0.03 mL hydrogen peroxide solution in a cuvette. The mixture was well shaken and placed in the spectrophotometer. Δt for absorbance increase by 0.1 was recorded and used in calculations. Peroxidase enzyme activity was expressed as peroxidase activity g $protein^{-1}$.

CAT (EC 1.11.3.6) activity was assayed in a method following [55]. Activity was determined by the decrease in the absorbance of the reaction mixture at 240 nm; 3 mL of the reaction mixture contains 50 mM (pH 7.0) of potassium phosphate buffer,

7.5 mM of H_2O_2, and 50 μL of the crude extract. Δt for the absorbance decrease from 0.45 to 0.40 was recorded and used in the calculations.

Total glutathione (GSH) content was determined spectrophotometrically [56]. Two milliliters of the leaf extract were added to 8 ml of phosphate buffer solution. One milliliter of 5-5′-dithiobis (2-nitrobenzoic acid) (DTNB) reagent was added. The optical density of the solution was measured at 412 nm. Standard curve was used to calculate the glutathione level.

The amount of anthocyanins was quantified spectrophotometrically according to [57]. After extraction of anthocyanin with methanol supplemented with 1% HCl, and centrifugation 20,000 × *g* for 20 min, the absorbance was measured at 535 and 650 nm. Anthocyanin values are reported as A535 – 0.25 (A650)/g fresh weights, taking into account the contribution of chlorophyll and its degradation products to absorption at 535 nm. Then, the results were also calculated as μg/g dry weight of leaves.

All spectrophotometric analyses were conducted at 25 °C with T60 UV–Vis spectrophotometer.

2.3 *Statistical Analyses*

The statistical analyses were performed with a *t*-test and multivariate ANOVA followed by Dunnett's test using IBM SPSS Statistics 20 software. $p \leq 0.05$* was considered statistically significant. Results were expressed as mean ± SE of three independent replicates.

3 Results

The presence of excess amounts of Ni and Pb in soil resulted in a marked reduction in the growth of broad bean plants, this being particularly evident at the highest Ni and Pb concentration (Fig. 1). Following the addition of 250, 500 mg Ni and 500, 1000 mg Pb kg^{-1} soil, shoot and root fresh weights were significantly decreased comparing to absolute control (plants grown in soil untreated with heavy metals and irrigated with tap water) ($F = 7.26$; $p = 0.006$*; Table 1). Shoot dry weight declined by around 53%, 62%, 60%, and 65.5%, while root dry weight decreased by approximately 53%, 65%, 65%, and 76%, respectively. By irrigation with MTW, shoot and root fresh weights of broad bean plants increased significantly comparing to respective controls (corresponding plants grown in soils treated with heavy metals and irrigated with tap water) ($F = 7.26$; $p = 0.006$*; Table 1). Similar increasing behavior was shown in dry weights (59%, 12%, 25%, and 71% in shoots and 88%, 66.6%, 66.6%, and 212.5% in roots, respectively, as compared to the respective controls; Fig. 1d).

Figure 2a shows that there was no significant change in the values of Chl (a + b) except the remarkable increase showed in plants grown in soil treated with 250 mg

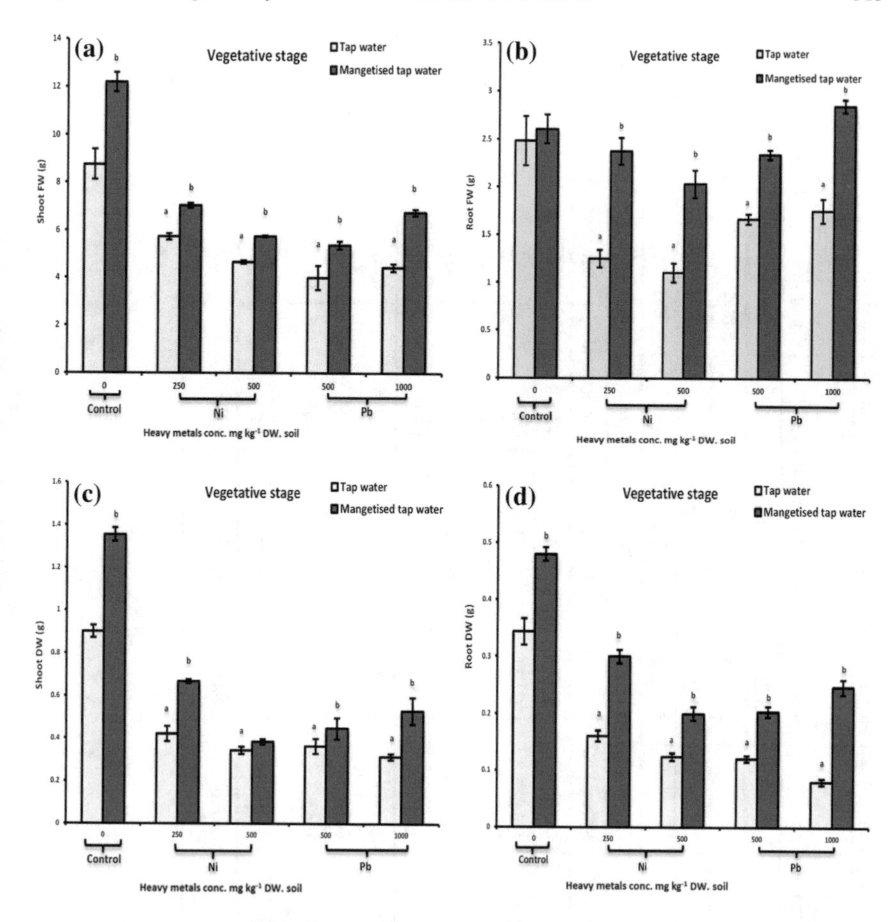

Fig. 1 Effect of magnetically treated water under heavy metal stress of 0, 250, 500 mg Ni kg^{-1} DW soil and 0, 500, 1000 mg Pb kg^{-1} DW soil on **a** shoot fresh weight, **b** root fresh weight, **c** shoot dry weight, and **d** root dry weight of *Vicia faba* L. at vegetative stage. Data represented as means ± SE of three replicates. a: Significant difference between non-magnetic treatments and absolute control at $p \leq 0.05$ using Dunnett's post hoc test statistic. b: Significant difference between magnetic and non-magnetic treatments using *t*-test for groups at $p \leq 0.05$

Ni kg^{-1}. Moreover, an additional enhancement on Chl (a + b) has been recorded under the interactive effect of 250 mg Ni and 500 mg Pb kg^{-1} soil with MTW by 19.8% and 23.7%, respectively, comparing to absolute control. On the other hand, treatments with Ni and Pb alone caused a slight decrease in carotenoid content when compared to the control. Interestingly, with irrigation by MTW, carotenoid contents were further decreased, and the minimum accumulation of carotenoids was noted in plants grown in soil treated with 500 mg Pb, which was by 44% less than that of absolute control. In contrast, carotenoid content was increased significantly in plants grown in soil treated with 1000 mg Pb and irrigated with MTW by 33.3% compared to plants grown in soil treated with 1000 mg Pb alone.

Table 1 Multivariate analysis assessing the effect of magnetically treated water on *Vicia faba* L. under elevated levels of Ni and Pb

Multivariate test: Pillai's trace test

Effect	Pillai's trace value	*F*-ratio	Hypothesis df	Error df	*Sign**
1. Metal type	0.986	8.256	17.000	2.000	*0.000**
2. Metal concentration	1.986	25.678	34.000	6.000	*0.000**
3. Magnetism	0.999	124.964	17.000	2.000	*0.008**
4. Metal type versus magnetism	0.998	56.332	17.000	2.000	*0.018**
5. Metal concentration versus magnetism	1.970	11.611	34.000	6.000	*0.003**

Source		df	Shoot FW	Root FW	Shoot DW	Root DW	Chl (a + b)	Carotenoids	Carbohydrates	Protein
1. Metal type ($n = 2$)	*F*-ratio	1	4.03	11.5	0.785	5.41	30.2	47.8	26.2	6.70
	p-value		*0.06*	*0.003**	*0.387**	*0.03**	*0.00**	*0.00**	*0.00**	*0.02**
2. Metal conc. ($n = 3$)	*F*-ratio	2	15.06	4.68	4.461	46.6	22.8	3.02	6.39	38.01
	p-value		*0.00**	*0.023**	*0.027**	*0.00**	*0.00**	*0.07*	*0.01**	*0.00**
3. Magnetism ($n = 2$)	*F*-ratio	1	57.7	87.5	18.045	45.19	4.49	6.09	26.8	187.2
	p-value		*0.00**	*0.00**	*0. 00**	*0.00**	*0.05**	*0.02**	*0.00**	*0.00**
4. Metal type versus magnetism ($n = 4$)	*F*-ratio	3	0.32	0.95	0.21	376	1.21	0.74	7.73	4.49
	p-value		*0.58*	*0.34*	*0.65*	*0.00**	*0.29*	*0.40*	*0.01**	*0.05**
5. Metal conc. versus magnetism ($n = 5$)	*F*-ratio	4	2.45	1.97	2.79	95.0	2.63	34.1	0.01	27.5
	p-value		*0.12*	*0.17*	*0.088*	*0.00**	*0.10*	*0.00**	*1.00*	*0.00**

(continued)

Table 1 (continued)

Source		df	Proline	H_2O_2	MDA	SOD	APX	POX	CAT	GSH	Anthocyanins
1. Metal type ($n = 2$)	*F*-ratio	1	4.91	72.973	24.804	0.274	1.383	16.400	0.063	0.192	24.475
	p-value		*0.04**	*0.000**	*0.000*	*0.607**	*0.255**	*0.001**	*0.804**	*0.666**	*0.000**
2. Metal conc. ($n = 3$)	*F*-ratio	2	1.91	68.711	12.448	12.348	12.348	1.754	79.791	82.741	1.258
	p-value		*0.18*	*0.000**	*0.000**	*0.000**	*0.201**	*0.000**	*0.000**	*0.000**	*0.308*
3. Magnetism ($n = 2$)	*F*-ratio	1	12.02	0.596	0.676	58.641	4.169	205.04	68.678	76.713	11.732
	p-value		*0.003**	*0.450*	*0.422*	*0.000**	*0.059*	*0.000**	*0.000**	*0.000**	*0.003**
4. Metal type versus magnetism ($n = 4$)	*F*-ratio	3	28.074	16.587	51.194	1.027	23.849	0.458	154.519	49.270	0.023
	p-value		*0.000**	*0.001**	*0.000**	*0.324*	*0.000**	*0.507*	*0.000**	*0.000**	*0.881*
5. Metal conc. versus magnetism ($n = 5$)	*F*-ratio	4	0.911	1.024	27.569	5.358	8.027	73.812	105.892	60.531	1.956
	p-value		*0.420*	*0.379*	*0.000**	*0.015**	*0.003**	*0.000**	*0.000**	*0.000**	*0.170*

Model carried out includes dependent parameters. Shoot FW, root FW, shoot DW, root DW, chlorophylls, carotenoids, total carbohydrates, total proteins, proline, hydrogen peroxide, lipid peroxidation, superoxide dismutase, ascorbic acid peroxidase, peroxidase, catalase, glutathione, and anthocyanins), and metal concentration and metal types as covariates

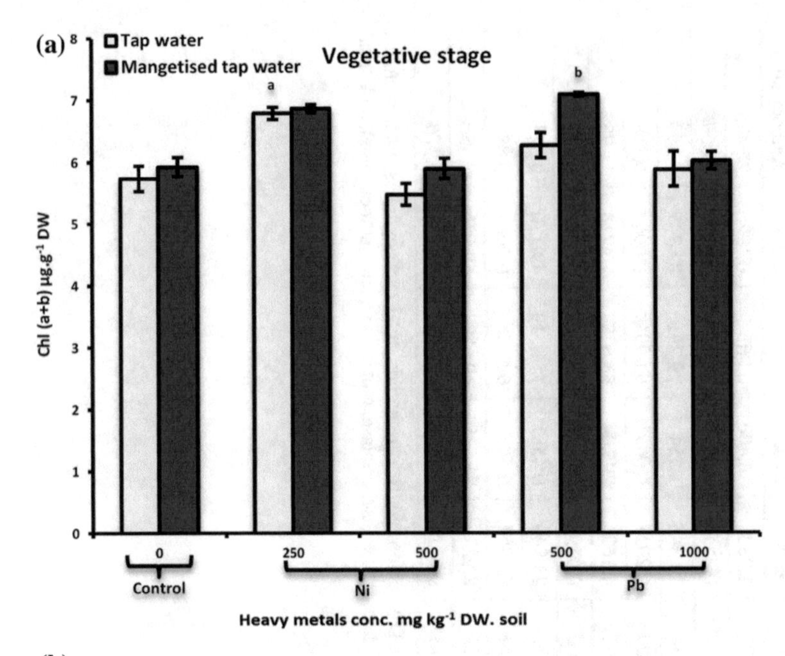

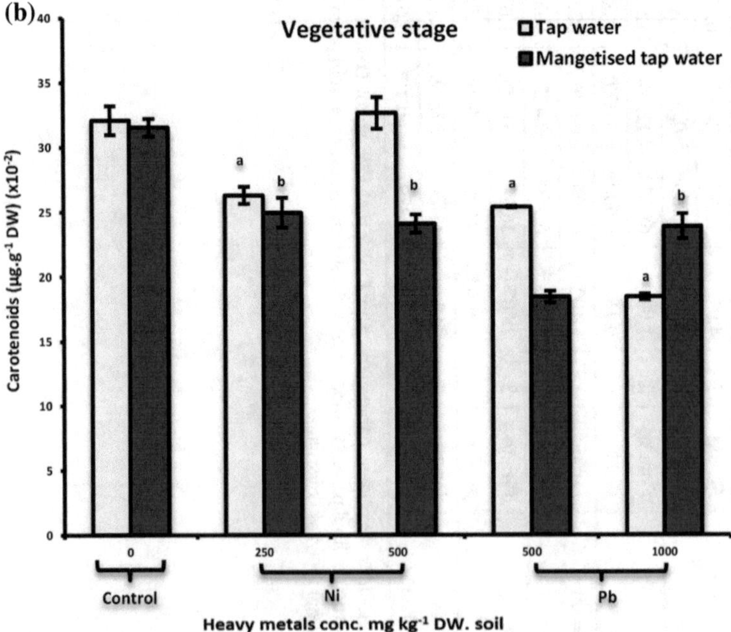

Fig. 2 Effect of magnetically treated water under heavy metal stress of 0, 250, 500 mg Ni kg^{-1} DW soil and 0, 500, 1000 mg Pb kg^{-1} DW soil on **a** Chl (a + b) and **b** carotenoids as (μg g^{-1} DW) of *Vicia faba* L. at vegetative stage. Data represented as means ± SE of three replicates. a: Significant difference between non-magnetic treatments and absolute control at $p \leq 0.05$ using Dunnett's post hoc test statistic. b: Significant difference between magnetic and non-magnetic treatments using *t*-test for groups at $p \leq 0.05$

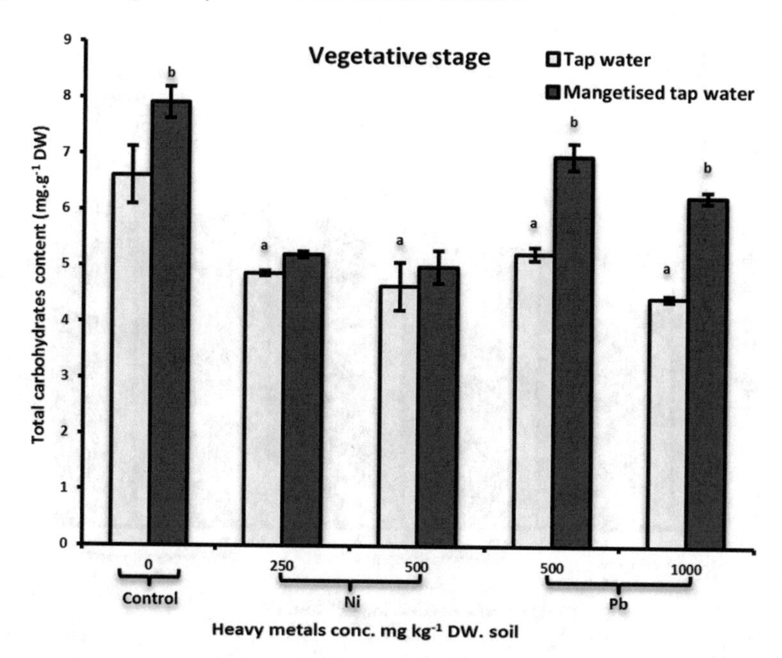

Fig. 3 Effect of magnetically treated water under heavy metal stress of 0, 250, 500 mg Ni kg^{-1} DW soil and 0, 500, 1000 mg Pb kg^{-1} DW soil on total carbohydrate content (µg g^{-1}DW) of *Vicia faba* L. at vegetative stage. Data represented as means ± SE of three replicates. a: Significant difference between non-magnetic treatments and absolute control at $p \leq 0.05$ using Dunnett's post hoc test statistic. b: Significant difference between magnetic and non-magnetic treatments using *t*-test for groups at $p \leq 0.05$

Ni and Pb additions were found to significantly decrease total carbohydrates in *Vicia faba* L. ($F = 26.151$; $p = 0.000$*; Table 1). However, irrigation with MTW increased total carbohydrates in plants unstressed with heavy metals ($p \leq 0.05$) by 20% and in plants stressed with 500 and 1000 mg Pb kg^{-1} soil by 33.5% and 41%, respectively, comparing to respective controls (Fig. 3).

A significant ($p \leq 0.05$) induction of protein content was observed by 11.3%, 32.4%, 46.5%, and 70.4% in plants grown in soil treated with 250, 500 mg Ni and 500, 1000 mg Pb, respectively, comparing to absolute control. Interestingly, irrigation with MTW significantly reduced total protein content in plants grown in soil untreated with heavy metals by 37%. Meanwhile in plants grown in soil treated with 250, 500 mg Ni and 500, 1000 mg Pb kg^{-1}, protein content is reduced by 29%, 7.4%, 15%, and 34%, respectively, compared to respective controls.

As shown in Fig. 4b, there was a progressive accumulation trend in proline content in Ni- and Pb-stressed plants compared to absolute control ($F = 4.909$; $p = 0.040$*; Table 1). However, under irrigation with MTW, proline content decreased significantly in plants grown in soils without heavy metals by 45% compared to absolute control and in plants grown in soils treated with 500, 1000 mg Pb kg^{-1} by 29% and 36%, respectively, comparing to respective controls. The results revealed that there

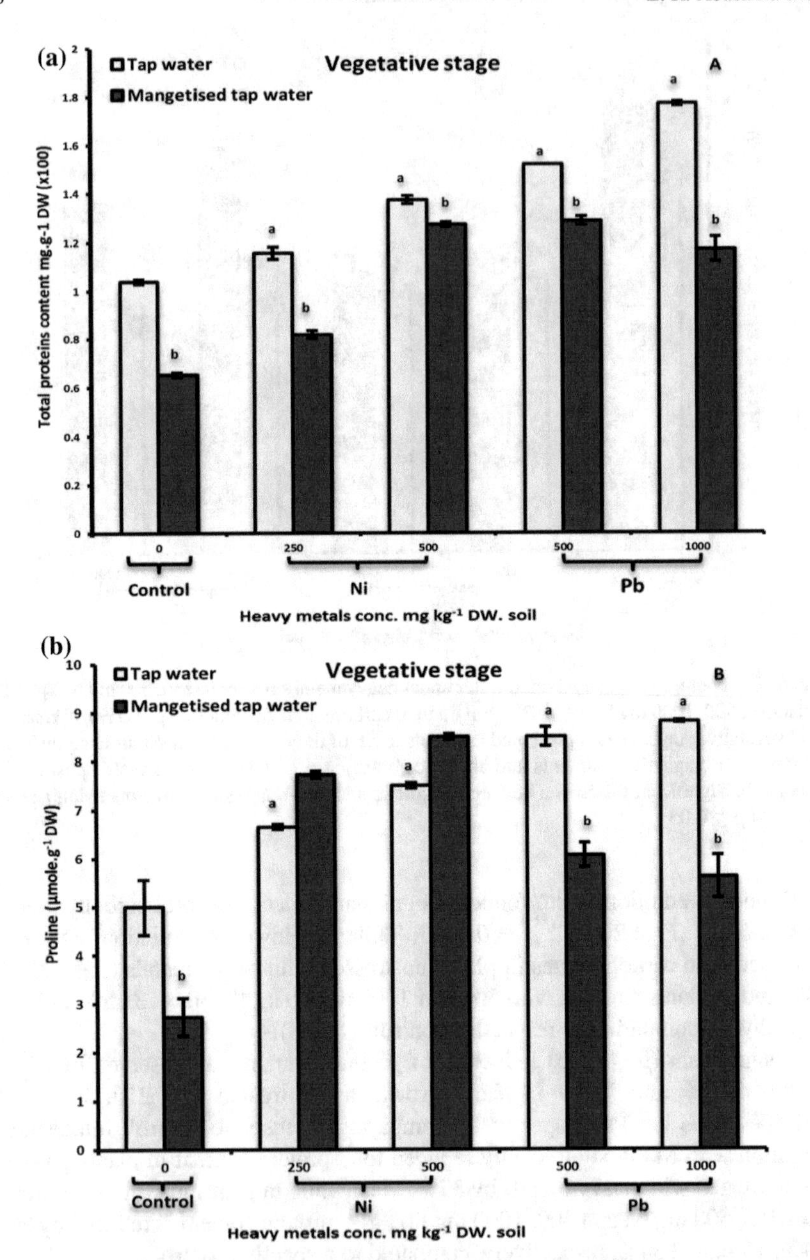

Fig. 4 Effect of magnetically treated water under heavy metal stress of 0, 250, 500 mg Ni kg^{-1} DW soil and 0, 500, 1000 mg Pb kg^{-1} DW soil on **a** total protein content ($\mu g\ g^{-1}$DW) and **b** proline ($\mu mol\ g^{-1}$ DW) of *Vicia faba* L. at vegetative stage. Data represented as means ± SE of three replicates. a: Significant difference between non-magnetic treatments and absolute control at $p \leq 0.05$ using Dunnett's post hoc test statistic. b: Significant difference between magnetic and non-magnetic treatments using *t*-test for groups at $p \leq 0.05$

was a remarkable increase in H_2O_2 level in *Vicia faba* L. with increasing heavy metal concentration in comparison with absolute control. However, under irrigation with MTW H_2O_2 content was further increased in plants grown in soils treated with 250 Ni kg^{-1} but decreased in plants subjected to 500 Pb kg^{-1} soil comparing to respective controls (Fig. 5). As shown in Fig. 6, there was an increasing trend in MDA values with increasing concentrations of heavy metals in relation to absolute control ($F = 12.448$; $p = 0.000$*; Table 1). Interestingly, MDA was sharply increased under irrigation with MTW by 125% in plants grown in soil untreated with heavy metals compared to absolute control. However, the maximum accumulation of MDA was noted in Pb-stressed plants irrigated with MTW, which was by 25% higher than that of Pb treatment alone.

The activity of SOD exhibited a significant decrease as the concentrations of heavy metals increased ($F = 12.348$; $p = 0.000$*; Table 1). The maximum decreasing percent was noticed in plants grown in soils treated with 1000 mg Pb by 42.6% comparing to absolute control. Meantime, irrigation with MTW increased SOD

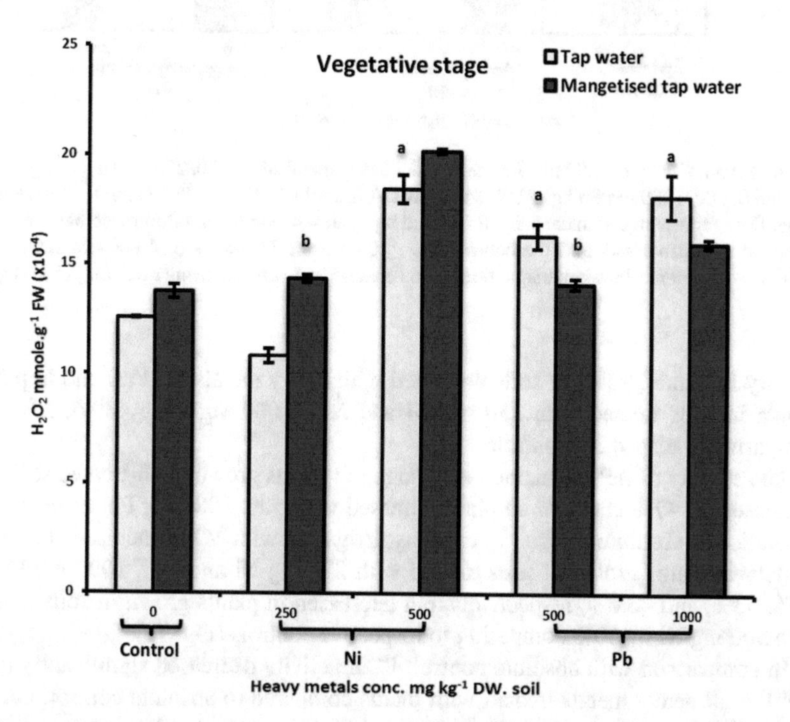

Fig. 5 Effect of magnetically treated water under heavy metal stress of 0, 250, 500 mg Ni kg^{-1} DW soil and 0, 500, 1000 mg Pb kg^{-1} DW soil on H_2O_2 (mmol g^{-1} FW) of *Vicia faba* L. at vegetative stage. Data represented as means ± SE of three replicates. a: Significant difference between non-magnetic treatments and absolute control at $p \leq 0.05$ using Dunnett's post hoc test statistic. b: Significant difference between magnetic and non-magnetic treatments using *t*-test for groups at $p \leq 0.05$

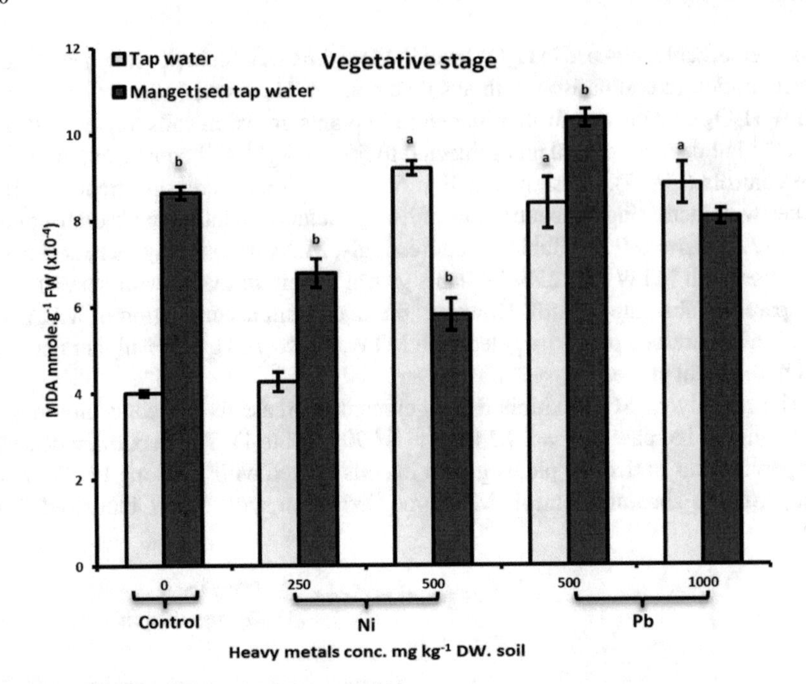

Fig. 6 Effect of magnetically treated water under heavy metal stress of 0, 250, 500 mg Ni kg^{-1} DW soil and 0, 500, 1000 mg Pb kg^{-1} DW soil on MDA (mmol g^{-1} FW) of *Vicia faba* L. at vegetative stage. Data represented as means ± SE of three replicates. a: Significant difference between non-magnetic treatments and absolute control at $p \leq 0.05$ using Dunnett's post hoc test statistic. b: Significant difference between magnetic and non-magnetic treatments using *t*-test for groups at $p \leq 0.05$

activity in plants grown in soils untreated with heavy metals by 45% and in plants grown in soils treated with 250 mg Ni and 500, 1000 mg Pb by 39.6, 27, 60% comparing to respective controls.

The activity of APX remained unchanged in plants grown in Ni-treated soils but decreased by 47% and 46% in plants stressed with 500, 1000 mg Pb, respectively, in relation to absolute control. In contrary, irrigation with MTW increased the APX activity in plants grown in soils treated with 250 mg Ni and 500, 1000 mg Pb by 56%, 93%, and 42.6%, respectively, but decreased in plants grown in soils treated with 500 mg Ni by 50% comparing to respective controls (Fig. 7b).

In comparison with absolute control, POX activity decreased significantly ($p \leq 0.05$) in all heavy metals treated with plants compared to absolute control. On the other hand, irrigation with MTW increased POX activity in plants grown in soil without heavy metals by 123% compared to absolute control and in plants grown in soils treated with 250, 500 mg Ni and 500 mg Pb kg^{-1} by 141, 146, 189%, respectively, but decreased POX activity in plants grown in soils treated with 1000 mg Pb kg^{-1} by 33% comparing to respective controls (Fig. 7c).

Figure 7d illustrates that catalase activity reduced significantly ($p \leq 0.05$) under the effect of heavy metal treatments by 40%, 48%, 74%, and 60.8% in plants grown in

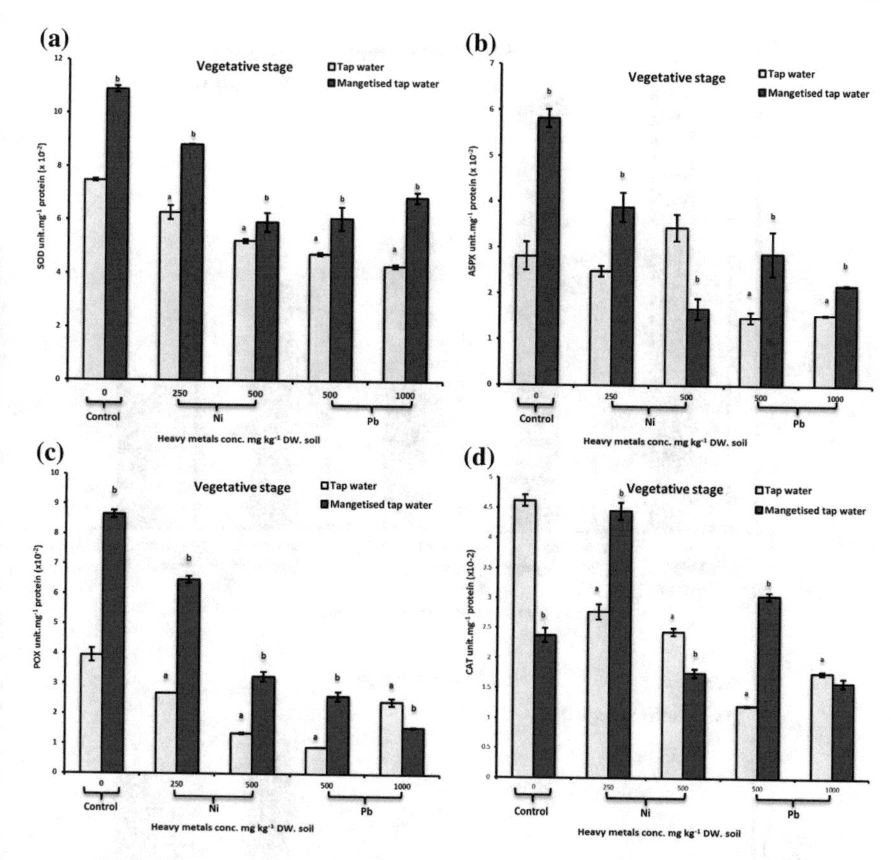

Fig. 7 Effect of magnetically treated water under heavy metal stress of 0, 250, 500 mg Ni kg^{-1} DW soil and 0, 500, 1000 mg Pb kg^{-1} DW soil on **a** superoxide dismutase (SOD); **b** ascorbate peroxidase (APX); **c** peroxidase activity (POX); and **d** catalase as (unit mg^{-1} protein) of *Vicia faba* L. at vegetative stage. Data represented as means ± SE of three replicates. a: Significant difference between non-magnetic treatments and absolute control at $p \leq 0.05$ using Dunnett's post hoc test statistic. b: Significant difference between magnetic and non-magnetic treatments using *t*-test for groups at $p \leq 0.05$

soils treated with 250, 500 mg Ni and 500, 1000 mg Pb kg^{-1}, respectively, comparing to absolute control. Meantime, lower catalase activity in plants grown under irrigation with MTW was found in most treatments except for a significant increase in plants grown in soils treated with low concentrations of heavy metals (i.e., 250 mg Ni and 500 mg Pb kg^{-1} soil by 61% and 150%, respectively) comparing to respective controls.

Comparing to control, GSH content significantly ($p \leq 0.05$) decreased in all plants treated with heavy metals compared to absolute control (Fig. 8a). However, irrigation with MTW significantly increased glutathione level in plants grown in soil without heavy metals compared to absolute control by 29% and in plants grown in soils treated with 250, 500 mg Ni and 500 mg Pb kg^{-1} by 74%, 76%, and 11%,

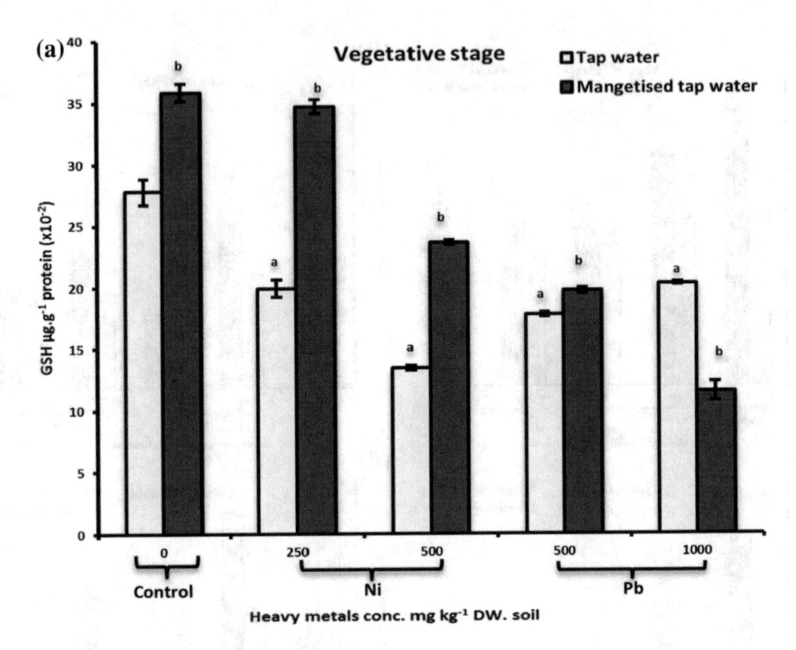

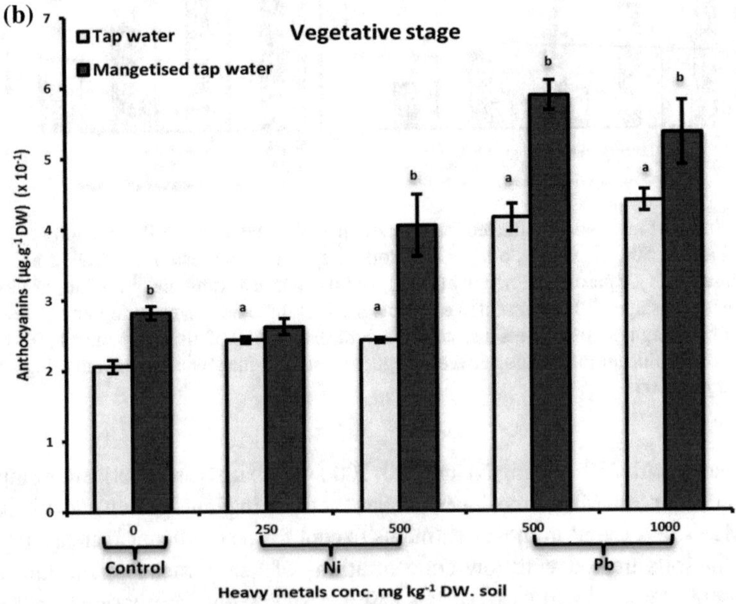

Fig. 8 Effect of magnetically treated water under heavy metal stress of 0, 250, 500 mg Ni kg^{-1} DW soil and 0, 500, 1000 mg Pb kg^{-1} DW soil on **a** glutathione content ($\mu g\ g^{-1}$ protein) and **b** anthocyanins ($\mu g\ g^{-1}$ DW) of *Vicia faba* L. at vegetative stage. Data represented as means ± SE of three replicates. a: Significant difference between non-magnetic treatments and absolute control at $p \leq 0.05$ using Dunnett's post hoc test statistic. b: Significant difference between magnetic and non-magnetic treatments using *t*-test for groups at $p \leq 0.05$

respectively, comparing to respective controls. Meantime, treatment with 1000 mg Pb resulted in more remarkable decrease in GSH by 43% when irrigated with MTW comparing to Pb treatment alone.

As shown in Fig. 8b, treatments with Ni and Pb caused a significant increase in anthocyanins comparing to absolute control ($F = 24.475$; $p \leq 0.000$*; Table 1). The maximal increase in anthocyanins was recorded in plants grown in soils treated with 500, 1000 mg Pb kg^{-1} by 101% and 113%, respectively. A marked stimulatory increase in anthocyanins was recorded as result of the interaction effect of heavy metals and irrigation with MTW in plants grown in soils treated with 500 mg Ni and 500, 1000 mg Pb kg^{-1} by 96%, 185%, and 160%, respectively, in comparison with absolute control. Similarly, anthocyanins were increased in plants grown in soils without heavy metals and irrigated with MTW by 37% comparing to absolute control (Fig. 8b).

4 Discussion

In the present study, the toxic effects of Ni and Pb on biochemistry and physiological processes in *Vicia faba* L. were evident from the reduced biomass in plants grown in soils treated with Ni and Pb. Since roots are the primary site for heavy metal exposure and toxicity, the root biomass was severely affected as reported by Jayasri and Suthindhiran [2] and Namdjoyan et al. [58]. High nickel concentrations have been found to affect the root cell membrane structure and reduce root water uptake, which leads to a decrease in plant water potential. Thus, physiological processes such as transpiration and respiration are negatively influenced and eventually end up with a reduction in plant growth [59]. Moreover, the interference with Pb toxicity and auxin-regulated cell elongation may explain the inhibition of plant growth [60, 61]. However, irrigation with MTW markedly increased fresh/dry weights of plants grown in soils without/with Ni and Pb. Similar stimulatory effect of MTW on snow pea, chickpea, flax, and lentil plants was reported [62, 63]. Yokatani et al. [64] explained that irrigation with magnetically treated water has a pronounced stimulatory effect on plant cell multiplication, growing, and development. The analysis of Bogatin et al. [65] also has shown that magnetized water led to increasing the number of crystallization centers and the change of the free gas content. For instance, degassing of water increases permeability in the soil which results in better permeability of irrigated water and dissociation of mineral fertilizers, owing to improving the conditions of root layers [36].

The criteria of photosynthetic pigments seemed to be interesting as heavy metal stress kept the values of chlorophylls Chl (a + b) around control. Moreover, a stimulation rather than inhibition was recorded in plants grown in soils treated with 250 mg Ni kg^{-1} comparing to absolute control. Our results are in agreement with [66] who reported that low concentrations of Pb and Ni resulted in less significant effect on Chl (a + b) of black gram seedlings, while higher concentrations significantly reduced chlorophyll contents. However, a significant reduction in carotenoids has been recorded in most treatments. The carotenoid content decrease in response to heavy metals indicates a severe effect on cell and its components, and heavy metals activate some mechanisms and degrade carotenoid pigments [67]. In this study, irrigation with MTW kept the values of Chl (a + b) content for most treatments more or less unchanged, while carotenoids showed an average decrease in all heavy metals treated with plants, except plants grown in soils treated with the high concentration of Pb, and there was a significant increase in carotenoid content comparing to respective controls. Campbell [68] declared that the magnetic field paramagnetically influenced chloroplasts similar to our findings for plants irrigated with MTW.

The decrease in total carbohydrate content was directly proportional to the applied concentration of Ni and Pb and may explain the massive biomass reduction in *Vicia faba* L. These results led to the conclusion that heavy metals may inhibit photosynthetic activity and increase partial utilization of carbohydrates in other metabolic pathways or stimulation of respiration rate [69]. On the other hand, irrigation with MTW increased total carbohydrate content in *Vicia faba* L. among all treatments comparing to respective controls, which may be the result of bioenergetics structural excitement causing cell pumping and enzymatic stimulation [70].

Total protein content in *Vicia faba* L. increased significantly with increasing Ni and Pb concentrations in treated soils (Fig. 4a); in confirmatory to our results, the exposure to Cu, Ni, Pb, and Zn led to an increase in the levels of chloroplast small HSP (smHSP) content in maize plants [71]. Furthermore, phytochelatins (PCs) include a family of small enzymatically synthesized peptides having a general structure of (γGlu-Cys)n-Gly, and these peptides are rapidly synthesized in response to toxic levels of heavy metals in all tested plants with Ag, As, As, Cd, Cu, Ca, Hg, Ni, Pb, Pd, Se, and ZN [72]. Hence, metal stress could induce synthesis of some proteins related to stress to overcome the toxic effects of heavy metals which consequently lead to an increase in total protein content [73]. Interestingly, reduction in total protein content under irrigation with MTW is noted comparing to respective controls. Such results are similar to findings of [74] who stated that different intensities of the magnetic field (0.5, 1.5 mT) caused a decrease in leaf protein in comparison with control. Reduction of protein content may be a cause of oxidative stress and produced free radicals by a magnetic field [75].

Interestingly, increasing Ni and Pb concentrations in soil induced a significant increase in proline content in *Vicia faba* L. Proline can act as a metal chelator and protein stabilizer [28, 76] and also helps in the reconstruction of chlorophylls when plants are exposed to heavy metals [77] and protecting cellular membranes by scavenging ROS [15, 28]. That evidence is indicated by unchanging values of Chl (a + b) and increasing protein content in the present study. Proline accumulation decreased in all plants when irrigated with MTW comparing to respective controls. Such dynamics in proline accumulation possibly count on the relative detoxification of ROS within plant cells, which can be explained that ROS production rates could be higher for being quenched or the amount of proline is not sufficient with respect to the ROS generated. The "caged reaction" theory explained the reduction of proline by oxidizing the free radicals that are enhanced during exposure to a magnetic field [78,

79]. Proline can be oxidized to various compounds that can protect plant tissue from potential damage [80, 81].

Hydrogen peroxide (H_2O_2) which acts as a signaling molecule serves a dual role in plant defense mechanism. It helps in increasing intolerance and stress acclimation at low concentration, while it acts as a ROS and induces cellular injury leading to cell death at higher concentrations [82, 83]. In the recent study, increasing heavy metal concentrations in soil induced a significant increase in H_2O_2 level in growing plants comparing to absolute control. Exposure of plants to redox inactive heavy metals also can induce ROS and result in oxidative stress through indirect mechanisms such as interaction with the antioxidant defense system, disruption of the electron transport chain, or induction of lipid peroxidation [84, 85]. Therefore, in our experiment, MDA content (an indicator of lipid peroxidation) was increased significantly in return. Kabal et al. [86] and Oves et al. [87] illustrated that plasma membrane is another target site for metal toxicity, in which metal can bind and disrupt membrane functions. Upon metal interaction, metal can induce qualitative and quantitative changes in the lipid composition of membranes, which in turn may alter the structure and functions of the membrane, and other important cellular processes.

However, irrigation with MTW kept the level of H_2O_2 unchanged in plants grown in soil without heavy metal contamination comparing to absolute control but showed an increase in plants grown in Ni-treated soil and, meanwhile, decreased in plants grown in Pb-treated soil comparing to respective controls. The increase in MDA level in plants grown in soil without heavy metals and soils treated with low concentrations of heavy metals are familiar to the results of [88–90] who explained this increases result of accumulation of H_2O_2 and other reactive species of oxygen in magnetic field exposed tobacco cells that may subsequently result in peroxidation of membrane lipids.

We have observed a significant reduction among defense system regarding SOD, POX, CAT, and APX activities as enzymatic and accumulation of GSH, and as non-enzymatic antioxidants in *Vicia faba* L. grown in soils treated with Ni and Pb treatments alone. Oveysi-Omran et al. [91] showed that the concentration of nickel in the maize plant nutrient solution causes a significant decline in the activity of antioxidant enzymes (catalase and peroxidase). Gallego et al. [92] explained it as the free oxygen radicals induced by heavy metal stress attack the antioxidant enzymes and inhibit their oxidative damages. Rastgoo and Alemzadeh [93] documented that a significant dose-dependent Pb caused a decrease in catalase activity at higher concentration and this indicates that the increase in metal concentration causes the inhibition of enzyme activity because in the high concentration of metal CAT is not properly able to protect the cell against ROS. Possible explanations for the decrease in SOD activity under heavy metal treatments may be linked to inactivation of the enzyme by the production of excess ROS and unspecific enzyme degradation [94].

Interestingly, there has been a reciprocal relationship between GSH level and total protein content (every treatment caused depletion in GSH content resulted consequently in increasing the total protein content). Phytochelatins participate in the defense against heavy metal toxicity, by sequestering metal ions into the vacuoles [95]. Its synthesis is catalyzed by phytochelatin synthase (PCS) using glutathione (GSH) as a substrate. A decay in GSH content in Populus × canescens roots [84] and *Oryza sativa* leaves [96] has been reported under cadmium stress as cadmium-induced depletion of GSH in favor of phytochelatin synthesis [97].

Unstressed *Vicia faba* plants (without heavy metal contamination) irrigated with MTW exhibited a marked decrease in catalase activity by 48% and an increase in SOD, APX, POX, and GSH activities by 45%, 107%, 123%, and 29%, respectively, comparing to absolute control. In this respect, it has been established that the primary action of magnetic field in biological systems is the induction of electrical charges and currents, hence ROS [98]. Antioxidant enzymes metabolize in order to have an effective defense against the over-production of toxic oxygen forms in plants exposed to environmental stresses, such as magnetic field [99]. Moreover, at the initial levels of Ni and Pb concentrations in soil under irrigation with MTW, enzymatic and non-enzymatic antioxidant activities were increased. However, at the high concentrations of heavy metals, we noticed a drop in CAT and APX activities in plants grown in soil treated with 500 mg Ni; meantime, POX and GSH were decreased as well in plants stressed with 1000 mg Pb. Therefore, at specific heavy metal levels, MTW could be inhibitory for one enzyme but promotive for the others. In our study, it could be explained from the over-production of ROS under the interactive effect of heavy metals and MTW; hence, activities of these ROS scavenging antioxidants or their production rate do not remain high enough to quench ROS. Such an increase and decrease in antioxidant activities in plants grown in soil contaminated with heavy metals were previously reported [28, 100, 101].

Anthocyanins are considered from the main pigments that are known to be involved in protecting plant organs from stresses and believed to increase the antioxidant response of plants to uphold the normal physiological status against biotic or abiotic stresses [102]. In the present experiment, a significant increase in anthocyanin content was observed in *Vicia faba* L. plants with increasing heavy metal concentrations in soils suggesting a possible defensive role of anthocyanins in Ni- and Pb-stressed plants. This was consistent with studies in several species of the *Brassica* genus, whereas the contents of anthocyanins were increased as a part of the defense mechanism to protect the photosynthetic apparatus from the stress induced by heavy metals [103]. However, anthocyanin content was further increased under irrigation with MTW in treated plants with heavy metals, indicating that irrigation with MTW could lead to efficient detoxification of Ni and Pb through the increased synthesis of a non-enzymatic antioxidant such as anthocyanins.

5 Conclusion

Broad bean subjected to soil contaminated with excess amounts of Ni and Pb exhibited growth impairment such as reduction in fresh/dry weight which was manifested by the reduction of total carbohydrate content, while proline and total protein contents were significantly increased. Oxidative stress damage showed from increasing

H_2O_2 and lipid peroxidation, anthocyanins and reduction in antioxidants activity. The reduction of fresh and dry weight was restored, and the reduction of total carbohydrates was prevented, while H_2O_2 and MDA accumulations were moderated with the irrigation of MTW. Moreover, anthocyanins and antioxidants showed a significant and average increase while proline and total protein contents were decreased in return. Interestingly, irrigation with MTW alone enhanced the growth parameters along with increasing total carbohydrates and exhibited a stimulatory effect on the production of anthocyanins and antioxidants; however, oxidative stress damage was clear from increasing MDA content and decreasing proline and total protein content. The clear protective activity of MTW in reducing Ni and Pb toxicity suggests their physiological relevance in increasing plant defensive mechanisms.

6 Recommendations

This study opens an interesting approach to explore the various actions that MTW seems to have in plants. Further studies on broader range of heavy metals and different magnetic field intensities will provide a clearer understanding of the machinery involved in the regulation of magnetically treated water involvement in reducing heavy metal toxicity in broad bean plants.

Acknowledgements We would like to express our appreciation to Agriculture Research Center for kindly providing the broad bean seeds and our deepest gratitude to Faculty of Agriculture, Suez Canal University, for the provision of their Research Farm facilities, land, and laboratories.

References

1. Awad YM, Vithanage M, Niazi NK, Rizwan M, Rinklebe J, Yang JE, Ok YS, Lee SS (2017) Potential toxicity of trace elements and nanomaterials to Chinese cabbage in arsenic- and lead-contaminated soil amended with biochars. Environ Geochem Health. https://doi.org/10.1007/s10653-017-9989-3
2. Jayasri MA, Suthindhiran K (2017) Effect of zinc and lead on the physiological and biochemical properties of aquatic plant *Lemna minor*: its potential role in phytoremediation. Appl Water Sci 7:1247–1253
3. Djingova R, Kuleff I (2000) Instrumental techniques for trace analysis. In: Vernet JP (ed) Trace elements: their distribution and effects in the environment. Elsevier, London, UK
4. Won HS, Lee BJ (2004) Nickel-binding properties of the C-terminal tail peptide of *Bacillus pasteurii* UreE. JBiochem 136:635–641
5. Benoit SL, Zbell AL, Maier RJ (2007) Nickel enzyme maturation in *Helicobacter hepaticus*: roles of accessory proteins in hydrogenase and urease activities. Microbiology 153:3748–3756
6. Ahmad MSA, Ashraf M (2011) Essential roles and hazardous effects of nickel in plants. Rev Environ Contam Toxicol 214:125–167
7. Rahman H, Sabreen S, Alam S, Kawai S (2005) Effects of nickel on growth and composition of metal micronutrients in barley plants grown in nutrient solution. J Plant Nutr 28:393–404

8. Kochian LV (1991) Mechanisms of micronutrient uptake and translocation in plants. In: Mortvedt JJ (ed) Micronutrients in agriculture. Soil Science Society of America, Madison, WI, pp 251–270
9. Seregin IV, Kozhevnikova AD (2006) Physiological role of nickel and its toxic effects on higher plants. Russ J Plant Physiol 53:257–277
10. Corley M, Mutiti S (2017) The effects of lead species and growth time on accumulation of lead in Chinese cabbage. Global Chall 1:1600020
11. Shahid M, Pinelli E, Pourrut B, Silvestre J, Dumat C (2011) Lead-induced genotoxicity to *Vicia faba* L. roots in relation with metal cell uptake and initial speciation. Ecotoxicol Environ Saf 74:78–84. https://doi.org/10.1016/j.ecoenv.2010.08.037
12. Kumar A, Prasad MNV, Sytar O (2012) Lead toxicity, defense strategies and associated indicative biomarkers in *Talinum triangulare* grown hydroponically. Chemosphere 89:1056–1065. https://doi.org/10.1016/j.chemosphere.2012.05.070
13. Konate A, He X, Zhang Z, Ma Y, Zhang P, Alugongo GM, Rui Y (2017) Magnetic (Fe_3O_4) nanoparticles reduce heavy metals uptake and mitigate their toxicity in wheat seedling. Sustainability 9:790. https://doi.org/10.3390/su9050790
14. Ashraf U, Kanu AS, Mo ZW, Hussain S, Anjum SA, Khan I, Abbas RN, Tang X (2015) Lead toxicity in rice; effects, mechanisms and mitigation strategies—a mini review. Environ Sci Pollut Res 22:18318–18332. https://doi.org/10.1007/s11356-0-15-5463-x
15. Sharma P, Dubey RS (2005) Lead toxicity in plants. Braz J Plant Physiol 17:35–52. https://doi.org/10.1590/S1677-04202005000100004
16. Clemens S (2006) Toxic metal accumulation, response to exposure and mechanism of tolerance in plants. Biochimie 88:1707–1719. https://doi.org/10.1016/j.biochi.2006.07.003
17. Pourrut B, Shahid M, Camille D, Peter W, Eric P (2011) Lead uptake, toxicity, and detoxification in plants. Rev Environ Contam Toxicol 213:113–136. https://doi.org/10.1007/978-1-4419-9860-6_4
18. Mishra A, Choudhary MA (1998) Amelioration of lead and mercury effects on germination and rice seedling growth by antioxidants. Biol Plant 41:469–473. https://doi.org/10.1023/a:1001871015773
19. Mittler R (2002) Oxidative stress, antioxidant and stress tolerance. Trends Plant Sci 7:M841–M851. https://doi.org/10.1016/S1360-1385(02)02312-9
20. Racchi ML (2013) Antioxidant defenses in plants with attention to *Prunus* and *Citrus spp*. Antioxidants 2:340–369. https://doi.org/10.3390/antiox2040340
21. Zhang FQ, Wang YS, Lou ZP, Dong JD (2007) Effect of heavy metal stress on antioxidative enzymes and lipid peroxidation in leaves and roots of two mangrove plant seedlings (*Kandelia candel* and *Bruguiera gymnorrhiza*). Chemosphere 67:44–50
22. Blokhina O, Virolainen E, Fagerstedt KV (2003) Antioxidants oxidative damage and oxygen deprivation stress: a review. Ann Bot 91:179–194
23. Noctor G, Foyer CH (1998) Ascorbate and glutathione: keeping active oxygen under control. Annu Rev Plant Physiol Plant Mol Biol 49:249–279
24. Wang S, Chu Z, Ren M, Jia R, Zhao C, Fei D, Su H, Fan X, Zhang X, Li Y, Wang Y, Ding X (2017) Identification of anthocyanin composition and functional analysis of an anthocyanin activator in *Solanum nigrum* fruits. Molecules 22:876. https://doi.org/10.3390/molecules22060876
25. Hayat S, Hayat Q, Alyemeni MN, Wani AS, Pichtel J, Ahmad A (2012) Role of proline under changing environments: a review. Plant Signal Behav 7:1456–1466
26. Chatterjee C, Dube BK, Sinha P, Srivastava P (2004) Detrimental effects of lead phytotoxicity on growth, yield, and metabolism of rice. Commun Soil Sci Plant Anal 35:255–265. https://doi.org/10.1081/css-120027648
27. Ali B, Xu X, Gill RA, Yang S, Ali S, Tahir M, Zhou WJ (2014) Promotive role of 5-aminolevulinic acid on mineral nutrients and antioxidative defense system under lead toxicity in *Brassica napus*. Ind Crop Prod 52:617–626. https://doi.org/10.1016/j.indcrop.2013.11.033
28. Ashraf U, Kanu AS, Deng Q, Mo Z, Pan S, Tian H, Tang X (2017) Lead (Pb) toxicity; physio-biochemical mechanisms, grain yield, quality, and Pb distribution proportions in scented rice. Front Plant Sci 8:259. https://doi.org/10.3389/fpls.2017.00259

29. Surendran U, Sandeep O, Joseph EJ (2016) The impacts of magnetic treatment of irrigation water on plant, water and soil characteristics. Agric Water Manag 178:21–29
30. Krishnaraj C, Yun S, Kumar VKA (2017) Effect of magnetized water (biotron) on seed germination of *Amaranthaceae* family. J Acad Ind Res 5(10):152–156
31. Pang XF, Deng B (2008) Investigation of changes in properties of water under the action of a magnetic field. Sci China Ser G: Phys Mech Astron 51:1621–1632
32. Chaplin M (2009) Theory vs experiment: what is the surface charge of water? Water 1:1–28
33. Ambashta RD, Sillanpää M (2010) Water purification using magnetic assistance: a review. J Hazard Mater 180:38–49
34. Lee S, Takeda M, Nishigaki K (2003) Jpn J Appl Phys Part 1 42:1828
35. Iwasaka M, Ueno S (1998) Structure of water molecules under 14 T magnetic field. J Appl Phys 83(11):6459–6461
36. McMahon CA (2009) Investigation of the quality of water treated by magnetic fields. B.E. courses ENG4111 and 4112 research project, Faculty of Engineering and Surveying, University of Southern Queensland
37. Su N, Wu C (2003) Effect of magnetic field treated water on mortar and concrete containing fly ash. Cem Concr Res 25:681–688
38. Morejon LP, Castro JC, Velazquez LG, Govea AP (2007) Simulation of *Pinus tropicalis* M. seeds by magnetically treated water. Int Agrophys 21:173–177
39. Moussa HR (2011) The impact of magnetic water application for improving common bean (*Phaseolus vulgaris* L.) production. New York Sci J 4(6):15–20
40. El Sayed HEA (2015) Impact of magnetic water irrigation for improve the growth, chemical composition and yield production of broad bean (*Vicia faba* L.). Plant Nat Sci 13(1):107–119
41. Bagherifard A, Ghasemnezhad A (2014) Effect of magnetic salinated water on some morphological and biochemical characteristics of artichoke (*Cynara scolymus* L.) leaves. J Med Plants Prod 2:161–170
42. Sadeghipour O (2015) Magnetized water alleviates drought damages by reducing oxidative stress and proline accumulation in mung bean (*Vigna radiata* L. Wilczek). Bull Environ Pharmacol Life Sci 4(8):62–69
43. Zeidan MS (2002) Effect of sowing dates and urea foliar application on growth and seed yield of determinate faba bean (*Vicia faba* L.) under Egyptian conditions. Egypt J Agron 24:93–102
44. Orcutt DM, Nilsen ET (2000) Physiology of plants under stress: soil and biotic factors. Wiley, New York 684 pp.
45. Karimi R, Chorom M, Solhi S, Solhi M, Safe A (2012) Potential of *Vicia faba* and *Brassica arvensis* for phytoextraction of soil contaminated with cadmium, lead and nickel. Afr J Agric Res 7(22):3293–3301
46. Metzner H, Rau H, Senger H (1965) Untersuchungen Zursynchronixier Barkelt Einzelner Pigment Mangel. Multantent Von Chlorella planta 65:186
47. Hedge JE, Hofretier BT (1962) In: Whistler RL, Be Miller JN (eds) Carbohydrate chemistry, vol 17. Academic Press, New York. Sadasivam S, Manickam A (eds) (1992) Biochemical methods: for agricultural sciences. Wiley Eastern Limited, New Delhi, pp 8–9
48. Bates LS, Waldeen RP, Teare ID (1973) Rapid determination of free proline for water stress studies. Plant Soil 39:205–207
49. Shi S, Wang G, Wang Y, Zhang L, Zhang L (2005) Protective effect of nitric oxide against oxidative stress under ultraviolet-B radiation. Nitric Oxide 13(1):1–9
50. Heath RL, Packer L (1968) Photoperoxidation in isolated chloroplasts. I. Kinetics and stoichiometry of fatty acid peroxidation. Arch Biochem Biophys 125:189–190
51. Bradford M (1976) A rapid and sensitive for the quantitation of microgram quantities of protein utilizing the principle of protein-dye binding. Anal Biochem 72:248–254
52. Beauchamp C, Fridovich I (1971) Superoxide dismutase: improved assays and an assay applicable to acrylamide gels. Anal Biochem 44:276–287
53. Nakano Y, Asada K (1980) Spinach chloroplasts scavenge hydrogen peroxide on illumination. Plant Cell Physiol 21:1295–1307

54. Malik CP, Singh MB (1980) In: Plant enzymology and histoenzymology, vol 41, no 2. Kalyani Publishers, New Delhi, p 53
55. Aebi HE (1983) Catalase. In: Bergmeyer HU (ed) Methods of enzymatic analysis, vol 3, 3rd edn. Verlag Chemie, Weinheim, pp 273–286
56. Griffith OW (1980) Potent and specific inhibition of glutathione synthesis by buthionine sulfoximine (s-n butyl homocysteine sulfoximine). J Biol Chem 254:7558–7560
57. Lange H, Shropshire WJ, Mohr H (1970) An analysis of phytochrome-mediated anthocyanin synthesis. Plant Physiol 47:649–655
58. Namdjoyan S, Kermanian H, Abolhasani SA, Modarres TS, Elyasi N (2017) Interactive effects of Salicylic acid and nitric oxide in alleviating zinc toxicity of Safflower (*Carthamus tinctorius* L.). Ecotoxicology 26. https://doi.org/10.1007/s10646-017-1806-3
59. Fuentes D, Disante KB, Valdecantos A, Cortina J, Ramón V (2007) Response of *Pinushalepensis* Mill. seedlings to biosolids enriched with Cu, Ni and Zn in three Mediterranean forest soils. Environ Pollut 145:316–323
60. Lane SD, Martin ES, Garrod JF (1978) Lead toxicity effects on indole-3-acetic acid induced cell elongation. Planta 144:79–84
61. Sengar RS, Gautam M, Sengar RS, Garg SK, Sengar K, Chaudhary R (2008) Lead stress effects on physiobiochemical activities of higher plants. Rev Environ Contam Toxicol 196:73–93
62. Abdul Qados AMS, Hozayn M (2010) Response of growth, yield, yield components and some chemical constituents of flax for irrigation with magnetized and tap water. World Appl Sci J 8(5):630–634
63. Grewal HS, Maheshwari BL (2011) Magnetic treatment of irrigation water and snow pea and chickpea seeds enhances early growth and nutrient contents of seedlings. Bioelectromagnetics 32:58–65
64. Yokatani KT, Hashimoto H, Yanagisawa M (2001) Growth of avena seedlings under a low magnetic field. Biol Sci Space 15:258–259
65. Bogatin J, Bondarenko N, Gak E, Rokhinson E, Ananyev I (1999) Magnetic treatment of irrigation water: experimental results and application conditions. Environ Sci Technol 33(8):1280–1285
66. Singh G, Agnihotri RK, Reshma RS, Ahmad M (2012) Effect of lead and nickel toxicity on chlorophyll and proline content of Urd (*Vigna mungo* L.) seedlings. Int J Plant Physiol Biochem 4(6):136–141
67. Parmar P, Dave B, Sudhir A, Panchal K, Subramanian RB (2013) Physiological, biochemical and molecular response of plants against heavy metals stress. Int J Curr Res 5(1):080–089
68. Campbell GS (1977) An introduction to environmental biophysics. Springer, New York, USA
69. Abu-Muriefah SS (2015) Effects of silicon on membrane characteristics, photosynthetic pigments, antioxidative ability, and mineral element contents of faba bean (*Vicia faba* L.) plants grown under Cd and Pb stress. Int J Adv Res Biol Sci 2(6):1–17
70. De Souza A, Garci D, Sueiro L, Gilart F, Porras E, Licea L (2006) Presowing magnetic treatments of tomato seeds increase the growth and yield of plants. Bioelectromagnetics 27:247–257
71. Heckathorn SA, Mueller JK, Laguidice S, Zhu B, Barrett T, Blair B, Dong Y (2004) Chloroplast small heat-shock proteins protect photosynthesis during heavy metal stress. Am J Bot 91(9):1312–1318
72. Maitani T, Kubota H, Sato K, Yamada T (1996) The composition of metals bound to class III metallothionein (phytochelatin and its desglycyl peptide) induced by various metals in root cultures of *Rubia tinctorum*. Plant Physiol 110:1145–1150
73. Doganlar ZB, Seher C, Telat Y (2012) Metal uptake and physiological changes in *Lemna gibba* exposed to manganese and nickel. Int J Biol. https://doi.org/10.5539/ijb.v4n3p148
74. Peyvandi M, Kazemi NKH, Arbabian S (2013) The effects of magnetic fields on growth and enzyme activities of *Helianthus annuus* L. seedlings. Iran J Plant Physiol 3(3):717–724
75. Aladjadjiyan A (2002) Study of the influence of magnetic field on some biological characteristics of *Zea mais*. J Cent Eur Agric 3(2):90–94

76. Mishra S, Dubey RS (2005) Heavy metal toxicity induced alterations in photosynthetic metabolism in plants. In: Pessarakli M (ed) Handbook of photosynthesis, 2nd edn. CRC Press, Taylor and Francis, New York, NY, USA, pp 845–863
77. Rastgoo L, Alemzadeh A, Afsharifar A (2011) Isolation of two novel isoforms encoding zinc- and copper-transporting P1BATPase from Gouan (*Aeluropus littoralis*). Plant Omics J 4(7):377–383
78. Scaiano JC, Cozens FL, McLean J (1994) Model of the rationalization of magnetic field effects *in vivo*. Application of radical-pair mechanism to biological systems. Photochem Photobiol 59:585–589
79. Parola A, Kost K, Katsir G, Monselise E, Cohen-Luria R (2006) Radical scavengers suppress low frequency EMF enhanced proliferation in cultured cells and stress effects in higher plants. Environmentalist 25:103–111
80. Stadtman ER (1993) Oxidation of free amino acids and amino acid residues in proteins by radiolysis catalyzed reactions. Annu Rev Biochem 62:797–821
81. Matysik J, Alia PSP, Bhalu B, Mohanty P (2002) Molecular mechanisms of quenching of reactive oxygen species by proline under stress in plants. Curr Sci 82:525–532
82. Stone JR, Yang S (2006) Hydrogen peroxide: a signaling messenger. Antioxid Redox Signal 8:243–270
83. Kaur G (2014) B-induced toxicity in plants: effect on growth, development and biochemical attributes. J Global Biosci 3(6):881–889
84. Schützendübel AandPolle A (2002) Plant responses to abiotic stresses: heavy metal-induced oxidative stress and protection by mycorrhization. J Exp Bot 53(372):1351–1365
85. Singh S, Parihar P, Singh R, Singh VP, Prasad SM (2016) Heavy metal tolerance in plants: role of transcriptomics, proteomics, metabolomics, and ionomics. Front Plant Sci 6:1143
86. Kabal K, Janicka-Russak M, Burzski M, Obus G (2008) Comparison of heavy metal effect on the proton pumps of plasma membrane and tonoplast in cucumber root cells. J Plant Physiol 165:278–288
87. Oves M, Saghir KM, Huda QA, Nadeen FM, Almeelbi T (2016) Heavy metals: biological importance and detoxification strategies. J Bioremed Biodeg 7:334. https://doi.org/10.4172/2155-6199.1000334
88. Sahebjamei H, Abdolmaleki P, Ghanati F (2007) Effects of magnetic field on the antioxidant enzyme activities of suspension—cultured tobacco cells. Bioelectromagnetics 28:42–47
89. Najafi S, Heidari R, Jamei R (2014) Photosynthetic characteristics, membrane lipid levels and protein content in the *Phaseolus vulgaris* L. (cv. Sadri) exposed to magnetic field and silver nanoparticles. Bull Environ Pharmacol Life Sci 3(2):72–76
90. Ghanati F, Morita A, Yokota H (2005) Effects of aluminum on the growth of tea plant and activation of antioxidant system. Plant Soil 276:133–141
91. Oveysi-Omran M, Darvizheh H, Zavareh M, Hatamzadeh A, Alibiglouei MH (2014) Effects of different concentrations of Ni on morphophysiological characteristics of maize (*Zea mays* L.) seedlings. Adv Environ Biol 8(12):664–672
92. Gallego SM, Benavides MP, Tomaro ML (1996) Effect of heavy metal ion excess on sunflower leaves: evidence for involvement of oxidative stress. Plant Sci 121:151–159
93. Rastgoo L, Alemzadeh A (2011) Biochemical responses of Gouan (*Aeluropus littoralis*) to heavy metals stress. Aust J Crop Sci 5(4):375–383
94. Filek M, Keskinen R, Hartikainen H, Szarejko I, Janiak A, Miszalski Z, Golda A (2008) The protective role of selenium in rape seedlings subjected to cadmium stress. Plant Physiol 165:833–844
95. Cobbett C, Goldsbrough P (2002) Phytochelatins and metallothioneins: roles in heavy metal detoxification and homeostasis. Annu Rev Plant Biol 53:159–182
96. Hsu YT, Kao CH (2004) Cadmium toxicity is reduced by nitric oxide in rice leaves. Plant Growth Regul 42:227–238
97. Grill E, Winnacker EL, Zenk MH (1985) Phytochelatins: the principal heavy metal complexing peptides of higher plants. Sci 230:674–676

98. Roy C, Repacholi M (2005) Electromagnetic field and health, a WHO perspective. World Health Organization, Geneva, Switzerland. Available from: http://www.world-aluminium.org/news/montreal/roy.htm
99. Sen Gupta SA, Webb RP, Holaday AS, Allen RD (1993) Overexpression of superoxide dismutase protects plants from oxidative stress. Plant Physiol 103:1067–1073
100. Guo B, Liang YC, Zhu YG, Zhao FJ (2007) Role of salicylic acid in alleviating oxidative damage in rice roots (*Oryza sativa*) subjected to cadmium stress. Environ Pollut 146:743–749
101. Anjum S, Tanveer M, Hussain S, Shahzad B, Ashraf U, Fahad S, Hassan W, Jan S, Saleem M, Khan I, Bajwa A, Wang L, Mahmood A, Samad RA, Tung S (2016) Osmoregulation and antioxidant production in maize under combined cadmium and arsenic stress. Environ Sci Pollut Res 23. https://doi.org/10.1007/s11356-016-6382-1
102. Neill SO, Gould KS (2003) Anthocyanins in leaves: light attenuators or antioxidants? Funct Plant Biol 30(8):865–873
103. Mourato M, Reis R, Martins LL (2012) Characterization of plant antioxidative system in response to abiotic stresses: a focus on heavy metal toxicity. In: Montanaro G, Dichio B (eds) Advances in selected plant physiology aspects. InTech, Vienna, Austria, pp 23–44

Irrigation with Magnetically Treated Water Induces Antioxidative Responses of *Vicia Faba* L. to Ni and Pb Stress at Harvest Stage

Amal H. Saleh, Ahmed I. Mohamed and Eman R. Abuslima

1 Introduction

Trace elements (TEs) are one of the main abiotic stresses and caused environmental pollution in recent decades [1–3]. Unlike other organic pollutants, heavy metals are not degradable and can be converted into harmless compounds via biological processes. The TEs persist for a long time in the environment and can enter into the food chain [4].

Toxicity of TEs can be manifested in many ways when plant cells accumulate them at high levels. The TEs can be divided into two groups: redox active (Fe, Cu, Cr, Co) and redox inactive (Cd, Zn, Ni, Al, Pb, etc.). The redox active-TEs are directly involved in the redox reaction in cells and result in the formation of O^{2-} and subsequently in H_2O_2 and OH^- production via the Haber–Weiss and Fenton reactions [5]. Exposure of plants to redox inactive-TEs also results in oxidative stress through the interaction with the antioxidant defense system, disruption of the electron transport chain, or induction of lipid peroxidation. The latter can be due to TEs-induced increase in lipoxygenase (LOX) activity [6].

Consequently, increasing the synthesis of reactive oxygen species (ROS) in biological systems is one of the immediate responses to different stress factors [7, 8]. Plants have complex ROS scavenging mechanisms at the molecular and cellular levels. These mechanisms inhibit or slow the oxidation of biomolecules and oxidative chain reactions which decrease the cellular oxidative damage and increase resistance to TEs [9]. Plants defense system includes enzymatic antioxidants (i.e., superoxide

A. H. Saleh · E. R. Abuslima (✉)
Department of Botany, Faculty of Science, Suez Canal University, 41522 Ismailia, Egypt
e-mail: eman_ramadan@science.suez.edu.eg

A. I. Mohamed
Department of Soil and Water, Faculty of Agriculture, Suez Canal University, 41522 Ismailia, Egypt

E.-S. E. Omran and A. M. Negm (eds.), *Technological and Modern Irrigation Environment in Egypt*, Springer Water,
https://doi.org/10.1007/978-3-030-30375-4_15

dismutase (SOD), guaiacol peroxidase (POX), catalase (CAT), and ascorbate peroxidase (APX) enzymes), ascorbate, glutathione (GSH), carotenoids, and proline [10].

Recently, many studies have been conducted to emphasize the importance of the magnetic fields (MF) as an eco-friendly choice to improve crops [11]. The applications of a magnetic field include the treatments of irrigation water, dry seeds, wet seeds, and seedlings [12].

Magnetically treated water (MTW) exhibited a marked significant increase in total protein, total amino acids, and proline contents in broad bean compared with control plants [13]. A significant increase in the activities of the antioxidant enzymes (CAT, POX, and SOD) has been observed in chickpea and common bean irrigated with MTW compared to irrigation with tap water [14–16]. On the other hand, magnetic field treatment decreased CAT activity in tobacco [17].

It was found that mungbean seedlings treated with 600 milliTesla (mT) magnetic field followed by cadmium stress; had low concentration of malondialdehyde, H_2O_2, and O^{2-}, and conductivity of electrolyte leakage compared to cadmium stress alone [18]. However, little is known whether magnetic field treatment of irrigation water could alleviate plants growth lesion induced by TEs stress and what is the protective mechanisms of MTW. The main purpose of this study is to explore the possibility if there is a protective mechanism against TEs stress such as Ni and Pb and to manifest the defensive mechanism of magnetically treated water (MTW).

2 Materials and Methods

2.1 Soil

Sandy soil was collected from the surface horizon (0–30 cm) of Agricultural Farm at the Faculty of Agriculture, Suez Canal University, Ismailia, Egypt. According to [19, 20], soil samples were spiked with liquid solutions of Ni (250 and 500 mg kg^{-1} of dry weight of soil) and Pb (500 and 1000 mg kg^{-1}) in the form of $Pb(NO_3)_2$ and $NiSO_4$, respectively, at the beginning of the experiment. The zero level of TEs treatment was considered as the control (C). The treatments were mixed uniformly with soil and incubated in the laboratory at the temperature range of 25–28 °C for a month. After incubation, 20 kg of soil from each group was placed in each plastic pot (50 cm in height and 30 cm in diameter) and fertilized with 50.0 mg N kg^{-1} dry soil as urea, 50.0 mg P kg^{-1} as diammonium phosphate, and 50.0 mg K kg^{-1} as potassium sulfate.

2.2 Plant Material and Growth Conditions

Seeds of *Vicia faba* were collected from Agricultural Research Center (ARC), Ismailia, Egypt. Next, seed surface was sterilized with 2.5% sodium hypochlorite for 15 min then washed exhaustively with deionized water. Seeds (six) of *Vicia faba* were grown in each pot. Broad bean samples (two plants) were collected after six weeks post sowing, while four other uniform plants were retained per pot and allowed to grow for 120 days from sowing. Each treatment group was divided into one subgroup irrigated with tap water, and the other subgroup irrigated with magnetically treated water (MOW). Magnetically treated tap water was originated with a magnetic field (1000 gauss) magnetron unit of 0.5-inch diameter. Plants were cultivated from November (2014) to March (2015) in an open field at the Agricultural Farm, Faculty of Agriculture, Suez Canal University. All pots were regularly irrigated every three days during November and December in 2014 while once a week during January, February, and March in 2015. To avoid water leakage, soil moisture was retained at 80% of field capacity.

2.2.1 Plant Chemical Analyses

Total Proteins

Total protein content was determined according to [21]. A 500 mg of fresh leaf sample was extracted in 5 mL phosphate buffer (0.1 M, pH 7.5). The supernatants were collected after centrifuging extracts. 100 μl of the extract was diluted to 1 ml phosphate buffer (pH 7.5), and then, a 5 ml of Bradford reagent was added. Absorbance was recorded photometrically at 595 nm using a standard of bovine serum albumin. Protein contents were calculated as mg g^{-1} DW. The remaining supernatant was used for further analysis of antioxidants enzymes activities.

Free Proline

For measuring proline, fresh leaf tissue (500 mg) was grounded in 10 ml of 3% sulfosalicylic acid according to the method described [22]. The extract was centrifuged at 10,000 × g for 10 min, and the supernatant transferred to a 1.5 ml tube. The extracted solution was reacted with glacial acetic acid and ninhydrin reagent (1.25 g ninhydrin in 30 mL glacial acetic and 20 mL 6 M H_3PO_4) in 1:1 (Vt/Vt) solution ratio and incubated at 95 °C for 1 h. The reaction was dismissed by placing the tube in an ice bath, and then, the mixture was mixed with 4 ml of toluene by vortexing for 5 min with a test tube mixer. The absorbance of the chromophore-containing toluene was measured at 520 nm against the blank (i.e., toluene). The proline content was estimated by using proline standard curve.

2.2.2 Oxidative Stress Measurements (H_2O_2, MDA, and EL)

Hydrogen peroxide in leaf tissues (200 mg) was extracted in 2 ml of 3% trichloroacetic and measured at 390 nm according to the modified method of Shi et al. [23]. Malondialdehyde (MDA) extracted from leaves tissues (300 mg) with 2 ml of 0.1% trichloroacetic acid and determined according to [24]. "The MDA concentration was determined by dividing the difference in absorbance (A532–A600) by its molar extinction coefficient (155 mM^{1} cm^{-1})"(http://www.scielo.br/pdf/bjpp/v17n4/a03v17n4.pdf). Electrolyte leakage (EL) in the cell sap from the leaf discs was determined after [25] using a Chemitrix type 700 portable conductivity meter.

2.2.3 Enzyme Assays

Enzymes were extracted from leaves of broad bean in a pre-chilled mortar in 20 mL chilled phosphate buffer (pH = 7.5). Extracts were then centrifuged at 6000 rpm for 20 min at 5 °C. Enzyme assays were conducted immediately following extraction.

Enzymatic Antioxidants

SOD (EC 1.15.1.1) was determined by measuring the inhibition of the photochemical reduction of nitro blue tetrazolium (NBT) as described method by [26]. The reaction mixture of enzyme was done in 3.0 mL buffer using 0.25 ml of 13 mM methionine, 80 μM NBT, and 0.1 mM EDTA. Next, 0.25 mL of 50 μM riboflavin was added in each tube. The tubes were shaken and then placed at 30 cm away from the light source for 20 min. The absorbance was measured at 560 nm. One unit of enzyme activity represents the amount of enzyme required for 50% inhibition of NBT reduction at 560 nm.

APX (EC 1.11.1.11) was determined following the method of [27]. The assay mix (2 mL) contained 25 mM potassium phosphate buffer (pH 7.0), 0.1 mM EDTA, 0.25 mM ascorbate, 1.0 mM H_2O_2 and 0.2 mL enzyme extract, and H_2O_2. The reaction was initiated by adding H_2O_2, and "oxidation of ascorbate was followed by the decrease in absorbance at 290 nm at 30 s interval for 5 min. One unit of APX activity is defined as the amount of enzyme" (http://eajournals.org/wp-content/uploads/Antioxidant-System-Responses-of-The-Cor) that oxidizes 1 μM of ascorbate per min at room temperature.

POX (EC 1.11.1.7) was determined following the described method by [28]. Specifically, the reaction was done in a cuvette by mixing 3 ml of phosphate buffer (0.1 M, pH 7), 0.05 mL of guaiacol solution, 0.1 mL of enzyme extract, and 0.03 mL of H_2O_2 solution. The mixture was homogenized by shaking. Δt for absorbance increase by 0.1 were recorded using the spectrophotometer and used in calculations. Peroxidase enzyme activity was expressed as peroxidase activity g-protein-1.

CAT (EC 1.11.3.6) activity was determined by absorbance decrease of the reaction mixture at 240 nm according to [29]. In particular, a 3 mL reaction mixture contains

50 mM (pH 7.0) of potassium phosphate buffer, and 7.5 mM of H_2O_2 was added to 50 μL of the crude extract. Δt for the absorbance decrease from 0.45 to 0.40 was recorded and used in the calculations.

Non-enzymatic Antioxidants

GSH content was measured spectrophotometrically following the method described by [30]. Two milliliters of the leaf extract were extracted by 8 ml of phosphate buffer solution. Additionally, one milliliter of 5-5′-Dithiobis (2-nitrobenzoic acid) (DTNB) reagent was added, and then, the optical density was measured at 412 nm. A standard curve was used to calculate the glutathione level.

The amount of anthocyanins was quantified spectrophotometrically following the procedure of [31]. Anthocyanin was extracted with methanol supplemented with 1% HCl. The absorbance was measured at 535 and 650 nm after centrifugation 20,000 × g for 20 min. Anthocyanins values are estimated as A535–0.25 (A650)/g fresh weights. We take into account the contribution of chlorophyll and its degradation products to absorption at 535 nm. Then, the results were also calculated as μg/g dry weight of leaves.

All spectrophotometric analyses were conducted at 25 °C with UV-VIS spectrophotometer (Model; T60, PG instruments limited, UK).

3 Statistical Analyses

The statistical analyses were performed to compare means of variables with a *t*-test and multivariate ANOVA followed by Dunnett's test using IBM SPSS Statistics 20 software. $P \leq 0.05^*$ was considered statistically significant. Results were expressed as mean ± SE of three independent replicates.

4 Results

4.1 Total Proteins

Sandy soils spiked with Ni, and Pb resulted in pronounced changes of the biochemical parameters of *Vicia faba*, which was manifested with high TEs concentrations applied. Total protein content was increased significantly by Ni and Pb treatments ($P = 0.000^*$; Table 1) comparing to absolute control. Furthermore, irrigation with MTW increased total protein content in plants untreated with TEs by 48% comparing to absolute control and by 51 and 34% in plants subjected to 250 mg Ni and 500 Pb mg kg^{-1} soil, respectively. In contrast, total protein content was decreased by

Table 1 Multivariate analysis assessing the effect of magnetically treated water on *V. faba* under elevated levels of Ni and Pb

Multivariate test: Pillai's trace test

Effect	Pillai's trace value	F-ratio	Hypothesis df	Error df	*Sign**
1. Metal concentration	1.962	33.299	22.0	14.0	*0.000**
2. Metal type	0.992	66.316	11.0	6.0	*0.000**
3. Magnetism	0.993	79.759	11.0	6.0	*0.000**
4. Metal type versus magnetism	0.995	109.361	11.0	6.0	*0.000**
5. Metal type versus metal concentration versus magnetism	1.953	26.689	22.0	14.0	*0.000**

Tests of between—subjects effects

Source	*df*	Protein	Proline	EL	MDA	H_2O_2	SOD	ASPX	POX	CAT	GLU	Anthocyanins
1. Metal conc. (n = 3)	F-ratio 2	237.4	29.56	5.907	9.06	7.26	124.5	299.1	66.13	120.8	188.5	3.943
	p-value	*0.000**	*0.000**	*0.012*	*0.002**	*0.006**	*0.000**	*0.000**	*0.000**	*0.000**	*0.000**	*0.000**
2. Metal type (n = 2)	F-ratio 1	263.64	19.55	6.201	15.01	3.764	366.2	504.9	257.2	7.795	35.45	36.114
	p-value	*0.000**	*0.000**	*0.024**	*0.001**	*0.070*	*0.000**	*0.000**	*0.000**	*0.013**	*0.000**	*0.000**
3. Magnetism (n = 2)	F-ratio 1	11.85	3.36	18.90	72.14	38.66	422.9	35.34	0.093	187.3	11.02	93.36
	p-value	*0.003**	*0.085*	*0.000**	*0.000**	*0.000**	*0.000**	*0.000**	*0.764*	*0.000**	*0.004**	*0.000**
4. Metal type versus magnetism ($n = 4$)	F-ratio 1	8.450	0.163	0.095	2.250	0.188	0.326	2.623	898.1	1.400	0.929	96.914
	p-value	*0.010**	*0.692*	*0.762*	*0.153*	*0.671*	*0.576*	*0.125*	*0.000**	*0.254*	*0.350*	*0.000**

(continued)

Table 1 (continued)

Tests of between—subjects effects												
Source	*df*	Protein	Proline	EL	MDA	H_2O_2	SOD	ASPX	POX	CAT	GLU	Anthocyanins
5. Metal type versus metal conc. versus magnetism ($n = 12$)	F-ratio 11	470.2	51.142	0.316	9.065	5.217	78.59	183.7	23.15	52.62	66.12	5.973
	p-value	*0.000**	*0.000**	*0.733*	*0.002**	*0.018**	*0.000**	*0.000**	*0.000**	*0.000**	*0.000**	*0.012**

Model carried out includes dependent parameters (total proteins, proline, electrolyte leakage, lipid peroxidation, hydrogen peroxide, superoxide dismutase, ascorbic acid peroxidase, peroxidase, catalase, glutathione, and anthocyanins), and metal concentration and metal types as covariates

34 and 49% in plants subjected to 500 mg Ni and 1000 mg Pb kg^{-1} soil and irrigated with MTW, respectively, comparing to corresponding controls (Fig. 1a).

4.2 Proline Content

Treatments with Ni and Pb showed a significant increase by 67 and 22% in proline level in plants subjected to high concentrations of 500 mg Ni and 1000 mg Pb kg^{-1} soil, respectively, (Fig. 1b). Likely, irrigation with MTW induced a significant accumulation of proline in plants non-subjected to TEs by 31% compared to absolute control and by 37.36% in plants subjected to 500 mg Pb kg^{-1} soil comparing to corresponding controls. Plants subjected to 500 mg Ni and 1000 mg Pb kg^{-1} soil and irrigated with MTW exhibited a reduction in proline content by 22 and 42% comparing to corresponding controls.

4.3 Hydrogen Peroxide (H_2O_2)

The results revealed that the level of H_2O_2 accumulated significantly by increasing concentrations of TEs ($p = 0.006$*; Table 1). Meantime, irrigating plants with MTW offset the accumulation of H_2O_2 by 30, 40, 18, and 40% in plants subjected to 250, 500 mg Ni and 500, 1000 mg Pb kg^{-1} soil, respectively, comparing to corresponding controls as shown in Fig. 2a.

4.4 Lipid Peroxidation (MDA)

Consequently, malondialdehyde (MDA) is generated, as a result of TEs exposure, which "is one of the final products of polyunsaturated fatty acids peroxidation in cells" (https://www.thefreelibrary.com/Role+of+external+proline+on+enhancing+defence+mec). The data in Fig. 2b illustrated that increasing concentrations of Ni and Pb induced lipid peroxidation in *Vicia faba* significantly ($P = 0.002$*; Table 1). However, irrigation with MTW significantly lowered MDA level in plants subjected to 250, 500 mg Ni and 1000 mg Pb kg^{-1} soil by 28%, 35%, and 15%, respectively, compared to corresponding controls. In plants subjected to 1000 mg Pb kg^{-1} soil, GSH content was increased by 18% compared to corresponding controls.

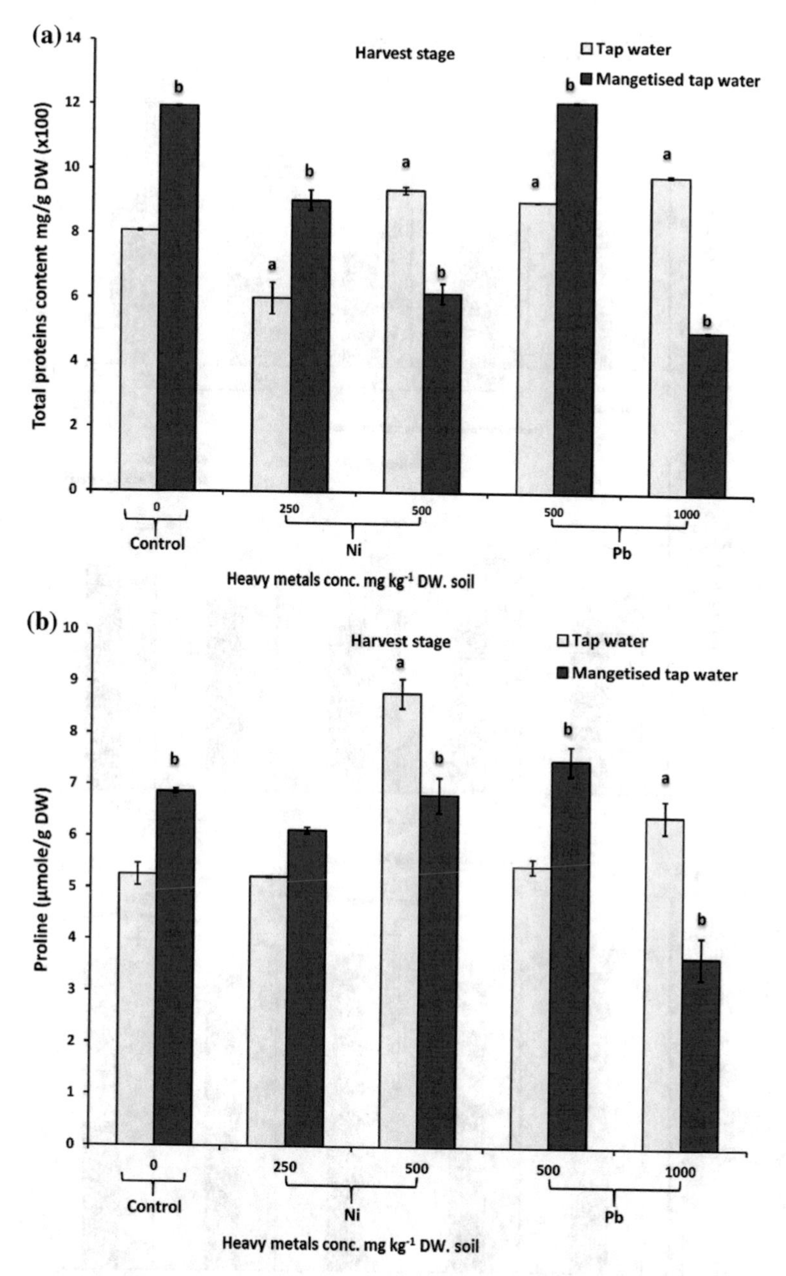

Fig. 1 Effect of magnetically treated water under trace elements stress of 0, 250, 500 mg Ni kg^{-1} soil and 0, 500, 1000 mg Pb Kg^{-1} soil on (**a**) total proteins content (mg/g DW) and (**b**) proline (μmole/g DW) of *Vicia faba* L. at harvest stage. Data represented as means ± SE ($n = 3$)

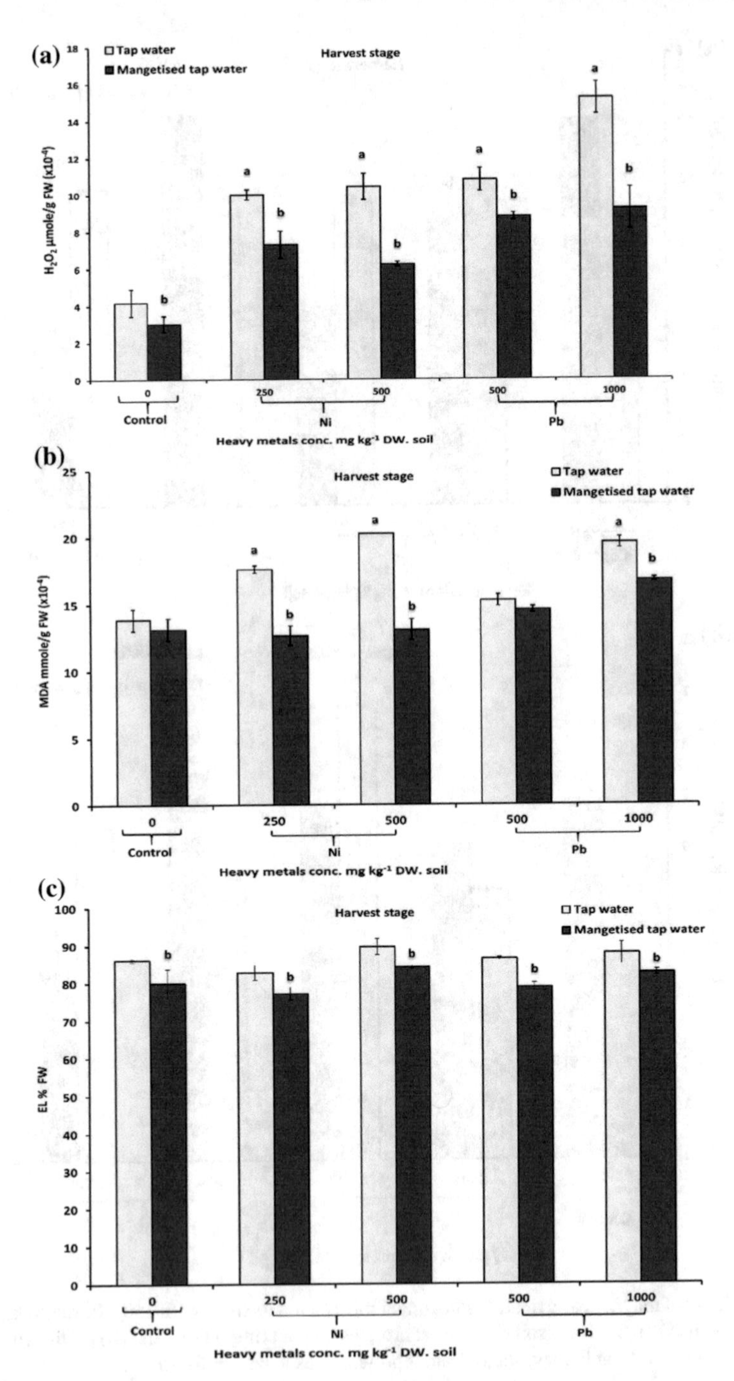

Fig. 2 Effect of magnetically treated water under trace elements stress of 0, 250, 500 mg Ni kg^{-1} DW soil and 0, 500, 1000 mg Pb Kg^{-1} DW soil on (**a**) H_2O_2 (mmole/g FW), (**b**) MDA (mmole/g FW), and (**c**) EL (%) of *Vicia faba* L. at harvest stage. Data represented as means ± SE ($n = 3$)

4.5 *Electrolyte Leakage (EL)*

There was no significant change in electrolyte leakage (%) in plants subjected to TEs comparing to absolute control. On the other hand, irrigation with MTW succeeded in reducing electrolyte leakage percent significantly in all the treatments ($P = 0.000*$; Table 1 and Fig. 2c).

4.6 *Antioxidant Enzymes Activity*

4.6.1 Enzymatic Antioxidants

Superoxide Dismutase

As shown in Fig. 3a, the activity of SOD showed a significant decrease in leaves of *Vicia faba* subjected to TEs. The highest reduction percent (~40%) recorded in plants subjected to 500 mg Pb kg^{-1} soil. Meantime, irrigation with MTW increased SOD by 34.6, 54.5, 27, and 100% in plants subjected to 250, 500 mg Ni and 500, 1000 mg Pb kg^{-1} soil, respectively, compared to corresponding controls, but decreased SOD significantly by 31% in plants non-subjected to TEs comparing to absolute control.

Ascorbic Acid Peroxidase

In comparison with control, ASPX increased significantly with increasing concentrations of TEs by 150, 395, and 419.6% in plants subjected to 500 mg Ni and 500, 1000 mg Pb kg^{-1} soil, respectively, comparing to absolute control. In contrary, irrigation with MTW decreased enzyme activity in plants subjected to TEs except those exposed to the highest concentration of Pb (1000 mg kg^{-1} soil), and ASPX were increased by 121% compared to corresponding control (see Fig. 3b).

Peroxidase

The POX activity decreased by 53 and 27% in plants subjected to 500 and 1000 mg Pb kg^{-1} soil, respectively, comparing to absolute control (Fig. 3c). Unlikely, irrigation with MTW decreased POX by 25 and 39% in plants subjected to 250, 500 mg Ni kg^{-1} soil while increased POX by 100% and 45% in plants subjected to 500, 1000 mg Pb kg^{-1} soil, respectively, comparing to corresponding controls. Similarly, plants non-subjected to TEs experienced an increase in POX activity by 33% when irrigated with MTW comparing to absolute control.

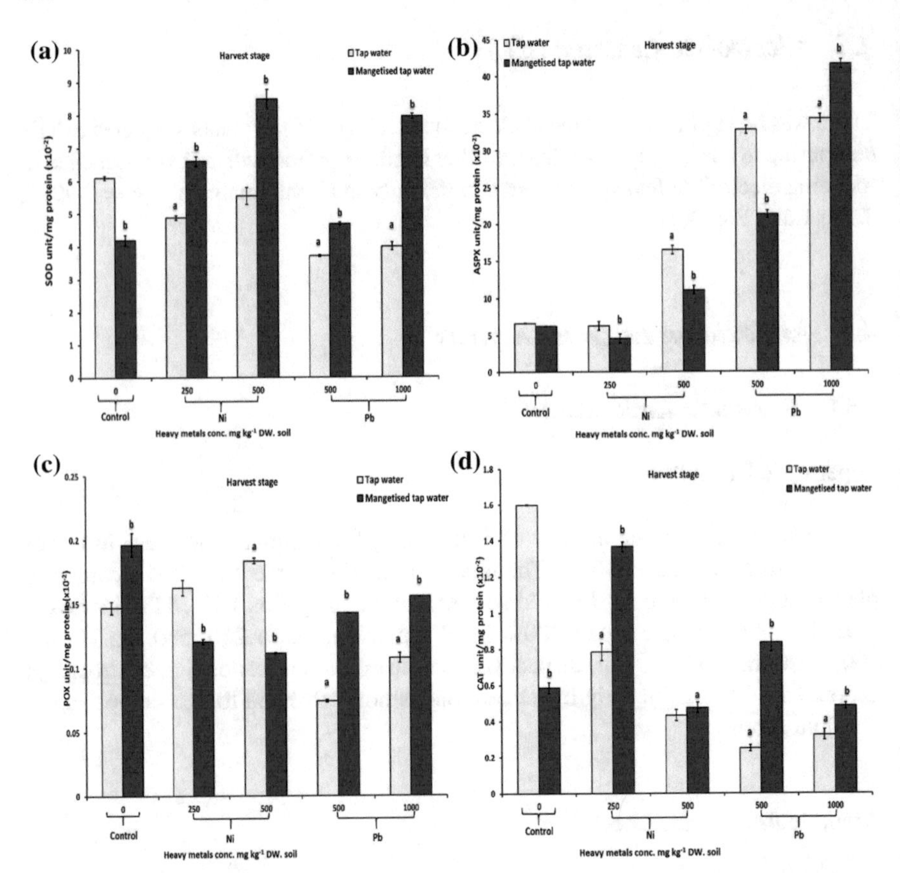

Fig. 3 Effect of magnetically treated water under trace elements stress of 0, 250, 500 mg Ni kg^{-1} DW soil and 0, 500, 1000 mg Pb Kg^{-1} DW soil on **a** SOD, **b** ASPX, **c** POX, and **d** CAT activities (unit/mg protein) of *Vicia faba* L. at harvest stage. Data represented as means ± SE ($n = 3$)

Catalase

The activity of CAT exhibited a significant reduction in plants subjected to TEs comparing to absolute control. The highest reduction percent (~84%) was recorded in plants subjected to 500 mg Pb kg^{-1} soil comparing to absolute control. On the other hand, irrigation with MTW increased CAT activity in most plants subjected to TEs on average, and the highest increase (~400%) was recorded with plants subjected to 500 mg Pb kg^{-1} soil, comparing to corresponding control (see Fig. 3d). Unlikely, a decrease of 63% in CAT activity was recorded in plants non-subjected to TEs and irrigated with MTW comparing to absolute control.

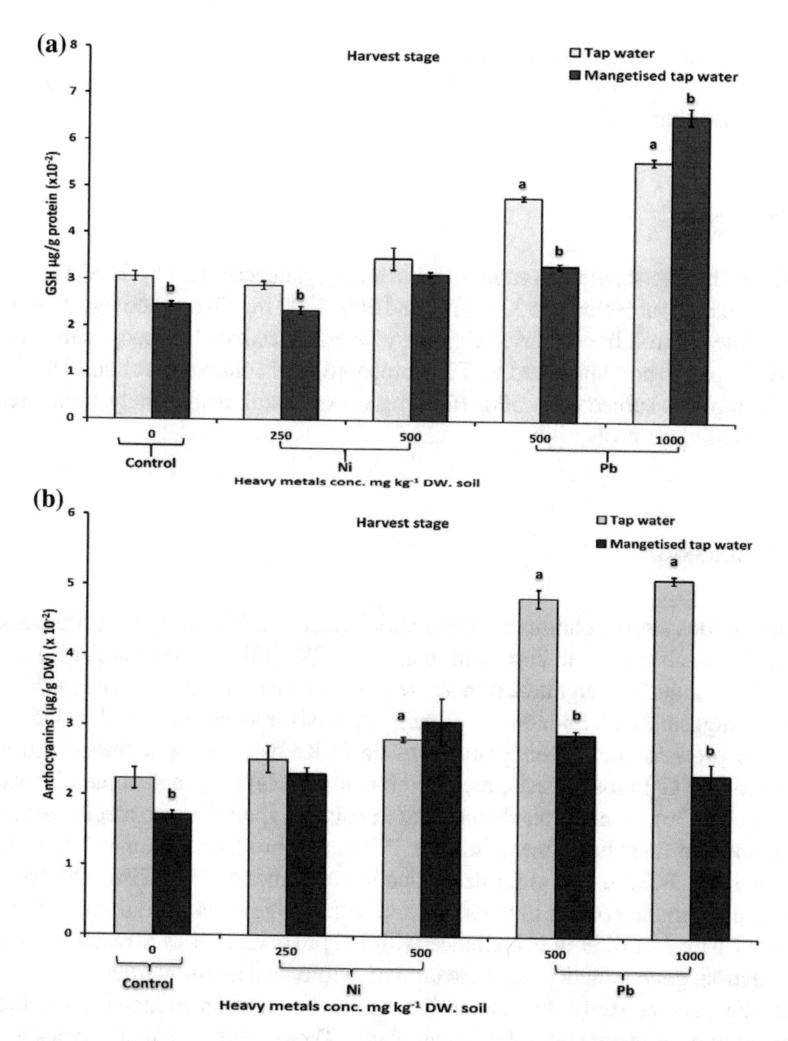

Fig. 4 Effect of magnetically treated water under trace elements stress of 0, 250, 500 mg Ni kg^{-1} DW soil and 0, 500, 1000 mg Pb Kg^{-1} DW soil on **a** GSH (µg/g protein), **b** anthocyanins (µg/g DW) of *Vicia faba* L. at harvest stage. Data represented as means ± SE ($n = 3$)

4.6.2 Non-enzymatic Antioxidants

Glutathione

Glutathione content increased by 55 and 81% in leaves of *Vicia faba* subjected to 500 and 1000 mg Pb kg^{-1} soil, respectively, compared to absolute control (Fig. 4a). Irrigation with MTW reduced glutathione content in plants non-subjected to TEs compared to absolute control by 19%, however; in plants subjected to 1000 mg Pb kg^{-1} soil, GSH content was increased again by 18% compared to corresponding controls.

Anthocyanins

As shown in Fig. 4b, there is an increase in anthocyanins content by 25.9, 116.5, and 129% in plants subjected to 500 mg Ni, and 500, 1000 mg Pb respectively, compared to absolute control. In contrary, irrigation with MTW reduced anthocyanins level by 23.3% in plants non-subjected to TEs compared to absolute control and by 40 and 54.5% in plants subjected to 500, 1000 mg Pb kg^{-1} soil, respectively, compared to corresponding controls.

5 Discussion

Under various stress conditions, plants may induce specific changes in protein synthesis that enable them to cope with that stress [32, 33]. The increase in TEs concentrations induced a significant increase in total protein content in *Vicia faba*. The authors suggest that TEs-induced protein synthesis may be attributed to induction of stress proteins such as enzymes involved in Krebs cycle, glutathione and phytochelatins (PCs) biosynthesis, metallothioneins (MTs) [34], and some heat shock proteins [35, 36], whereas plants must have evolved hypertolerance mechanisms that accommodate their high metal content. Plants can produce cysteine-rich peptides such as GSH, PCs, or (MTs) for detoxification or homeostasis of TEs. However, the decrease in protein content in plants treated with the high concentrations of TEs and irrigated with MTW is in accordance with [37] explained it as a result of protein degradation, gene mutation, or indication of cytological abnormalities.

In the present study, the increasing TEs concentration induced a significant increase in proline content in faba bean plants. These results are in accordance with Singh et al. [38] who reported that Pb and Ni significantly increased proline content of black gram as compared to control. [39] suggested that TEs-induced proline accumulation in plants is not directly emanated from TEs stress, but "water balance disorder, which occurs as a result of metal excess" (https://www.hindawi.com/journals/tswj/2015/756120/), is responsible for the induction of proline. In this regard, proline functions as an osmoregulator or osmoprotectant. Our results are in line with [40] who interpreted that free proline under Pb and Cd stress works an antioxidant and reduces free radical damage, which consequently increases GSH content in cells.

In the current study, proline content reduced significantly under the effect of MTW which is similar to the results of [41]. The reduction of proline under the effect of the magnetic field by oxidizing the free radicals that are enhanced during exposure to a magnetic field [42, 43] is called "caged reaction" theory. Proline consumption in "caged reaction" oxidizes proline to various compounds [44], and this process can protect plant tissue from potential damage [45]. In addition, the proline pathway could be shifted by magnetic field exposure, through oxidation of proline to glutamate or forming glutamic acid g-semialdehyde [45].

Hydrogen peroxide is one of the ROS, upon reaction with lipids, proteins, pigments, and nucleic acids results in lipid peroxidation, membrane damage, and alterations in enzymatic machinery, thereby ultimately affecting cell viability [46] and causing cell death [47]. In the present study, TEs treatments induced a significant increase in H_2O_2 level, which explains the significant increase in MDA content. The plasma membrane is the first functional part of the plant cell that which comes in contact with toxic metals where, hence, ions alter the membrane fluidity and structural conformation of membrane-bound enzymes (e.g., ATPase) and their activity [48, 49] and cause an increase in MDA content. There was a remarkable decrease in H_2O_2 level, MDA content, and EL values in all TEs-stressed plants irrigated with MTW comparing to corresponding controls. The authors suggest that irrigation with MTW reduces free radicals, oxidative stress and maintains the membrane integrity as confirmed by the results of [18].

Elevated TEs concentrations induced a highly significant reduction in most of the antioxidants (i.e., SOD, CAT, and POX) comparing to absolute control. In most cases, the activity of enzymes from the antioxidant system increases as an attempt to combat the ROS, but when the content of the TEs "in the plant tissue or the generation of ROS is too high, enzymatic activity may be reduced" [50, 51]. "The lower enzymatic activity may be caused by the oxidation of the thiol groups from the enzymes due to the increase in the generation of H_2O_2" [52]. Gallego et al. [53] explained it as the free oxygen radicals induced by TEs stress attack the antioxidant enzymes and inhibit their oxidative damages.

The role of GSH in the antioxidant defense system provides a strong basis for its use as a stress marker. It was demonstrated that increasing concentration of GSH is correlated with oxidative stress tolerance in plants during metal stress [54]. *Arabidopsis* plants with low concentrations of glutathione were susceptible to even low concentrations of Cd [55]. Moreover, increased glutathione biosynthesis plays a role in Pb-detoxification by their ability to detoxify ROS within involving in H_2O_2-scavenging mechanism [56, 57].

The effect of irrigation with MTW varied accordingly on antioxidants activity which is in line with the results of [17, 58] who recorded a similar fluctuation in antioxidant activities under the effect of magnetic field. In this respect, it has been established that the primary action of the magnetic field in biological systems is the induction of electrical charges and currents [59]. Hence, a magnetic field is known to affect radical pair recombination of paramagnetic molecules, and depending on their spin orientations, radical recombination, or diffusion, the formation of free radicals (e.g., oxygen radicals) may take place [60, 61]. Increasing the concentration of free radicals is potentially damaging and can initiate chain reactions to form new free radicals, and antioxidant defense mechanism acts as a protection from deleterious effects of these radicals [57, 62].

In our study, a substantial, significant positive correlation was observed between increasing soil TEs concentrations and anthocyanins content of *Vicia faba* plants.

Cellular anthocyanins level can increase when a plant encounters various abiotic stresses [63, 64]. As a general rule, anthocyanins are considered light attenuators and antioxidants. "It is believed that under stress situations, their main function is the quenching of the reactive oxygen species generated by stress" [65]. Hale et al. [66] proposed that anthocyanins might have an active protective role in *Brassica* plants against Mo toxicity by complexing this metal and sequestering it into the vacuoles. However, the reduction of anthocyanins content under the effect of irrigation with MTW may be explained from the reduction in ROS and oxidative stress. Neverthless, metal–phytochelatin complexes can be transported into the vacuole thus sequestering the metals away from sensitive enzymes [67]. This system seems to provide plants with a moderate level of resistance to TEs, and hence, anthocyanins content decreases.

6 Conclusions

In conclusion, it was found that TEs exposure leads to oxidative damage of *Vicia faba* which was displayed from increasing H_2O_2 level and MDA, proline and total proteins contents while decreasing activity of most enzymatic antioxidants. In the meantime, the difference in the effect of MTW on *Vicia faba* was related to the presence or the absence of TEs and their concentrations; however, it was clear that using tap water treated with 1000 gauss magnet can reverse the effect of Ni and Pb by decreasing H_2O_2 level and MDA, proline and total proteins contents while increasing antioxidant enzymes activity comparing to TEs-treated plants irrigated with normal tap water.

7 Recommendations

Irrigation with MTW increased *Vicia faba* L. plant biomass and yield, and therefore, it considered a recommendable environmentally technique that can be used to improve yield qualities. The current results prove that irrigation with MTW (1000 gauss) can alleviate toxic effects induced by low concentrations of Ni and Pb in *Vicia faba* L. plant and increase its productivity, but it cannot be effective against the high concentrations. Therefore, further investigation is needed on different MF intensities and different kinds of heavy metals.

Acknowledgements We want to express our appreciation to Agriculture Research Center for kindly providing the broad bean seeds. Our deepest gratitude to Faculty of Agriculture, Suez Canal University, for provision of their Research Farm facilities, land, and laboratories.

References

1. Gisbert C, Ros R, Deharo A, Walker DJ, Pilarbernal M, Serrano R, Navarro-Avino J (2003) A plant genetically modified that accumulates Pb is especially promising for phytoremediation. Biochem Biophys Res Commun 303:440–445
2. Castro R, Caetano L, Ferreira G, Padilha P, Saeki M, Zara L, Martines M, Castro G (2011) Banana peel applied to the solid phase extraction of copper and lead from river water. Chem Res 50(6):3446–3451
3. Awad YM, Vithanage M, Niazi NK, Rizwan M, Rinklebe J, Yang JE, Ok YS, Lee SS (2017) Potential toxicity of trace elements and nanomaterials to Chinese cabbage in arsenic-and lead-contaminated soil amended with biochars. Environ Geochem Health. https://doi.org/10.1007/s10653-017-9989-3
4. Chiban M, Soudani A, Sinan F, Tahrouch S, Persin M (2011) Characterization and application of dried plants to remove heavy Metals, nitrate, and phosphate ions from industrial wastewaters. Clean—Soil, Air, Water 39(4):376–383
5. Schützendübel A, Polle A (2002) Plant responses to abiotic stresses: heavy metal-induced oxidative stress and protection by mycorrhization. J Exp Bot 53(372):1351–1365
6. Gill M (2014) Heavy metal stress in plants: a review. Int J Adv Res 2(6):1043–1055
7. Singh S, Sinha S (2005) Accumulation of metals and its effects in *Brassica juncea* (L.) Czern (cv. Rohini) grown on various amendments of tannery waste. Ecotoxicol Environ Saf 62:118–127
8. Shi K, Gao Z, Shi TQ, Song P, Ren LJ, Huang H, Ji XJ (2017) Reactive oxygen species-mediated cellular stress response and lipid accumulation in oleaginous microorganisms: the state of the art and future perspectives. Front Microbiol 8:793. https://doi.org/10.3389/fmicb.2017.00793
9. Michalak A (2006) Phenolic compounds and their antioxidant activity in plants growing under heavy metal stress. Pol J Environ Stud 15(4):523–530
10. He J, Wang Y, Ding H, Ge C (2016) Epibrassinolide confers zinc stress tolerance by regulating antioxidant enzyme responses, osmolytes, and hormonal balance in solanum melongena seedlings. Braz J Bot 39(1):295–303
11. Sharma A, Shekhawat S (2017) An overview of magnetic field exposure on germination behavior of seeds under saline stress. Int J Pharm Bio Sci 8(1):713–716
12. Dhawi F (2014) Why magnetic fields are used to enhance a plant's growth and productivity? Annu Rev Plant Biol 4(6):886–896
13. El Sayed HEA (2015) Impact of magnetic water irrigation for improve the growth, chemical composition and yield production of broad bean (*Vicia faba* L.). Plant Nat Sci 13(1):107–119
14. Pintilie M, Oprica L, Surleac M, Dragut IC, Creanga DE, Artenie VE (2006) Enzyme activity in plants treated with magnetic liquid. Rom J Phys 51:239–244
15. Hozayn M, Qados AMS Abdul (2010) Irrigation with magnetized water enhances growth, chemical constituent and yield of chickpea (*Cicer arietinum* L.). Agric Biol J N Am 1(4):671–676
16. Moussa HR (2011) The impact of magnetic water application for improving common bean (*Phaseolus vulgaris* L.) production. New York Sci J 4(6):15–20
17. Hassan S, Parviz A, Faezeh G (2007) Effects of magnetic field on the antioxidant enzyme activities of suspension–cultured tobacco cells. Bioelectromagnetics 28:42–47
18. Chen YP, Li R, He JM (2012) Magnetic field can alleviate toxicological effect induced by cadmium in mungbean seedlings. Ecotoxicol 20:760–769
19. Orcutt DM, Nilsen ET (2000) Physiology of plants under stress: soil and biotic factors. Wiley, New York, p 684
20. Karimi R, Chorom M, Solhi S, Solhi M, Safe A (2012) Potential of *Vicia faba* and *Brassica arvensis* for phytoextraction of soil contaminated with cadmium, lead and nickel. Afr J Agri Res 7(22):3293–3301

21. Bradford M (1976) A rapid and sensitive for the quantitation of microgram quantities of protein utilizing the principle of protein-dye binding. Anal Biochem 72:248–254
22. Bates LS, Waldeen RP, Teare ID (1973) Rapid determination of free proline for water stress studies. Plant Soil 39:205–207
23. Shi S, Wang G, Wang Y, Zhang L (2005) Protective effect of nitric oxide against oxidative stress under ultraviolet-B radiation. Nitric Oxide-Biol Ch 13(1):1–9
24. Heath RL, Packer L (1968) Photoperoxidation in isolated chloroplasts. I. Kinetics and stoichiometry of fatty acid peroxidation. Arch Biochem Biophys 125:189–190
25. Dionisio-Sese ML, Tobita S (1998) Antioxidant responses of rice seedlings to salinity stress. Plant Sci 135:1–9
26. Beauchamp C, Fridovich I (1971) Superoxide dismutase: improved assays and an assay applicable to acrylamide gels. Anal Biochem 44:276–287
27. Nakano Y, Asada K (1980) Spinach chloroplasts scavenge hydrogen peroxide on illumination. Plant Cell Physiol 21:1295–1307
28. Malik CP, Singh MB (1980) Plant enzymology and histoenzymology. Kalyani Publishers, New Delhi 41(2):53
29. Aebi HE (1983) Catalase. In: Bergmeyer HU (ed) Methods of enzymatic analysis, vol 3, 3rd edn. Verlag Chemie, Weinhem, pp 273–286
30. Griffith OW (1980) Potent and specific inhibition of glutathione synthesis by buthionine sulfoximine (s-n butyl homocysteine sulfoximine). J Biol Chem 254:7558–7560
31. Lange H, Shropshire WJ, Mohr H (1970) An analysis of phytochrome-mediated anthocyanin synthesis. Plant Physiol 47:649–655
32. Santos I, almeida JM, Salema R (1998) Plants of *Zea mays* L. developed under enhanced UV-b radiation. I. Some ultrastructural and biochemical aspects. J Plant Physiol 141:450–456
33. Rodziewicz P, Swarcewicz B, Chmielewska K, Wojakowska A, Maciej SM (2014) Influence of abiotic stresses on plant proteome and metabolome changes. Acta Physiol Plant 36(1):1–19
34. Robinson NJ, Tommey AM, Kuske C, Jackson PJ (1993) Plant metallothioneins. Biochem J 295:1–10
35. Verma S, Dubey RS (2003) Lead toxicity induces lipid peroxidation and alters the activity of antioxidant enzymes in growing rice plants. Plant Sci 164:645–655
36. Mishra S, Agrawal SB (2006) Interactive effects between supplemental ultraviolet-B radiation and heavy metals on the growth and biochemical characteristics of *Spinacia oleracea* L. Braz J Plant Physiol 18(2):307–314
37. Shabrangi A, Majd A, Sheidai M (2011) Effects of extremely low frequency electromagnetic fields on growth, cytogenetic, protein content and antioxidant system of *Zea mays* L. Afr J Biotech 10(46):9362–9369
38. Singh G, Agnihotri RK, Reshma RS, Ahmad M (2012) Effect of lead and nickel toxicity on chlorophyll and proline content of Urd (*Vigna mungo* L.) seedlings. Int J Plant Physiol Biochem 4(6):136–141
39. Clemens S (2006) Toxic metal accumulation, responses to exposure and mechanisms of tolerance in plants. Biochimie 88(11):1707–1719
40. Bayçu G, Rognes SE, Özden H, Gören Sağlam N, Csatari I, Szabo S (2017) Abiotic stress effects on the antioxidative response profile of Albizia julibrissin Durazz. (Fabaceae). Braz J Bot 40(1):21–32
41. Alfaidi MA, Al-Toukhy AA, Al-Zahrani HS, Howladar MM (2017) Effect of irrigation by magnetized sea water on guinea grass (panicum maximum) leaf content of chlorophyll a, b, carotenoids, pigments, protein & proline. Adv Environ Biol 11(1):73–83
42. Scaiano JC, Cozens FL, McLean J (1994) Model of the rationalization of magnetic field effects *in vivo*. Application of radical-pair mechanism to biological systems. Photochem Photobiol 59:585–589
43. Parola A, Kost K, Katsir G, Monselise E, Cohen-Luria R (2006) Radical scavengers suppress low frequency EMF enhanced proliferation in cultured cells and stress effects in higher plants. Environmentalist 25:103–111

44. Stadtman ER (1993) Oxidation of free amino acids and amino acid residues in proteins by radiolysis catalyzed reactions. Annu Rev Biochem 62:797–821
45. Matysik J, Alia PSP, Bhalu B, Mohanty P (2002) Molecular mechanisms of quenching of reactive oxygen species by proline under stress in plants. Curr Sci 82:525–532
46. Dixit V, Pandey V, Shyam R (2001) Differential antioxidative responses to cadmium in roots and leaves of pea (*Pisum sativum* L. cv.Azad). J Exp Bot 52(358):1101–1109
47. Stone JR, Yang S (2006) Hydrogen peroxide: a signaling messenger. Antioxid Redox Signal 8:243–270
48. Ros R, Cooke DT, Burden RS, James CS (1990) Effect of herbicide MCPA and the heavy metals, cadmium and nickel, on the lipid composition, Mg-ATPase activity and fluidity of plasma membrane from rice, *Oryza sativa* (cv Bhatia) shoots. J Exp Bot 41:457–462
49. Ros R, Morales A, Segura J, Picazo I (1992) In vivo and in vitro effects of nickel and cadmium on the plasma ATPase from rice (*Oryza sativa* L.) shoots and roots. Plant Sci 83:1–6
50. Hu Z, Xie Y, Jin G, Fu J (2015) Growth responses of two tall fescue cultivars to Pb stress and their metal accumulation characteristics. Ecotoxicol 24(3):563–572
51. Rabêlo FHS, Borgo L (2016) Changes caused by heavy metals in micronutrient content and antioxidant system of forage grasses used for phytoremediation: an overview. Ciência Rural 46(8):1368–1375
52. Gill SS, Tuteja N (2010) Reactive oxygen species and antioxidant machinery in abiotic stress tolerance in crop plants. Plant Physiol Biochem 48:909–930
53. Gallego SM, Benavides MP, Tomaro ML (1996) Effect of heavy metal ion excess on sunflower leaves: Evidence for involvement of oxidative stress. Plant Sci 121:151–159
54. Freeman JL, Person MW, Nieman K, Albrecht C, Peer W, Pickering IJ, Salt DE (2004) Increased glutathione biosynthesis plays a role in nickel tolerance in *Thalspi* Nickel hyperaccumulators. Plant Cell 16:2176–2191
55. Xiang C, Werner BL, Christensen EM, Oliver DJ (2001) The biological functions of glutathione revisited in arabidopsis transgenic plants with altered glutathione levels. Plant Physiol 126(2):564–574
56. Piechalak A, Tomaszewska B, Baralkiewicz D, Malecka A (2002) Accumulation and detoxification of lead ions in legumes. Phytochemistry 60(2):153–162
57. Ashraf U, Kanu AS, Deng Q, Mo Z, Pan S, Tian H, Tang S (2017) Lead (Pb) toxicity; physio-biochemical mechanisms, grain yield, quality, and Pb distribution proportions in scented rice. Front Plant Sci 8:259. https://doi.org/10.3389/fpls.2017.00259
58. Shabrangi A, Majd A (2009) Effect of magnetic fields on growth and antioxidant systems in agricultural plants. In: Prog. electromagnetic research symposium proceedings, Beijing, China, pp 1142–1147
59. Roy C, Repacholi M (2005) Electromagnetic field and Health, A WHO perspective. World Health Organization, Geneva, Switzerland, Available from: http://www.world-aluminium.org/news/montreal/roy.htm
60. Kula A, Sobczak A, Kuska R (2002) Effects of electromagnetic field on free-radical processes in steelworkers. Part I: magnetic field influence on the antioxidant activity in red blood cells and plasma. J Occup Health 44:226–229
61. Sobczak A, Kula B, Dancii A (2002) Effects of electromagnetic field on free-radical processes in steelworkers. Part II: Magnetic field influence on vitamin A, E and selenium concentrations in plasma. J Occup Health 44:230–233
62. Wang HY, Zeng XB, Guo SY, Li ZT (2008) Effect of magnetic field on the antioxidant defense system of recirculation-cultured *Chlorella vulgaris*. Bioelectromagnetics 29:39–46
63. Dixon RA, Xie DY, Sharma SB (2005) Proanthocyanidins: a final frontier in flavonoid research. New Phytol 165:9–28
64. Grotewold E (2006) The genetics and biochemistry of floral pigments. Annu Rev Plant Biol 57:761–780
65. Neill SO, Gould KS (2003) Anthocyanins in leaves: light attenuators or antioxidants? Funct Plant Biol 30(8):865–873

66. Hale KL, McGrath SP, Lombi E, Stack SM, Terry N, Pickering IJ, George GN, Pilon-Smits EA (2001) Molybdenum sequestration in Brassica species. A role for anthocyanins? Plant Physiol 126:1391–1402
67. Rauser WE (1990) Phytochelatins. Annu Rev Biochem 59:61–86

Conclusions

Update, Conclusions, and Recommendations to "Technological and Modern Irrigation Environment in Egypt: Best Management Practices and Evaluation"

El-Sayed E. Omran and Abdelazim M. Negm

1 Introduction

Agriculture in Egypt is currently a major source in the Egyptian economy. Agricultural production in Egypt depends on irrigation. The future of agriculture in Egypt faces many challenges. The most important of these challenges is a water deficit, environmental pollution, food shortages, and others. A key challenge of water resources management in Egypt is the imbalance between increasing water demand and limited supply. The improvement of irrigation systems and better control over water by farmers is therefore crucial to the drive to raise productivity in the current context of scarce water resources. New technologies might be a significant contribution toward helping us reach that goal. The latest technologies are to improve agricultural procedures and the use of modern irrigation technologies. The development of irrigation systems, crop varieties more drought tolerant, as well as intercropping systems represented important strategies for the decision maker and played a significant role in increasing productivity. Next to all this, biotechnology offers a better way to enhance crop productivity and therefore bridge the gap between food production and consumption. Therefore, achieving agricultural sustainability requires close cooperation between government, non-governmental organizations, and farmers groups. Therefore, this book brings together contributions to technological and modern irrigation environment from an agricultural point of view from experts in the field. It

E.-S. E. Omran (✉)
Soil and Water Department, Faculty of Agriculture, Suez Canal University, Ismailia 41522, Egypt
e-mail: ee.omran@gmail.com

Institute of African Research and Studies and Nile Basin Countries, Aswan University, Aswan, Egypt

A. M. Negm
Water and Water Structures Engineering Department, Faculty of Engineering, Zagazig University, Zagazig 44519, Egypt
e-mail: amnegm@zu.edu.eg; amnegm85@yahoo.com

E.-S. E. Omran and A. M. Negm (eds.), *Technological and Modern Irrigation Environment in Egypt*, Springer Water,
https://doi.org/10.1007/978-3-030-30375-4_16

is devoted to a wide variety of modern irrigation issues. The book presents state-of-the-art knowledge that can be effectively used for solving a variety of problems in modern irrigation as well as the latest developments in the research area. The focus of the book is placed on water sensing and information technologies, automated irrigation technologies, and improve irrigation efficiency. The aim was to establish current knowledge, describe the various applications, present recent advances, and discuss possibilities for research and adoption of the techniques and interdisciplinary collaboration.

Therefore, this chapter presents general conclusions of the technological and modern irrigation environment and its importance for Egypt and the researchers. In designing sustainable agricultural production systems, it is necessary to give due consideration to various resources used in modern irrigation environment in Egypt, which render the resultant agricultural system unsustainable. Therefore, the book intends to improve and address the following main theme:

- Irrigation Practice: Problems and evaluation.
- Smart irrigation technology.
- Irrigation management.
- Irrigation system design.
- Water reuse and treatment.

The next section presents a brief of the important findings of some of the recent (updated) published studies on the technological and modern irrigation environment in Egypt, then the main conclusions of the book chapters in addition to the main recommendations for researchers and decision makers. The update, conclusions, and recommendations presented in this chapter come from the data presented in this book.

2 Update

The following are the major update for the book project based on the main book theme.

2.1 Update on Irrigation Practice: Problems and Evaluation

Three approaches shed light on the main problems and irrigation evaluation in Egypt. The first approach deals with water resources, type, and common problems in Egypt. "Water availability issues in Egypt are rapidly assuming alarming proportions. By the year 2020, Egypt will be consuming 20% more water than it has. With its loosening grip on the Nile, water scarcity could endanger the country's stability and regional dominance" [1] and (https://www.ecomena.org/egypt-water/). The development of water resources is expected to be very poor. There is excessive growth of population.

Furthermore, agriculture activities in Egypt consumed about 80% of the Nile water budget. Therefore, potential scarcity might occur in Egypt; especially, there is a critical argument due to the buildup of the Grand Ethiopian Renaissance Dam (GERD). Clearly, the construction of the GERD will negatively affect the situation of natural water resources in Egypt [2]. Filling the GERD by 74 billion m^3 of Nile water may lead to change in the demographic map of Egypt.

The second approach calls for an evaluation of irrigation schemes and irrigation systems. Irrigation systems may or may not be well designed and properly used. Improvement of water management on the farm may conserve water, soil, and labor and may increase crops yields. Most modifications suggested for irrigation systems improvement requires only simple changes in management practices. "In order to assess and evaluate irrigation performance, several methods should be developed and used. The main methods used to evaluate irrigation system performance are the fuzzy set theory, direct measurements for indicators, analysis hierarchy process (AHP), and remote sensing (RS). All these methods should be identified and discussed in the future. Although they offer a wide range of choices for characterizing all aspects of performance evaluation, however, still there is no agreed approach that could be offered to assess the performance of different irrigation systems [3]. It could be concluded that the selection of the evaluation framework and method largely depends on the nature of the irrigation system and the purpose of evaluation" (https://www.sciencedirect.com/science/article/pii/S037837741830129X).

The third approach is the evolution of irrigation system, tools, and technologies. A study of irrigation history in Egypt underscores the ancient origins of agriculture. The modern world is in debt to this great civilization, which contributed several of our basic agricultural innovations, especially cultivation and irrigation technology, and the horticultural art. Ancient Egyptian agriculture is also shown to be the mother of science (https://www.coursehero.com/file/p2meluh/Plantings-became-ordered-and-set-in-stra).

2.2 Update on Smart Irrigation Technology

Three potential approaches were identified for using new and modern technology to increase sustainable agriculture in Egypt. A first potential approach is the smart sensing system for precision agriculture. A key challenge facing the geospatial community specifically has been dealing with the large extents of spatial data being managed and accessed, particularly remote sensing (RS) data. "Internet of things" (IoT) concept was introduced [4], in which present reality items are associated with an embedded system including electronics and sensors through which the data can be transferred reliably. Cloud computing has been utilized for agricultural data storage [5, 6]. Many workers have adopted these new technologies (e.g., smartphones) to aid their work [7, 8]. The smartphones application for agricultural sensing offers significant benefits to the PA. Agricultural IoT can be viewed as a network of sensors, cameras, and devices, which will work toward a common goal of helping a farmer, does his job in an intelligent manner. Through sensor networks, agriculture can be connected to the IoT, which permits us to make connections among agronomists, farmers, and crops regardless of their geographical differences. Spatial data integrated with IoT and cloud computing will bring our agriculture world online [9]. The main goal was to make fully utilized the new technologies in PA fields, which can diminish the farmer problems. The need to provide the decision makers with fast, reliable, and up-to-date information, which will help farmers make the right decisions, has become a requirement for PA users and decision making. Therefore, how emerging technology can provide real-time data to support the PA process is the most imperative issue that needs investigation.

The second potential way is the development of recent information and data on irrigation technology and management. Irrigation technology improvement and related developments in water management depending on offer an opportunity for agriculture to mitigate water shortage more efficiently for the allocated water [10]. Irrigation improvement schemes require data collection and evaluation, planning and design of the scheme, and implementation. Collection data and its evaluation involve collecting information from soil surveys, water analyses, and topographic surveys [11]. A detailed assessment and specific information for the experimental areas for irrigation schemes are required to (i) update information and data on irrigation technology, (ii) promote intensive and sustainable irrigated agriculture, (iii) improve food security and (iv) support decision making in Egypt.

The third potential approach is the medicinal plants in a hydroponic system under water deficit conditions as a way to save water. Under a deficient irrigation system, as a system for providing water with a degree of potential stress that produces minimal effects on the crop and sometimes increases the active substances in medicinal plants. Deficit irrigation can achieve greater economic benefit than increasing the yield per unit of water for a particular medicinal plant. Aromatic plants can also be produced in most hydroponic systems with increasing rates of growth and percentages of volatile oils and reduction of the growing period. After considering all aspects, hydroponic culture can have good results in the production of medicinal plants with a high content of biologically active substances. This type of agricultural systems can be a means of growing medicinal plants for an economic purpose. In addition to economic and chemical advantages, hydroponic systems for the cultivation of medicinal plants help protect wild plants and the variety of species that can be found in the country. One of the main targets of the Egyptian government is to encourage the use of ecologically friendly agriculture as hydroponic systems [12].

2.3 Update on Irrigation Management

Four approaches were identified for managing irrigation as a potential way for sustainable agriculture. First, accurate estimation of crop coefficients for better irrigation water management in Egypt is needed. Proper irrigation water management under the prevailing conditions of water scarcity in Egypt required the knowledge of exact values of water consumptive use of the cultivated crops. This can be attained by the accurate estimation of the values of ETo and crop Kc. Earlier studies compared different ETo equations for their accuracy revealed that the Penman–Monteith equation is the most accurate because it fits detailed theoretical base and its accommodation of small periods [13]. It was reported that the Kc is affected by all the factors that influence soil water status, for instance, the irrigation method and frequency, the weather factors, the soil characteristics, and the agronomic techniques that affect crop growth [14]. Consequently, the reported values of crop coefficients in the literature can vary significantly from the actual measured values in a location, if growing conditions differ from those where the said coefficients were experimentally obtained. Although the calculation of ETo can be easily implemented, the calculation of Kc required more efforts to be calculated.

Second, vermicomposting (VC) influences agriculture in Egypt. There is a nice prospect for vermicomposting adaptation by municipal waste systems in country operation [15]. Some of the waste could be used in the generation of organic fertilizers, animal fodder, production of food, energy production, and other resolutions. Vermiculture is also an important process to transform organic waste into vermicomposting. In suitable environments such as Nile Valley, earthworms played a significant role in sustainable agriculture. Vermicomposting affects soil properties. The addition of VC to the soil increases water-holding capacity and by maintaining evaporation losses to a minimum as good adsorbent of atmospheric moisture eventually helps in maintaining the ecology of the hydrologic cycle. Increasing water-holding capacity is one of the soil erosion control measures that influences soil productivity in both managed and natural ecosystems [14].

Finally, irrigation water use efficiency and economic water productivity of different plants under Egyptian conditions get more attention. In Egypt, fixed water assets led to the synergy of government, farmers, and scientific communities' efforts to increase irrigation water use efficiency IWUE of most plants, where cultivation is the major user of water. The efficiency of irrigation water use may also be seen in a plant physiological sense and in particular as a comparison of the yield or economic return of an irrigated crop or pasture to the total amount of water transpired by the crop or pasture. In fact, in recent literature (e.g., [16]), this is commonly referred to as irrigation water productivity and not WUE.

2.4 Update Irrigation System Design

Three distinct technologies are identified for irrigation system design. First, improving the performance of surface irrigation system by designing pipes for water conveyance and on-farm distribution is important. Agricultural irrigation future has many challenges, such as global warming, the low efficiency with which water resources have been used for irrigation, and 40% or more of the water diverted for irrigation is wasted through either deep percolation or surface runoff. These losses often represent certain foregone opportunities for water because they delay the arrival of water at downstream diversions, and this produces low-quality water. The big problem in the future is the growth of alternative demands for water, such as urban and industrial needs [17]. In the future, agriculture irrigation will undoubtedly face the problem of maximizing efficiency. Irrigation efficiency is greatly dependent on the type and design of water conveyance and distribution systems. Designing of economic pipe diameter is important to minimize cost, water loss, and land requirement.

The second technology is the micro-sprinkler irrigation of orchard. The orchard sprinkler is a small spinner or impact sprinkler designed to cover the inter-space between adjacent trees; there is little or no overlap between sprinklers. Orchard sprinklers are designed to be operated at pressures between 10 and 30 psi, and typically, the diameter of coverage is between 5 and 10 m. They are located under the tree canopies to provide approximately uniform volumes of water for each tree [18]. Water should be applied fairly even to areas to be wetted even though some soil around each tree may receive little or no irrigation. The individual sprinklers can be supplied by hoses and periodically for each position [19]. Improvements in irrigation water management and irrigation systems save water, nutrients, chemicals, time, and money. For high yield and quality, management of the irrigation system should establish to provide the required amount of water at the optimum timing. The main issue for the field crops growing under orchard trees is how to improve the distribution uniformity. At the same time, high distributing water and controlling the irrigation system are very significant to the orchard. Automated irrigation and fertilization system will collaborate with the soil moisture sensors to provide an efficient means of irrigation scheduling [20].

The third is the drip irrigation technology. The drip irrigation system is suitable for a wide variety of horticultural and agronomic crops. In many respects, it applies to those crops presently under surface drip irrigation. The advantages of the subsurface drip irrigation system are negligible interference with farming activity, elimination of mechanical damage to laterals, decreased weed infestation, elimination of runoff and evaporation from the soil surface may be zero, and improved uptake of nutrition elements by the roots, notably phosphorous. The disadvantages are high costs for burying the laterals into the soil, plugging hazard by intruding roots and sucked-in soil particles, inconvenience in monitoring the performance of drippers and laterals, and strict maintenance is mandatory [21]. A mobile drip irrigation system provides certain advantages over other move systems [22].

2.5 Update Water Reuse and Treatment

Two techniques were identified to treat water irrigation with a magnetic field. The first technique is to use irrigation with magnetically treated water induces antioxidative responses of *Vicia faba* L. to Ni and Pb stress at harvest stage. Many studies have been conducted to emphasize the importance of the magnetic fields (MF) as an eco-friendly choice to improve crops [23]. The applications of a magnetic field include the treatments of irrigation water, dry seeds, wet seeds, and seedlings [24]. Magnetically treated water (MTW) exhibited a marked significant increase in total protein, total amino acids, and proline contents in broad bean compared with control plants [25]. A significant increase in the activities of the antioxidant enzymes (CAT, POX, and SOD) has been observed in chickpea and common bean irrigated with MTW compared to irrigation with tap water. On the other hand, magnetic field treatment decreased CAT activity in tobacco [26].

The second technique is to use irrigation with magnetically treated water to enhance plant growth and rehabilitates the toxicity of nickel and lead. Modern agricultural efforts are currently in search of efficient and eco-friendly production technology for improving crop productivity without harming the environment, and magnetic water treatment is one such area [27]. Water in liquid form can be affected by magnetic fields which assist its purification [28]. It enhances tolerance to abiotic stresses. Under stress conditions, the magnetized water treatment caused more activity of antioxidant enzymes and free proline in mung bean plant compared to ordinary water [29].

3 Conclusions

Throughout the current book project, the editorial' teams were able to reach several conclusions, which have been drawn from this book. Besides methodological insights, the chapter originates key lessons from the cases in the book, in particular, the promising characteristics of both the technological and modern irrigation environment in Egypt. These conclusions are important to increase sustainable food supply in Egypt. These are discussed in the following in no particular order.

3.1 Irrigation Practice: Problems and Evaluation

It could be concluded clearly that Egypt may suffer from water shortage through the next few years due to increasing the present and future water demands. In addition, this chapter discusses the dispute between The Nile Basin countries on The Nile Water; especially, the problem of Grand Ethiopian Renaissance Dam (GERD). More than 80% of water resources in Egypt are consumed in irrigation for agricultural soil. The major irrigation system used in Egypt is surface irrigation with high water loss due to evaporation and seepage. Egyptian experts predict a water reduction of about 11–19 billion m^3 owing to the construction of that Ethiopian dam.

Consequently, about one-third of the total agriculture land in Egypt might be subjected to drought. Therefore, attention should be paid to overcome these problems through the development of irrigation systems, reuse of agricultural wastewater, desalination of seawater, and managing the discharge and usage of groundwater. Furthermore, applications of advanced irrigation systems should be preceded by studying the major characteristics of the water resource to ensure its safe use.

In Egypt, irrigation is dependent upon the Nile River and has been systematically utilized since the predynastic era. In the time of the political unification of Egypt under the Pharaohs, basin irrigation was supplemented with the growth of crops year-round, specifically cereal grains. Although the river floods in summer, it carries a much volume of water for the rest of the year. Egypt's dry and warm climate supports more than one crop every year, so farmers started growing multiple crops. The Aswan Dam was built in 1902 to be completed twice in 1912 and the second in 1933. In order to better regulate the extra water, the government built a series of other barrages: Asyut, Zifta, Esna, Edfina, and Nag Hammadi, as well as the structure of the Delta Bridge, which replaced Al-Qanatir Al-Khairiya. First utilizing buckets and then shadoof after the Middle Kingdom, Egyptians lifted water from the river to irrigate their crops. The Archimedes Screw (Tanbour) and waterwheel (saqia—in Arabic) were introduced to Egypt by the second century. These water-lifting devices were by animal-powered and meant a drastic increase in the water amount that could be poured onto the fields during the summer when the water volume in the Nile was at its lowest.

3.2 Smart Irrigation Technology

To deal with smart irrigation technology, two objective are identified to mitigate climate change impact. The first is to assess the status of smart sensors, cloud computing, and IoT development to support PA. Some conclusions from the systematic review can be drawn. Proximal soil sensing allows measuring many of soil-plant properties in situ. Examples include portable X-ray, visible/near-infrared spectroscopy, digital camera, smartphone, multi-stripe laser triangulation scanning, ground-penetrating radar, and electromagnetic induction sensor. There is an availability of a number of smartphone applications for target farmers. Most of these applications are easily accessible if target users have access to smartphones. Direct estimation of soil properties in the field is possible and can be measured with accuracy levels suitable for soil and plant monitoring requirements. Second, this chapter proposed a system that integrates the Internet of things (IoT) with cloud computing and sensors, which are crucial to building smart precision agriculture. The Cloud is "a collection of platforms and infrastructures on which data are stored and processed, allowing farmers to retrieve and upload their data for a specific application," at any site with available Internet access. Hence, the Cloud is a pool of resources accessible via the Internet. Combining the Cloud, IoT, and sensors is crucial for the system. Four layers are proposed in the system, which are sensor layer, transmission layer, Cloud services layer, and application layer. Benefits and possible limitations of the proposed system are identified.

The required new knowledge, data, and information for irrigation, evapotranspiration concepts, and computer software are vital to calculating evapotranspiration of crops and irrigation requirements. Furthermore, a detailed assessment and specific information for the experimental areas for irrigation schemes are necessary to update information and data on irrigation technology to promote intensive and sustainable irrigated agriculture and improve food security and support decision making in Egypt.

Medicinal plants cultivated on a commercial scale to satisfy the large demand for natural remedies and to provide elemental natural materials for medicines, flavors, and cosmetics. Because of the rapid consumption of water resources and the increased stroke of climate change, ordinary irrigation practices can no longer be continued in many areas, especially in the Middle East. Therefore, the production of medicinal plants in hydroponics under greenhouse conditions advances faster. Some hydroponics systems are matching other water control methods to save water further. Water management is not the only advantage of hydroponics, but it can also get rid of the use of pesticides, fertilizers, and herbicide and gives more control to the farmers, and they can plant their medicinal plants any time of the year. Under the system of deficit irrigation, as a potential technique for saving water, it may advance water utilization and affect plant growth and may increase active constituents. In the situation of remodeling water productivity, there is an increasing interest of medicinal plants under the hydroponic systems with the application of deficit irrigation, whereby water supply is reduced diminished below maximum levels, and delicate stress is allowed with minimum effects on yield and sometimes increased the biosynthesis of secondary metabolites. Deficit irrigation can advance greater economic benefit than boost yields per unit of water for a given plant. Farmers are more willing to use water more efficiently, and more water-efficient cash-crop selection helps optimize returns.

3.3 Irrigation Management

Proper irrigation water management under the prevailing conditions of water scarcity in Egypt required the knowledge of exact values of water consumptive use of the cultivated crops. This can be attained by the accurate estimation of the values of ETo and crop Kc. Although the calculation of ETo can be easily implemented, the calculation of Kc required more efforts to be calculated. This chapter provided the date and the values of the Kc for 33 crops (field crops, fruit trees, and vegetable crops) to contribute to irrigation water management in Egypt.

Egypt is a country in which solid waste is up-surging abruptly. The agricultural and municipal solid waste generated through urban and rural areas, respectively. To combat the problem of waste generation and GHG emissions through waste burning, vermicomposting is one of the best methods. The organic waste conversion into value-added products plays a significant role in balancing the circular economy by making money from organic waste. Vermicomposting covers many aspects like solid waste reduction as biofertilizers have positive effects on plants growth. It also reduces GHG emissions. So, from every point of view, vermicomposting is a better option to manage Egypt waste generation. Moreover, more aspects need to be considered for vermicomposting adoption like substrate composition, earthworm species, and pre-composting.

Egypt *characterizes by plenty of space cultivable soil, but water scarcity is a limiting factor in reclamation processes. Consequently, improving IWUE and EWP is most important than increasing yield for each land unit. The equations that calculate IWUE and EWP did not estimate the retained in the plant. It should be increased farmers' and decision makers' understanding of the scientific communities' results and applied it, especially in concerning of agronomic management effects on IWUE and EWP across various plant types, genotypes, soil properties, climate conditions, irrigation management (systems, rates, frequencies, and timing), soil amendments, plant practices,* etc. *Irrigation management (supplemental irrigation, irrigation systems, rates, and* frequencies), tillage methods, *organic amendments, hydrogel application,* grafting technique, cultivar selection, crop rotation, as well as *spraying of* abscisic acid, proline, silicate, salicylic, auxin and cytokinin, etc., *are the fundamental practices have a high potential for maximizing the benefits of water unit.*

3.4 Irrigation System Design

Egypt is highly vulnerable to climate change, which increases the water demand and causes a loss of crops. Thus, one of the main challenges facing the sequential government during the previous decades was to enhance the agriculture sector by increasing the efficiency of water use. "To achieve high performance in the surface irrigation system, it must be designed to irrigate uniformly, with the ability to apply the right depth at the right time. Properly designed, installed, maintained, and managed irrigation system greatly reduces the volume of irrigation water and hence saves energy and money. Besides, it improves crop yield and quality" (https://link.springer.com/chapter/10.1007/978-1-4419-7637-6_3). Developed water distribution systems are more effective, efficient, and far better than the conventional system, and developed water distribution systems at different locations in the country are running in good condition without any major constraint. This developed distribution system could be widely adopted as a model in the field for increasing agricultural production in Egypt. Among all the water distribution systems, the PVC pipeline system is the most suitable. Improved and efficient water management practices can help to maintain farm profitability in an era of increasingly limited and more costly water supplies.

Good management of the micro-sprinkler irrigation system provides great water and soil conservation and reduces applied water requirement. Using micro-sprinkler irrigation system improves the existing water productivity, water use, efficiency, economic returns, increases yield and great control of applied water through improving DC and SE for the field crops growing under orchard area.

Drip or trickle irrigation is the most efficient method of irrigating. The drip irrigation system works by applying water slowly, directly and at frequent intervals to the soil through mechanical devices called emitters or drippers, localized at selected points along the water delivery line. In addition, the drip irrigation system has been used to deliver the fertilizers and pesticides with irrigation water. The different types of drip irrigation comprise surface, subsurface, bubbler, micro-sprinkler, mobile drip irrigation, and ultra-low drip irrigation. All drip irrigation systems consist of a pumping unit, fertigation unit, filtration unit, a control head, main and sub-main pipes, laterals, and emitters.

3.5 Water Reuse and Treatment

Trace elements (TEs) exposure leads to oxidative damage of *Vicia faba,* which was displayed from increasing H_2O_2 level, proline, and total proteins contents while decreasing activity of most enzymatic antioxidants. The difference in the effect of MTW on *Vicia faba* was related to the presence or the absence of TEs and their concentrations. It was clear that using tap water treated with 1000 gauss magnet could reverse the effect of Ni and Pb by decreasing H_2O_2 level and MDA, proline, and total proteins contents while increasing antioxidant enzymes activity comparing to TEs-treated plants irrigated with normal tap water. Contents of H_2O_2, lipid peroxidation, and electrolyte leakage were declined in all TEs-subjected plants irrigated with irrigation with magnetically treated water (MTW) compared to corresponding controls. Antioxidants content is increased, while anthocyanins decreased comparing to corresponding controls. Irrigation with MTW (1000 gauss) could reduce the toxicity of Ni and Pb in *Vicia faba.*

Magnetically treated water (MTW) significantly enhanced the morphological parameters, decreased proline, and total protein content in all treatments. The criteria of H_2O_2 and the lipid peroxidation showed a fluctuation trend among the different concentrations of heavy metals. Moreover, MTW caused a significant increase in most antioxidants activities in plants grown in soil with/without heavy metals. The present study revealed a novel role of irrigation with MTW in the adaptation of broad bean plants to toxic effects of Ni and Pb based on plant growth, physiological and biochemical criteria.

4 Recommendations

Throughout this book project, the editorial' teams noted some areas that could be explored to further improvement. Based on the authors' findings and conclusions, this section offers a set of recommendations providing suggestions for future researchers and decision making. The following recommendations are mainly obtained from the chapters presented in this volume:

1. For an irrigation scheme evaluation, the values of performance indicators should be applied in order to evaluate the performance level concerning the irrigation scheme. For their calculation, performance indicators call upon a certain number

of parameters that have to be measured in the areas of irrigation schemes for the progressive elaboration. In addition to evaluation of system performance in the field, which indicates the location and magnitude of water losses, such a study would require a thorough knowledge of system costs, and the relation between water and crop production in the area studied.

2. Development and recent data provide Egypt with a planning tool for rational exploitation of developing and manage recent information and data concerning soil, crops, and water resources. This planning is intended to lead to an increase in crop production for local consumption, as well as promote the production of high-value crops. The planning tool will support decision making with direct regard to:

   - Establishing agro-ecological zones (AEZ) maps.
   - Identifying water availability and selection of potential irrigation areas.
   - Identifying criteria for crop selection and estimated water requirements.
   - Identification of the most favorable regions to develop irrigation management practices.
   - Giving priority to irrigation water distribution.
   - Organization and control of irrigation supply management.
   - Developing irrigated agriculture in small, medium, and large-scale projects on old and new lands.
   - Further, decision making will be made with regard to:
     - Supporting the national policy options for irrigation water distribution.
     - Upgrading the Egyptian agricultural production.
     - Recommending options for water harvesting and storage in the coastal regions.
     - Identifying energy requirements for irrigation systems.

3. Various studies have been done to understand the vermicomposting process by using surface species, which only confined to use surface feeder species. Hence, to have a holistic understanding of the impact of vermicomposting produced by the earthworms deep research is needed to study the vermicomposting quality considering factors species type, temperature, substrate, and field conditions to improve soil quality. The variations were observed in different field conditions for agriculture management systems. Based on various types of feedstocks availability, more studies should be concentrated on local species found in that particular area, because they perform better in the local environment. Moreover, these species having the adaptability to varying physiochemical properties of the feedstocks are needed. The research also needs to focus on the particular process involved in an increment of nutrient availability to the plants. There are many aspects, which are still unknown regarding vermicomposting technology. Therefore, more experiments needed to unravel different scientific hiccups affecting the field studies linked with management strategies.

4. All efforts should come together to improve the agricultural management that reflects on the increase both of IWUE and EWP, especially in arid and semi-arid regions, then decreasing the demand for water in the agricultural sector.
5. Evaluation helps to identify problems and the measures required to correct them. No single indicator is satisfactory for all descriptive purpose. In general, a set of indices are used for evaluating the performance of a surface irrigation system. Of course, adequate monitoring and evaluation of performance are needed to improve water management practices in order to achieve an increase in overall efficiency.
6. Increase production while decreasing water use and costs is considered one of the most important contemporary issues today. In order for agriculture to continue to feed the Egyptian people, farmers have to learn to utilize from existing water supplies as far as possible in new lands. Many farmers are converting to micro-irrigation for their trees and field crops because of the potential to save water and increase efficiency. Micro-irrigation system can save water by applying less water exactly where it is needed. In order to maximize yield and profits, farmers have to apply excellent irrigation practices as well as consistent maintenance of their micro-irrigation system. Micro-sprinkler irrigation system is a very common method for irrigating tree crops in Egypt.
7. Irrigation with MTW increased plant biomass and yield; therefore, it considered a recommendable environmentally technique that can be used to improve yield qualities. Irrigation with MTW can alleviate toxic effects induced by low concentrations of Ni and Pb in *Vicia faba* L. plant and increase its productivity, but it cannot be effective against the high concentrations. Therefore, further investigation is needed on different MF intensities and different kinds of heavy metals.
8. It is important to explore the various actions that MTW seems to have in plants. Further studies on a broader range of heavy metals and different magnetic field intensities will provide a clearer understanding of the machinery involved in the regulation of magnetically treated water involvement in reducing heavy metals toxicity in broad bean plants.

Acknowledgements Abdelazim Negm acknowledges the partial support of the Science and Technology Development Fund (STDF) of Egypt in the framework of the grant No. 30771 for the project titled "A Novel Standalone Solar-Driven Agriculture Greenhouse-Desalination System: That Grows Its Energy And Irrigation Water" via the Newton-Mosharafa funding scheme.

References

1. Dakkak A (2017) Egypt's water crisis—recipe for disaster. Environment, Middle East, Pollution, Water

2. Omran E, Negm A (2018) Environmental impacts of the GERD project on Egypt's Aswan high dam lake and mitigation and adaptation options. In: The handbook of environmental chemistry. Springer, Berlin, Heidelberg
3. Elshaikh AE, Xiyun J, Shi-hong Y (2018) Performance evaluation of irrigation projects: theories, methods, and techniques. Agric Water Manag 203:87–96
4. Bian F et al (eds) (2013) Geo-informatics in resource management and sustainable ecosystem. In: International symposium, GRMSE 2013, Wuhan, China, November 8–10 2013, proceedings, Part 2, 2013
5. Prasad S, Peddoju S, Ghosh D (2013) AgroMobile: a cloud-based framework for agriculturists on mobile platform. Int J Adv Sci Technol 59:41–52
6. Channe H, Kothari S, Kadam D (2015) Multidisciplinary model for smart agriculture using internet-of-things (IOT), sensors, cloud-computing, mobile-computing & big-data analysis. Int J Comput Technol Appl 6(3):374–382
7. Mosa ASM, Yoo I, Sheets L (2012) A systematic review of healthcare applications for smartphones. BMC Med Inform Decis Mak 12(1):67
8. Habib MA et al (2014) Smartphone-based solutions for fall detection and prevention: challenges and open issues. Sensors 14(4):7181–7208
9. Duan Y-e (2011) Design of intelligent agriculture management information system based on IOT. In: International conference on intelligent computation technology and automation (ICICTA), vol. 1, pp 1045–1049
10. Chóliz J, Cristina S (2019) Uncertainty in irrigation technology: insights from a CGE approach. Water 11(3):617
11. National Water Research Center (2013) Designing local framework for integrated water resources management project. 1st technical report, pp 10–15
12. Rorabaugh P (2017) Introduction to hydroponics and controlled environment agriculture. Hydroponics page. Controlled Environment Agriculture Center, University of Arizona
13. Valipour M (2014) Analysis of potential evapotranspiration using limited weather data. Appl Water Sci
14. Munnoli PM, Saroj B (2011) Water-holding capacity of earthworms' vermicompost made of sugar industry waste (press mud) in mono-and polyculture vermireactors. Environmentalist 31:394–400
15. Mahmoud M, Yahia E (2011) Vermiculture in Egypt: current development and future potential. Ph.D. Agro Industry and Infrastructure Officer Food and Agriculture Organization (FAO/UN), Cairo, Egypt
16. Han X et al (2018) Effects of crop planting structure adjustment on water use efficiency in the irrigation area of Hei River Basin. Water 10:1305
17. Cosgrove W, Loucks DP (2015) Water management: current and future challenges and research directions. Water Resour Res 51(6):4823–4839
18. USDA D.o.A. (2013) Natural resources conservation service *Microirrigation.* In: National engineering handbook, Part 623 (Chapter 7), pp 19–191
19. Yuan S et al (2017) Optimization of movable irrigation system and performance assessment of distribution uniformity under varying conditions. Int J Agric Biol Eng 10(1):72–79
20. Godin R, Broner I (2013) Micro-sprinkler irrigation for Orchards. Crop Series, Irrigation, Fact Sheet No. 4.703
21. Waller P, Yitayew M (2016) Irrigation and drainage engineering. Springer International Publishing, Springer, Switzerland, Heidelberg, pp 1–18
22. Fayed M (2016) Design and evaluation of mobile drip irrigation system. Ph.D. The Faculty of Agriculture, Al-Azhar University, Cairo, Egypt
23. Sharma A, Shekhawat S (2017) An overview of magnetic field exposure on germination behavior of seeds under saline stress. Int J Pharm Bio Sci 8(1):713–716
24. Dhawi F (2014) Why magnetic fields are used to enhance a plant's growth and productivity? Annu Rev Plant Biol 4(6):886–896
25. El Sayed H (2015) Impact of magnetic water irrigation for improve the growth, chemical composition and yield production of broad bean (Vicia faba L.). Plant Nat Sci 13(1):107–119

26. Moussa H (2011) The impact of magnetic water application for improving common bean (Phaseolus vulgaris L.) production. New York Sci J 4(6):15–20
27. Krishnaraj C, Soon-Il YS, Kumar V (2017) Effect of magnetized water (Biotron) on seed germination of amaranthaceae family. J Acad Indus Res 5(10):152–156
28. Ambashta R, Sillanpää M (2010) Water purification using magnetic assistance: a Review. J Hazard Mater 180:38–49
29. Bagherifard A, Ghasemnezhad A (2014) Effect of magnetic Salinated water on some morphological and biochemical characteristics of artichoke (CynarascolymusL.) leaves. J Med Plants Prod 2:161–170

Printed in the United States
by Baker & Taylor Publisher Services